Analytics for the Sharing
Engineering and Busine

Emanuele Crisostomi · Bissan Ghaddar ·
Florian Häusler · Joe Naoum-Sawaya ·
Giovanni Russo · Robert Shorten
Editors

Analytics for the Sharing Economy: Mathematics, Engineering and Business Perspectives

Editors
Emanuele Crisostomi
DESTEC
University of Pisa
Pisa, Italy

Florian Häusler
Moovel Group GmbH
Stuttgart, Germany

Giovanni Russo
Department of Information & Electrical
Engineering and Applied Mathematics
University of Salerno
Fisciano, Salerno, Italy

Bissan Ghaddar
Ivey Business School
University of Western Ontario
London, Canada

Joe Naoum-Sawaya
Ivey Business School
London, ON, Canada

Robert Shorten
Dyson School of Design Engineering
Imperial College London
London, UK

ISBN 978-3-030-35034-5   ISBN 978-3-030-35032-1 (eBook)
https://doi.org/10.1007/978-3-030-35032-1

© Springer Nature Switzerland AG 2020
This work is subject to copyright. All rights are reserved by the Publisher, whether the whole or part of the material is concerned, specifically the rights of translation, reprinting, reuse of illustrations, recitation, broadcasting, reproduction on microfilms or in any other physical way, and transmission or information storage and retrieval, electronic adaptation, computer software, or by similar or dissimilar methodology now known or hereafter developed.
The use of general descriptive names, registered names, trademarks, service marks, etc. in this publication does not imply, even in the absence of a specific statement, that such names are exempt from the relevant protective laws and regulations and therefore free for general use.
The publisher, the authors and the editors are safe to assume that the advice and information in this book are believed to be true and accurate at the date of publication. Neither the publisher nor the authors or the editors give a warranty, expressed or implied, with respect to the material contained herein or for any errors or omissions that may have been made. The publisher remains neutral with regard to jurisdictional claims in published maps and institutional affiliations.

This Springer imprint is published by the registered company Springer Nature Switzerland AG
The registered company address is: Gewerbestrasse 11, 6330 Cham, Switzerland

Contents

Introduction 1
Emanuele Crisostomi, Bissan Ghaddar, Florian Häusler,
Joe Naoum-Sawaya, Giovanni Russo and Robert Shorten

The Mathematics of Sharing

Optimization Methods: An Applications-Oriented Primer 5
Antonio Frangioni and Laura Galli

Networked Systems Theory: Distributed Algorithms for Optimal
Cooperation of Dynamical Systems 25
Lucia Pallottino

On Distributed Generalized Nash Equilibrium Seeking 39
Sergio Grammatico

Queueing Theory in the Context of Shared Resources 51
Christopher King

Enablers for Collaborative Consumption

Advances in Cloud Computing, Wireless Communications
and the Internet of Things 71
Gopika Premsankar and Mario Di Francesco

Distributed Ledger Technologies and the Collaborative Economy 95
Pietro Ferraro and Daniel Conway

Sharing Economy: A Business Perspective 109
Paolo Roma

Distributed Algorithms for Internet-of-Things-Enabled Prosumer
Markets: A Control Theoretic Perspective 125
Syed Eqbal Alam, Robert Shorten, Fabian Wirth and Jia Yuan Yu

Negotiation Approaches for Sharing Systems 151
Pierre Pinson, Fabio Moret, Thomas Baroche
and Athanasios Papakonstantinou

Behaviour Change for the Sharing Economy 173
Léa Deleris and Pól Mac Aonghusa

Platforms and New Use Cases

Sharing Data in Automotive Applications 191
Joachim Taiber

On Parsing Shared Information: An Application from the Connected Car Domain 205
Rodrigo Ordóñez-Hurtado, Giovanni Russo, Sam Sinnott
and Robert Shorten

Mobility on Demand in the United States 227
Susan Shaheen and Adam Cohen

Data-Driven Rebalancing Methods for Bike-Share Systems 255
Daniel Freund, Ashkan Norouzi-Fard, Alice Paul, Carter Wang,
Shane G. Henderson and David B. Shmoys

Peer-to-Peer Energy Trading 279
Thomas Morstyn and Malcolm D. McCulloch

Healthcare and the Sharing Economy 301
Jad Bitar

Industry 4.0 309
Edoardo Calia and Davide D'Aprile

Industry 4.0 and The Sharing Economy 335
Robert Shorten, John Oliver, Deirdre Clayton, Ammar Malik
and Hugo Lhachemi

Conclusion 347
Emanuele Crisostomi, Bissan Ghaddar, Florian Häusler,
Joe Naoum-Sawaya, Giovanni Russo and Robert Shorten

Introduction

Emanuele Crisostomi, Bissan Ghaddar, Florian Häusler, Joe Naoum-Sawaya, Giovanni Russo and Robert Shorten

This edited volume grew out of our shared interest in one of the most disruptive, and paradoxically, oldest ideas, currently shaping businesses across the globe: collaborative consumption. Collaborative consumption, or the sharing economy as it is also known, refers to businesses that are based on sharing resources and services, as opposed to traditional ownership-based models. Driven by widespread connectivity, new technology bricks such as blockchain and 3D printing, and more informed consumers, this new paradigm is causing a wave of disruption across industries, and is leading to empowerment of citizens and communities in ways that were unimaginable just a short time ago. Prominent examples are the peer-to-peer lodging company Airbnb, transportation network companies such as Uber, and various online platforms which facilitate the exchange of services among users. This new class of business models is fundamentally different from the current standard approaches. A key component of the sharing economy is the rise of the prosumer; that is the

---

E. Crisostomi (✉)
University of Pisa, Pisa, Italy
e-mail: emanuele.crisostomi@unipi.it

B. Ghaddar · J. Naoum-Sawaya
Ivey Business School, University of Western Ontario, London, Canada
e-mail: bghaddar@ivey.ca
e-mail: jnaoum-sawaya@ivey.ca

F. Häusler
Moovel, Stuttgart, Germany
e-mail: florian.haeusler@gmail.com

G. Russo
Department of Information & Electrical Engineering and Applied Mathematics,
University of Salerno, Fisciano, Salerno, Italy
e-mail: giovarusso@unisa.it

R. Shorten
Dyson School of Design Engineering, Imperial College London, London, UK
e-mail: r.shorten@imperial.ac.uk

© Springer Nature Switzerland AG 2020
E. Crisostomi et al. (eds.), *Analytics for the Sharing Economy:
Mathematics, Engineering and Business Perspectives*,
https://doi.org/10.1007/978-3-030-35032-1_1

traditional consumer is replaced by a prosumer who both produces and consumes resources. This can make sharing economies both resilient and efficient, but not without significant challenges. For example, issues of fairness, dimensioning, and social compliance, and increasingly the orchestration of both humans and machines, are all issues that go to the heart of the design of sharing economy models. However, notwithstanding these challenges, market forces driving the emergence of a shared economy are irresistible, and many large corporations are moving quickly to develop new shared services. Roughly speaking, opportunities around shared products give rise to a number of sharing models [1]. First, services based on opportunistic sharing of resources exploit the availability of idle resources and are facilitated by ubiquitous mobile connectivity. Second, in federated sharing, sharing efficiency is achieved by aggregating the demand of many users and their available resources. Finally, driven by the success of existing shared services, a number of large companies are already exploring ways to design products with the specific objective of these products being shared.

Our objective in this book is to provide a somewhat holistic overview of sharing economy systems. The design of sharing economy products is driving not only new and innovative business models, but also innovation in technology, in business, and in the creation of new services. It is with this background in mind that we have organised this edited volume. First we give an overview of some of the mathematics that is of use in designing shared services; then we speak about some of the technologies that are enabling this revolution; and finally, we present a number of use cases that illustrate some of the services that are emerging in this area.

Finally, while this book has grown out of our common interest in this new area, we are very grateful to the Danish ForskEL Energy Collective project (grant no. 2016-1-12530) and to the Science Foundation Ireland Analytics for the Sharing Economy Project (grant 16/IA/4610), both of which fostered and enabled some of the collaboration that gave rise to this book.

Reference

1. Crisostomi E, Shorten R, Štüdli S, Wirth F (2018) Electric and plug-in hybrid electric networks: optimization and control. CRC Press, Taylor & Francis Group, Automation and Control Engineering Series

The Mathematics of Sharing

Emanuele Crisostomi, Bissan Ghaddar, Florian Häusler,
Joe Naoum-Sawaya, Giovanni Russo and Robert Shorten

1 Introduction

In this first part of the book we give an overview of some of the mathematics that are typically used to design shared services. Indeed, independently on the specific context of interest, sharing economy systems are usually characterized by a number of common features. For instance, such systems are usually designed to be operated in a *distributed* way as a (typically large) set of agents/actors that interact with each other through some form of (possibly stochastic and asynchronous) communication to access a shared good. Consider, for example, trading fashion through a clothes

---

E. Crisostomi
University of Pisa, Pisa, Italy
e-mail: emanuele.crisostomi@unipi.it

B. Ghaddar · J. Naoum-Sawaya
Ivery Business School, Canada
e-mail: bghaddar@ivey.ca
e-mail: jnaoum-sawaya@ivey.ca

F. Häusler
Moovel, Germany
e-mail: florian.haeusler@gmail.com

G. Russo
Department of Information & Electrical Engineering and Applied Mathematics,
University of Salerno, Fisciano, Salerno, Italy
e-mail: giovarusso@unisa.it

R. Shorten
Dyson School of Design Engineering, Imperial College London, London, UK
e-mail: r.shorten@imperial.ac.uk

swapping platform such as Kleiderkreisel[1] where users interact with each other and share information through the platform to buy, sell, or trade clothes.

The goal of this part is to provide a primer on some of the key methodological tools, typically used when designing a sharing economy system. These methodological tools are then employed in the remaining parts of this book. We start with Chap. 2, where it is shown how several problems arising in the context of this book can be recast as an optimization problem. In particular, an introduction is offered on the main theoretical and software tools that are nowadays available to system designers to solve the kind of optimization problems arising in sharing economy systems. A simplified bike sharing use-case is used throughout the chapter to guide the reader through the discussion of modeling and algorithmic issues. In Chap. 3 we build upon these optimization tools to consider the case where the systems are distributed. In this chapter certain distributed optimization algorithms are surveyed and it is shown how the problem of optimally managing a shared resource between a set of users can be recast as a (distributed) optimization problem. Within a sharing economy system, the users requiring access to the shared good can either cooperate or compete with each other. A collection of inter-dependent decision making problems from agents each having their own individual aim can be recast as a *game*. Certain tools from Game Theory and Nash equilibria, relevant for sharing economy systems, are surveyed in Chap. 4. Finally, in Chap. 5, a primer is given on queuing theory. Indeed, in many sharing economy applications, it may happen that the request for a given resource exceeds the available supply. Queuing theory offers a way for designing protocols regulating access to the resource and the chapter links the technical concepts to a set of concrete examples.

---

[1]https://www.european-business.com/kleiderkreisel-gmbh/portrait/.

Optimization Methods: An Applications-Oriented Primer

Antonio Frangioni and Laura Galli

Abstract Effectively sharing resources requires solving complex decision problems. This requires constructing a mathematical model of the underlying system, and then applying appropriate mathematical methods to find an optimal solution of the model, which is ultimately translated into actual decisions. The development of mathematical tools for solving optimization problems dates back to Newton and Leibniz, but it has tremendously accelerated since the advent of digital computers. Today, optimization is an inter-disciplinary subject, lying at the interface between management science, computer science, mathematics and engineering. This chapter offers an introduction to the main theoretical and software tools that are nowadays available to practitioners to solve the kind of optimization problems that are more likely to be encountered in the context of this book. Using, as a case study, a simplified version of the *bike sharing problem*, we guide the reader through the discussion of modelling and algorithmic issues, concentrating on methods for solving optimization problems to proven optimality.

Keywords Optimization methods · Combinatorial optimization · Integer programming · Bike sharing

1 Introduction

Effectively sharing resources is a complex task. A benefit of having dedicated resources (say, a personal car) for a given task is that there is no need to take complex decisions about their use, as they are always available. The obvious drawback is that they can be heavily under-utilized, thereby substantially decreasing their value while consuming other valuable resources (say, scarce parking space). Allowing a resource to be shared may dramatically increase its social and economic value, but

A. Frangioni · L. Galli (✉)
Dipartimento di Informatica, Università di Pisa, Largo B. Pontecorvo 3, 56127 Pisa, Italy
e-mail: laura.galli@unipi.it
e-mail: antonio.frangioni@unipi.it

it also opens up a number of complex issues regarding its fair and efficient management that need to be solved for the system to deliver its potential benefits. Besides the many technological aspects (say, self-driving capabilities, accurate navigation, reliable communication, ...) that are crucial to make sharing possible, the ability of taking optimal decisions about highly complex systems (say, a large fleet of self-driving cars) is also a fundamental pre-requisite for the sharing economy to thrive.

Many of the decision procedures can be cast under the form of *optimization problems*, where one *objective function* depending on the *decision variables* has to be optimized (minimize the effort or maximize the benefit) along all possible states of the system, typically represented via *constraints*. Problems of this kind are pervasive and can be found e.g. in design of structures and trajectories [9], production planning [12], resource allocation [13], scheduling [10], control [11], and many others. The development of mathematical and software tools for the solution of these problems is a highly inter-disciplinary branch of applied mathematics lying at the interface with diverse other subjects, including (but not limited to) management science, computer science, and engineering, and is often generally referred to as *Operations Research* (OR). At its heart, OR is based on mathematical techniques that can be traced back to Newton and Leibniz, and that have seen a tremendous development since the second half of last century together with the meteoric rise of availability of computational power. This is indeed relevant, as optimization problems are generally "difficult". Without going into the mathematical details, suffices here to say that the worst-case computational cost of the best known algorithms for solving most optimization problems grows extremely rapidly (exponentially) with the size of the problem, and not just in theory: the increase in complexity often shows off in practice. This motivated an enormous effort to develop efficient (as much as possible) algorithms for different classes of problems, exploiting all available *structure* that each one has. The result is a complex panorama of many different *classes* of optimization problems: for instance, it is customary to distinguish between *continuous optimization* where decision variables can take, say, real values, *combinatorial optimization* where solutions belong to some combinatorial space (sets, paths, trees, ...) (a.k.a. *discrete optimization* since usually decision variables are restricted to only attain discrete values), and *functional optimization* (a.k.a. *optimal control*) where solutions are functions. Each of these classes typically requires different mathematical tools to be addressed (say, analysis for continuous and functional optimization and combinatorics for combinatorial optimization), although in practice the boundaries are blurred, and efficient optimization methods typically borrow from several different areas of mathematics and computer science.

Using mathematical optimization in practice is therefore a complex task that requires navigating nontrivial trade-offs. The process typically starts by constructing a *mathematical model* trying to capture the "essence" of one's practical system to be managed. However, the model should attain the right compromise between correctly representing the reality at hand, and admitting "sufficiently efficient" algorithms that can provide the desired solutions in a time compatible with the constraints that the actual use case imposes (which can be rather tight). Therefore, a crucial decision while developing a mathematical optimization model is what *class* of optimization problems one selects to express the model into, since this dictates what (mathematical, hence algorithmic and ultimately software) tools one has at disposal to actually solve it, and therefore the chances to obtain a practical system. Indeed, it is one of the most important cultural legacies of the last century—from Gödel's incompleteness theorems to the discovery of non-computable functions down to complexity theory—that just being able to write a mathematical model of one's system does not imply that solutions can actually be found. Thus, using optimization methods requires in principle some understanding of the several "core techniques" that have been developed to tackle different kind of problems, such as linear, non-linear, combinatorial and stochastic optimization, calculus of variations, optimal control, game theory, and several others.

However, for problems in contexts like the one of this book, the choices are typically limited. In fact, the problems are usually large-scale (thousands to hundreds of thousands of decision variables), with rather complex constraints encompassing both discrete and continuous decisions. In order to have any chance of solving problems of this type, in most cases one has to restrict to the class of *Mixed Integer Linear Programs* (MILP), which only allows linear relationships between decision variables, thereby forcing to use somehow un-natural constructs for modelling certain relationships. However, the class is expressive enough to allow to represent very many systems, and the vast majority of optimization models in the applications of interest for this book (logistics, transportation, telecommunications, ...) are customarily written as MILP. Furthermore, many software tools are currently available for solving this class of problems. Which does not mean that MILP are not "difficult" problems: in general, the complexity of algorithms for their (exact) solution is exponential. However, due to more than 70 years of continuous research by a large community, the current status of solution algorithms for MILP is such that for several practical problems, just writing the MILP model and passing it to a solver may be enough to obtain solutions with a "reasonable" computational effort. Although there is no guarantee that this will happen for any specific application, the effort for testing this approach is low enough that most often this is the first step that should be attempted (unless perhaps if the problem is extremely-large scale, or needs to be solved in extremely short time, or has many complex nonlinear relationships); with any luck, it may also be the only step that is needed.

In this chapter, after an introduction about general optimization models in Sect. 2 we will therefore concentrate on MILP. The concepts will be illustrated by means of a case study inspired by a bike sharing system, described in Sect. 3. In Sect. 4 we will concentrate on a crucial and nontrivial aspect, which is that there are many possible ways to *formulate* the same system as a MILP model, and that this choice is actually crucial for the effectiveness of the solution algorithms. Finally, in Sect. 5 we will provide pointers for further reading on the subject.

2 Optimization Problems

A quite general optimization problem is

$$\text{(P)} \quad \min\{\, f(x) \,:\, G(x) \leq 0 \,,\, x_i \in \mathbb{Z} \ \ i \in I \,\} \,,$$

where $x = [\, x_1, \ldots, x_n \,]$ is the (*finite*) vector of *decision variables*. When not otherwise specified, each x_i can attain real values, i.e., $x \in \mathbb{R}^n$; this already restricts (P) to a finite-dimensional space, thereby excluding problems where x_i may be, say, itself a function. The *objective function* $f(x)$ and the (*finite number of explicit*) *constraints* $G(x) = [\, g_j(x) \,]_{j=1,\ldots,m}$ are usually assumed to be real-valued functions, i.e., $f : \mathbb{R}^n \to \mathbb{R}$ and $G : \mathbb{R}^n \to \mathbb{R}^m$. For f, this implies that the decision-maker only has one objective, which is often not true in practice as there could be any number of—possibly, contrasting—measures to be optimized. Yet, this is necessary to univocally define the concept of *optimal solution*; with a vector-valued $f(x) = [\, f_h(x) \,]_{h=1,\ldots,k} : \mathbb{R}^n \to \mathbb{R}^k$, the lack of a complete order in $\mathbb{R}^k$ (for $k > 1$) means that one has rather to resort to the concept of *nondominated* (a.k.a. *Pareto-optimal*) solution, but this would mean that (P) does not identify *one* solution that can be automatically taken as the answer without human oversight, as it is often necessary to do. Thus, in presence of multiple objective functions it is customary to either form a weighted sum ($f(x) = \sum_{h=1,\ldots,k} \alpha_h f_h(x)$, a.k.a. *weighting*) or define thresholds on the least acceptable value for all objectives save one (*budgeting*), and fall back to the single-objective case. Even so, f and G cannot clearly be just "any" function, as there are functions that simply cannot be computed; in general, the assumption is that they are "easy" to compute, most often algebraic functions. Yet, even restricting (P) to algebraic functions is not enough to tame its complexity: even with simple polynomials, the problem may be undecidable (provably, for some instances there cannot be any solution algorithm). This justifies why it is customary to work with *classes of optimization problems* corresponding to restricting the shape of f and G. Indeed, by far the most common class of optimization problems for the applications of interest here is that of *Mixed-Integer Linear Programs* (MILP), where both f and G are *linear*; that is,

$$\text{(MILP)} \quad \min\{\, cx \,:\, Ax \leq b \,,\, x_i \in \mathbb{Z} \ \ i \in I \,\} \,,$$

where $c \in \mathbb{R}^n$ is the *cost* (row) *vector*, $A \in \mathbb{R}^{m \times n}$ is the *constraint matrix*, $b \in \mathbb{R}^m$ is the (column) vector of *right-hand sides*. This finally justifies why, besides the explicit constraints, (P) also has *integrality constraints* on a (possibly empty) subset $I \subseteq N = \{\, 1, \ldots, n \,\}$ of the variables; although these could be expressed in an algebraic way, doing so would require "complex" functions.

Even with such a dramatic restriction to the set of available functions, MILP is a "hard" problem. However, at least it is now "clear where the hardness comes from": with no integrality constraints ($I = \emptyset$) the problem becomes a *Linear Program* (LP), which is instead "easy". Formally, this means that there are solution algorithms whose complexity grows at most as a polynomial (although, not a very low degree one) in the size of the problem. In practice, this means that LPs with up to hundreds of thousands of variables and constraints can routinely be solved with standard approaches on standard PCs, and that LPs with up to billions of variables can be approached with appropriate techniques on HPC systems. As we shall see in Sect. 4, this is a crucial property upon which all the available solution approaches to MILP rely. Indeed, the fact that LP admit "efficient" algorithms is the main reason why MILP is the most widely used class of optimization problems: although MILPs are in general "hard", formulating one's problem in this shape allows to exploit the huge amount of research, and the many available actual software products, dedicated to their solution. The advances in those approaches over the last decades have been such that many practical problems, despite being theoretically "hard", can nowadays be routinely solved by just applying off-the-shelf technologies once they have been *properly* formulated as MILPs.

Yet, doing so is not straightforward, for two reasons. The first is that while MILP is a quite "expressive" class, in the sense that an enormous number of different relevant problems can be written as such, the severe restriction of being only able to use linear functions requires getting familiar with a number of *modelling tricks*, whereby the limitations of the framework are sidestepped by clever contraptions. The second, and perhaps more important, is that there are usually very *many different MILP formulations* of a given problem, which are typically *not* equivalent in terms of computational cost. Quite on the contrary, writing the "right" MILP formulation of the problem can easily be the deciding factor in being able to solve it, with orders-of-magnitude differences between apparently similar formulations being not uncommon. All in all, there is no systematic way to formulate an optimization problem as a MILP, and devising a good model is often more of an art than of a science. However, some general guidelines can be provided, which is what we will attempt in the rest of this chapter. In particular, the next section, using one specific case study, will focus on the process of writing a MILP formulation out of the "informal" description of one's situation. Then, in Sect. 4 we will very briefly summarize the main concept underlying the most effective (least ineffective) solution methods for MILP, as doing so allows to pinpoint the characteristics that a "good formulation" should ideally possess.

However, to conclude this section it is important to remark that MILP is not the most general class of optimization problems for which (relatively) efficient solution tools are available. Indeed, it is possible to "slightly" enlarge the set of functions available to express f and G while keeping a solution cost comparable with that of a MILP of the same size. For instance, this is possible with Mixed-Integer Quadratic Programs (MIQP) that have *convex quadratic functions* of the form $f(x) = x^T Q x + qx$, with Q a positive semidefinite matrix. More in general, *Second-Order Cone constraints* of the form $x_n \geq \sqrt{\sum_{i=1}^{n-1} x_i^2}$ can be used as well; a surprising number of different nonlinear functions can be formulated as such, and the best current software tools can usually handle Mixed-Integer Second-Order Cone Programs (MI-SOCP) efficiently. Even Mixed-Integer Nonlinear Programs (MINLP) with general *convex functions* can be tackled with similar approaches, although with a somehow lower efficiency. Finally, there are tools capable of solving (or attempting to) MINLP where f and G are general algebraic or transcendental functions; however, their efficiency in practice can easily be orders of magnitude lower than those for MILP problems of comparable size. All in all, the best course of action when devising a mathematical model is most often to start with a MILP one, unless perhaps if the practical problem have either complex nonlinear relationships that cannot reasonably be simplified, or simple nonlinear relationships that can be expressed within the classes of functions that make MINLPs "not too much more difficult" than MILPs.

3 Case Study: The Bike Sharing Problem

Bike Sharing (BS) is an urban mobility service that makes public bicycles available for shared use. The bicycles are located at some stations distributed across the urban area. Customers can take a bicycle from one of the stations, use it for a journey, then leave it at a possibly different station, paying according to the time of usage. This is an important service in *green logistics*, in that on one hand it helps to decrease traffic and CO_2 emissions, and on the other hand it offers a partial solution to the so-called "last mile problem" related to proximity travels.

The cost of operating a BS system has several components: the setup costs (i.e., buying and installing stations and bikes), the cost of the back-end system to operate the equipment, and daily operating costs like maintenance, insurance, and the cost of redistributing bikes among the stations. Ideally, this being a "system" one should optimize all its aspects simultaneously; however, in practice this is usually impossible. For once, different costs are incurred at different times: setup costs are paid once, before the system can start operating, and the consequences of the corresponding *strategic decisions* affect the day-to-day operations for a long period. Instead, *operating decisions* such as how many bikes should be moved from one station to another can only be taken after the particular situation of a given day is known. Taking all these decisions in one step is impossible, due both to lack of data about the future and to the fact that that the corresponding optimization problem would be so huge as to being intractable. Therefore, it is crucial to devise the right trade-off between system model tractability and accuracy.

In practice, optimization of a complex system is usually decomposed into independent sub-problems, where independence comes either from considering different time horizons (strategic vs. operational decisions), or from purposely ignoring present but sufficiently "weak" dependencies, say by approximating the effect of the choices in one subproblem on the others' feasibility and cost. Besides making the problems easier to solve, this usually leads to the discovery that each of the sub-problems has strong similarities with other known problems, coming from similar or even completely different settings. This is why the OR literature often focuses on classes of "abstract" models (say, Vehicle Routing models, Location models, ...) that represent common mathematical structures that appear in many different real-world contexts, possibly as sub-problems or with variants. Once one's problem has been identified, it is therefore usually possible to find similar cases in the literature, which can greatly help in devising a good (MILP) model.

Our case study gives rise to several optimization sub-problems such as: bike station location and fleet dimensioning, allocation of bikes to stations, re-balancing incentives setting, and bike repositioning. We will concentrate on the latter, that represents one of the main daily operating costs, with a consistent impact on the budget. In fact, even if in the early morning all the bike stations meet the desired level of occupation, due to the users' travel behaviour in the evening some stations are typically full and others are empty. Repositioning is crucial in order to offer a good service all day long, and is usually done by means of capacitated vehicles, based at a central depot, that pick up some bicycles from the stations where the level of occupation is too high and deliver them to those stations where the level is too low. The depot also keeps a buffer of bicycles to allow a more flexible redistribution. Driving the vehicles around the urban area is expensive, thus one needs to decide how to route the vehicles to perform the redistribution at minimum cost. This optimization problem is known as the *Bike sharing Re-balancing Problem* (BRP), and has recently received considerable attention from both the OR community and practitioners in the area. We will only consider the *static version* in which the occupancy of each station is considered fixed, as this corresponds to the heavier re-balancing operations performed at night when the system is closed or demand is very low. A *dynamic version* exists where real-time usage information is taken into account to update the redistribution plan even as the vehicles are performing it, but this requires more complex modelling techniques to represent uncertainty of future events that are out of the scope of this simple treatment.

3.1 BRP: System Model

The first step to develop a mathematical model is to define the so-called *system model*, that describes the involved entities, their relations and the corresponding parameters. A system model is usually defined already as much as possible in terms of mathematical objects, and significant decisions are made as to which entities and relations are significant, which ones are ignored, and which relations are simplified.

In the BRP case, the system consists of the depot, the vehicles, the stations, and the urban road network that connects them. This can be described using a complete directed graph $G = (N, A)$, where the set of nodes $N = \{0, 1, \ldots, n\}$ contains the depot, node 0, and the stations, nodes $N' = \{1, 2, \ldots, n\}$. An arc $(i, j) \in A$ represents the shortest path in the actual city road network connecting the locations corresponding to the two nodes i and j, with the associated traveling cost c_{ij}. Since the BS system is located in an urban area, where one-way streets often strongly affect the feasible paths, arcs (i, j) and (j, i) may have different costs. Each station node i has a request q_i, which can be either positive or negative: if $q_i > 0$ then i is a pickup node, where q_i bikes must be removed, while if $q_i < 0$ then i is a delivery node, where $-q_i$ bikes must be supplied. The requests are computed as the difference between the number of bikes present at station i when performing the redistribution and the desired number of bikes in the station in the final configuration. This is a typical example of a link between two decision phases: the desired number of bikes is established during the planning phase, where re-distribution costs are estimated without taking into account detailed routing choices (e.g., by some average), and then taken as a constraint in the operational phase. A fleet of m identical vehicles, each with capacity Q (bikes), is located at the depot. The bikes removed from pickup nodes can either go to a delivery node or back to the depot, and similarly for those supplied to delivery nodes.

The problem is to decide how to route at most m vehicles through the graph, with the *objective* of minimizing the total traveling cost while satisfying the following *constraints*:

1. each vehicle performs a route that starts and ends at the depot;
2. each station is visited exactly once, which implies that its request is completely fulfilled by the one vehicle visiting it;
3. each vehicle starts from the depot with some initial load, comprised between 0 (empty) and Q (full), and the vehicle load at each step of the route, corresponding to the sum of requests of the visited stations plus the initial load, is never negative or greater than Q.

In setting the system model, some important decisions have already been taken. For instance, it might be possible to serve a station with more than one vehicle, each satisfying a fraction of the demand. This would enlarge the space of feasible BRP solutions, allowing more flexibility in the redistribution plan at the cost of complicating the operational procedures; we assume that the operator has decided against it, but a somehow different model could be developed to check the economical impact of the choice and perform what-if analysis. Also, we allow flow of bikes on the depot to be either positive or a negative, which is useful to model cases in which some bikes have to be brought back to the depot for maintenance, and then put in operations again. Finally, it is assumed that there is enough time available to perform all the operations (although some routes can be "long") and that stations can be serviced at any time; this allows to completely ignore the exact moment at which each operation is performed, that may instead be a relevant information in other variants of the problem (e.g., the dynamic one).

3.2 BRP: MILP Formulation

The second step is developing a MILP formulation. As it often happens, the model is closely related to a widely studied class of (difficult) combinatorial problems known in the literature as *Capacitated Vehicle Routing Problem* (CVRP) [2, 3]. In particular, BRP is a *Pickup and Delivery Vehicle Routing Problem* (PDVRP). Thus, to define a

MILP formulation one can use as a guide those that have already been proposed in the literature.

A MILP model of BRP uses three types of *decision variables*:

- *binary arc* variables $x_{ij} \in \{0, 1\}$ taking value 1 if arc (i, j) is used by a vehicle (irrespectively to which one), and 0 otherwise;
- non-negative *continuous arc flow* variables f_{ij} representing the current load of the vehicle traveling along arc (i, j), if any;
- *integer arc* variables y_{ij}, representing the position of arc (i, j), if used, in the corresponding route.

The MILP model is the following:

$$\min \sum_{(ij) \in A} c_{ij} x_{ij} \tag{1}$$

$$\sum_{(ij) \in A} x_{ij} = 1 \qquad i \in N' \tag{2}$$

$$\sum_{(ji) \in A} x_{ij} = 1 \qquad i \in N' \tag{3}$$

$$\sum_{(0j) \in A} x_{0j} \leq m \tag{4}$$

$$\sum_{(0j) \in A} x_{0j} - \sum_{(j0) \in A} x_{j0} = 0 \tag{5}$$

$$\sum_{(ij) \in A} f_{ij} - \sum_{(ji) \in A} f_{ji} = q_i \qquad i \in N' \tag{6}$$

$$\underline{f}_{ij} x_{ij} \leq f_{ij} \leq \bar{f}_{ij} x_{ij} \qquad (i, j) \in A \tag{7}$$

$$\sum_{(ji) \in A} y_{ji} - \sum_{(ij) \in A} y_{ij} = 1 \qquad i \in N' \tag{8}$$

$$x_{ij} \leq y_{ij} \leq (n+1) x_{ij} \qquad (i, j) \in A \tag{9}$$

$$x_{ij} \in \{0, 1\}, \quad y_{ij} \in \mathbb{N} \qquad (i, j) \in A \tag{10}$$

The objective function (1) minimizes the traveling cost. Constraints (2)–(3) ensure that each station node has exactly one incoming and one outgoing arc, while (4) and (5) ensure, respectively, that no more than m routes (= vehicles) leave the depot, and that each leaving vehicle eventually returns. Next, *flow conservation* constraints (6) ensure that f_{ij} properly account for the number of bikes on the (single) vehicle traversing arc (i, j). Then, (7) serve a double purpose. On one hand, when $x_{ij} = 1$, i.e., a vehicle is traversing arc (i, j), they guarantee that the load on a vehicle is feasible by imposing appropriate upper and lower bounds on it (for instance $\underline{f}_{ij} = \max\{0, q_i, -q_j\}$ and $\bar{f}_{ij} = \min\{Q, Q+q_i, Q-q_j\}$, whose validity can be easily verified with some reflection). On the other hand, when $x_{ij} = 0$ they guarantee that $f_{ij} = 0$; hence, bikes can only "flow" on the graph following vehicles, as it is logically required. One may believe that these constraints (plus (10) dictating the binary nature of x_{ij}) be enough to correctly model the problem, but this is not so because they do not forbid *sub-tours*, i.e., closed oriented loops that *do not include the depot*. To picture this, consider two nodes i and j having $q_i = -q_j > 0$: it would be therefore possible to set $x_{ij} = x_{ji} = 1$, $f_{ij} = q_i$ and $f_{ji} = 0$, i.e., have a vehicle "appearing" at i, carrying all required bikes to j, getting back to i and "disappearing" there. Constraints (8)–(9), which are a version of the so-called *Miller-Tucker-Zemlin*

(MTZ) subtour elimination constraints originally devised for the *Traveling Salesman Problem* (TSP) [1], avoid sub-tours (of any length). The idea is to associate an ordering to the vertices of each route by assigning an integer "position" variable y_{ij} to each arc in the route. Starting from the depot, the value of the position variables increases by one as we move to the next arc along the route. Clearly, since a route is cyclic, we can impose an ordering for all the vertices in the route except from the depot: indeed, (8) is not imposed for $i = 0$. Note that the maximum number of arcs in a route is $n + 1$—a single route that visits all the vertices—yielding the upper bound on the position variables.

Admittedly, devising such a formulation is not a trivial task. A number of *formulation tricks* have been used, such as extensive use of *flow conservation constraints* ((6), (8) and in fact even (2)–(5)) and constraints imposing "logical" conditions, such as "$x_{ij} = 0 \implies y_{ij} = 0, x_{ij} = 1 \implies y_{ij} \in [\,1\,,\,n+1\,]$" ((9), and similarly for (7)). Getting used to all the devious tricks necessary to devise a MILP formulation requires some experience; however, as already recalled, most problems one encounters have very likely already been modeled, possibly with some variations, and therefore help is usually available. For instance, the MTZ constraints for the PDVRP are more complex than those for the CVRP where all goods originate from the depot; actually, in that case, the flow constraints (6) are sufficient to avoid sub-tours, as one can easily see considering our counter-example above. Yet, the more complex version had already been devised far before that BRP was ever conceived.

The advantage of formulating one's problem as a MILP is that, once this is done, efficient (as much as possible) software tools are available for (attempting to) solving it.

3.3 Software Tools

Due to the huge number of practical applications leading to MILP models, there is no shortage of software tools for solving these problems. Actually, even forming the coefficient matrix A and the right-hand-side vector b corresponding to a MILP is itself a nontrivial and error-prone task, especially considering that practical applications may originate even considerably more complex MILP models than (1)–(10). This is why software tools known as *algebraic modelling languages* are available to perform it, allowing to describe one's model in a way that is very similar to how equations are written in a document; these equations are then processed to automatically construct the data of the problem (A, b, c, I). Several of these tools exist, both commercial ones such as AMPL [14] and GAMS [15], or open-source ones like Coliop [16] and ZIMPL [17]. Also, *modelling systems* are available that implement similar functionalities within many different general-purpose programming languages, such as C++ (FlopC++ [18], COIN-Reharse [31]), python (PuLP [19], Pyomo [20]), Julia (JuMP [21]), Matlab (YALMIP [22]) and others. As an alternative, it is possible to construct the model by either writing a file with proper format (e.g., MPS [23] or LP [24]), or directly calling the in-memory API of the desired solver.

However the modelling phase is performed, the problem can then be solved using any of the several available solvers, again both commercial ones like Cplex [25], GuRoBi [26], MOSEK [27] and others, or open-source ones like Cbc [28] or SCIP [29]. The effectiveness of the solver should be expected to vary considerably; although all more or less based on the same ideas, quickly summarised in the next Section, implementation details may make an enormous difference on specific instances. Indeed, MILPs are "difficult", and therefore one should in principle expect the problems to take a long time to solve; a large set of sophisticated algorithmic techniques has been developed to try to improve on this, whose implementation differs between different solvers. In general one should expect commercial solvers to be more efficient than open-source ones (which justify the high licensing fees they usually require), but exceptions are not unheard-of. Indeed, all solvers have a large set of *algorithmic parameters* controlling their behaviour, whose appropriate tuning can make a very substantial difference.

Above all, however, *choosing the right formulation* is usually the most important factor dictating how efficiently the problem will ultimately be solved. To be able to at least introduce all this aspects, a very quick recap of the algorithmic techniques that are employed is necessary.

4 Algorithmic Approaches and "Good Formulations"

There are very many different algorithms for "hard" problems. In many practical applications, recognising the fact that efficiently finding *provably* optimal solutions is difficult, it is usual to resort to *heuristics*, i.e., algorithms that strive to efficiently provide "good" solutions, but give no guarantees on their quality. Most often heuristics are developed for a much more specific class of problems than MILP (say, PDVRP), although general frameworks for specific classes of heuristic approaches have been developed, such as *local search*, *tabu search*, *simulated annealing*, *genetic algorithms* and many others, and general-purpose solvers exist (e.g. LocalSolver [30]). In some cases, *approximation algorithms* are available that provide explicit *a-priori* bounds on the quality of the solution obtained within a given computational effort, but again these strongly depend on the specific problem, and are not available for general MILP. In this section we will rater concentrate on *exact* algorithms, that guarantee to find an optimal solution, albeit possibly at a computational cost that may grow exponentially fast with the size of the problem. In particular we will (briefly) discuss the *Branch-and-Bound* (B&B) approach underlying all the general-purpose solvers alluded to above, with its crucial variants known as *Branch-and-Cut* and *Branch-and-Price*.

These approaches are inherently based on the concept of *relaxation* to provide *bounds* on the optimal value of the problem. In general, a relaxation of an optimization problem (P) (assuming minimization) is another optimization problem (P′) such that: (i) its feasible set is larger than that of (P), and (ii) its objective function has the same or smaller value than that of (P) on the latter's feasible solutions. By denoting as $\nu(\cdot)$ the optimal value of an optimization problem, one clearly has $\nu(P') \leq \nu(P)$. The question then arises of how to construct interesting relaxations. However, as anticipated, the answer is trivial for (MILP), as one can simply use its *continuous relaxation*

$$\text{(LP)} \quad \min\{\, cx \,:\, Ax \leq b \,\} \ ,$$

i.e., the LP obtained by simply dropping the integrality constraints. It is immediate to realise that $\nu(\text{LP}) \leq \nu(\text{MILP})$, which may provide *optimality conditions* for a feasible solution $\bar{x}$ of (MILP). Indeed, consider the (very fortunate) case in which the optimal solution $\bar{x}$ of (LP) satisfies the integrality constraints: it is immediate to realise that, in this case, $c\bar{x} = \nu(\text{LP}) \leq \nu(\text{MILP}) \leq c\bar{x}$. Thus, exploiting the efficient available LP algorithms we could, in principle, be able to not only find an optimal solution of the "difficult" (MILP), but also have a *certificate of optimality* proving it without any doubt. This actually hinges on being able to prove that $\bar{x}$ is optimal for (LP) in the first place, which typically relies on *duality* and ultimately *convexity* of the problem, but we are not going to delve deeper in these concepts.

Unfortunately, in general $\nu(\text{LP}) < \nu(\text{MILP})$. Thus, besides *bounding* by relaxations some other mechanism is required to ensure that the lower bound is increased and eventually reaches $\nu(\text{MILP})$. Unfortunately, the only (practical) general ways that have been devised for ensuring this are *branching*, a.k.a. (partial) *enumeration*, and *cutting*. Both are, for the best variants known so far, processes that may require an exponential number of iterations to achieve the desired results. Yet, they are also the best (least worse) general-purpose available algorithms for MILP, hence what is used in practice. We will now give a brief recount of their basic principles.

4.1 Branch-and-Bound

Branch-and-Bound (B&B) uses a "divide and conquer" approach to explore the set X of feasible solutions of (MILP), as described in Algorithm 1. The idea is to partition X into a finite collection of subsets $X_1, \ldots, X_k$ and solve separately each one of the subproblems restricted to each X_i; one then compares the optimal solutions to the subproblems, and chooses the best one. However, since the subproblems are usually almost as difficult as the original problem, it is typically necessary to solve each of them by recursively iterating the same approach, that is, splitting them into further subproblems. This is the *branching* part of the method, which leads to a *tree of subproblems* T. Because X is typically exponentially large, the B&B uses *lower bounds* on the optimal value of each subproblem to try to avoid exploring parts of it. These are obtained by solving the continuous relaxation. Actually, the corresponding (LP) may have no solution, which immediately implies that X_i has no integer solution as well; any LP solver can efficiently detect it, customarily returning ∞ as the optimal value and therefore allowing to *fathom* the corresponding node of T by *infeasibility*. Otherwise, a continuous optimal solution $\bar{x}_i$ is found that may occasionally be integer, in which case we can possibly have found a better *upper bound* on $v(\text{MILP})$ than previously known, updating the *incumbent value* $\bar{z}$ (the cost of the best solution found thus far). The essence of the algorithm lies in the following observation: if the optimal value $\underline{z}_i = c\bar{x}_i$ of the continuous relaxation of X_i satisfies $\underline{z}_i \geq \bar{z}$, then this subproblem need not be considered further, since the optimal solution of X_i is no better than the best feasible solution encountered thus far. In this case, the corresponding node of T can be *fathomed by the bound*; it is immediate to realize that this surely happens if $\bar{x}_i$ is integer, since then $c\bar{x}_i \geq \bar{z}$. Otherwise, *branching* has to occur, i.e., the current X_i is further split. Provided that the total number of subproblems that can be generated is finite, the B&B algorithm clearly terminates in a finite (albeit, very possibly, exponentially large) number of iterations having correctly identified the optimal value $\bar{z} = v(\text{MILP})$. The *incumbent solution* having produced the incumbent value $\bar{z}$ is usually conserved as well, and at termination it is guaranteed to be an optimal solution. Finiteness of the branching operation is usually trivial. For instance, if all variables are binary, the typical branching consists in selecting one variable in $\bar{x}_i$ that has a fractional value and creating two subproblems, in one of which the variable is fixed to 0, while in the other it is fixed to 1; a slightly more general version is easily devised for integer variables. We remark in passing that branching for nonconvex MINLPs is a considerably more sophisticated process, justifying the higher practical cost of the latter w.r.t. MILPs.

Algorithm 1: Branch-and-Bound

Input: (MILP), subproblem-selection-rule, LP-relaxation, branching rule
Output: *Optimal value* $\bar{z}$

1 Initialize $T = \{X\}, \bar{z} = \infty$;
2 **while** $T \neq \emptyset$ **do**
3     $X_i \leftarrow$ subproblem-selection-rule(T) // select an active subproblem
4     $(\bar{x}_i, \underline{z}_i) \leftarrow$ LP-relaxation(X_i) // solve continuous relaxation ;
5   **if** $\underline{z}_i = \infty$ **then**
6   delete X_i // subproblem X_i infeasible
7   **else**
8   **if** $\bar{x}_i$ *is feasible for (MILP)* **then**
9            $\bar{z} = \min\{\bar{z}, c\bar{x}_i\}$ // new feasible solution found
10   **if** $\underline{z}_i \geq \bar{z}$ **then**
11   delete X_i // subproblem X_i fathomed by bound
12   **else**
13           $T \leftarrow$ branching rule(X_i)
14   // break X_i into further subproblems and add them to T ;
15   **end**
16   **end**
17 **end**

Many aspects of the practical implementation of a B&B are potentially crucial, such as the exact choice of the branching operation, the selection of the next active subproblem, and the details of the solution of the continuous relaxation. Also, MILP solvers usually employ *general-purpose heuristics* which try to generate feasible integer solutions, say by "cleverly" rounding the optimal continuous solutions $\bar{x}_i$. However, by far the most important factor dictating the effectiveness of a B&B is how often fathoming of a node (line 11) happens. This is clearly related to *tightness* of the lower bound, i.e., how close the optimal value $\underline{z}_i$ of the continuous relaxation is to the actual optimal value of X_i (although, of course, the quality of the upper bound $\bar{z}$ also is a factor). As a rule of thumb, MILPs for which $\underline{z}_i$ is within a very few percentage points off the real value can be solved efficiently via the B&B, whereas if the *relative gap* is, say, above 10% then a huge number of nodes will be enumerated.

Crucially, this does not depend on the B&B solver, but rather on the *quality of the formulation* that the user has provided it. It is therefore important to explicitly define what the "quality" of a formulation is.

4.2 Strong, "Large" Formulations

For LP, a "good" formulation is one that has a *small number* of variables and constraints, because the computational cost of its solution depends (polynomially) on these. The situation for MILP is drastically different.

The relevant mathematical concepts are pictorially illustrated in Fig. 1, depicting a MILP in two integer variables. Three different formulations are represented, as *polyhedra*. The three formulations are *equivalent* in a MILP sense, in that they define the same feasible region (white points, intersection between the $\mathbb{Z}^2$ lattice and each polyhedron). However, the polyhedra have different "size"; intuitively, the larger the polyhedra, the smallest the value of the corresponding continuous relaxation, and therefore the worse the gap, although of course the "direction" of the objective function also has an impact. Among the three, "the best" polyhedron (shaded area with dotted edges) is depicted. This is the smallest polyhedron representing a correct formulation, a.k.a. the *convex hull* of all feasible integer solutions. Without entering into details of the definition, the convex hull providers the "perfect" formulation: all its *vertices* are integer-valued (a.k.a. *integrality property*), and it can be seen that this implies that solving the continuous relaxation solves the problem to provable opti-

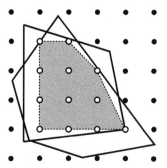

Fig. 1 2D illustration

mality. Unfortunately, most MILP formulations do *not* have the integrality property. However, it is now clear that the *quality* of a MILP formulation can be generally judged by how close it is to defining the convex hull of the integer solutions.

Unfortunately, in most cases such formulations are necessarily "very large". For all "difficult" problems, the "perfect" formulation includes an *exponential* number of different constraints (say, $m \in O(2^n)$). In fact, to define "strong" formulations it is usually necessary to use families of constraints that are exponential in size.

Our BRP problem provides a fitting example of this phenomenon: to impose connectivity of the solution, an alternative to (8)–(9) are the *Subtour Elimination Constraints* (SEC)

$$\sum_{i \in S, j \in S} x_{ij} \leq |S| - 1 \quad S \subset N' \,, \quad 2 \leq |S| \leq n - 2 \,. \tag{11}$$

Clearly, a sub-tour touching a subset $S \subset N'$ of stations must contain at least $|S|$ arcs, as our example illustrated: (11) forbids this, so that all sub-tours must necessarily include the depot. If (11) are added to the formulation, the y_{ij} variables and the corresponding constraints (8)–(9) are no longer needed; this saves $|A|$ variables, but at the cost of *exponentially many* constraints. Yet, the formulation using (11) is well-known to be usually much *stronger* than that using MTZ, and therefore in principle preferable. The issue, of course, is that even for graphs with relatively few nodes writing down an exponentially large set of constraints would not be possible, and anyway it would lead to a very inefficient solution algorithm due to the enormous cost of solving the corresponding LP. Yet, formulations of this kind can be effectively used.

The idea is to only add constraints that are "needed". More precisely, one can initialize a *Restricted Master Problem* (RMP) containing only a small subset of the original constraints (say, only those corresponding to all possible subsets S of size two). The corresponding LP is solved, and the solution $\bar{x}_i$ is checked: if it satisfies *all* the original constraints one stops, having clearly solved the continuous relaxation as if all the constraints had been there, otherwise one or more violated constraints are added to the RMP and the process is iterated. In practice this scheme is usually quite effective, since although several LPs are solved instead of only one, their size is tiny if compared to that that the complete one would have. However, for this scheme to make sense, a "smart" way is needed to verify whether or not $\bar{x}_i$ satisfies all the constraints (11), and if not find at least a violated one. Indeed, if this was done by enumerating them all, then it would have an exponential cost, which would rapidly become unbearable as the size of the graph grows. Fortunately, again, this can be done. Without going into details, the idea is to recast the question under the form of an *optimization problem*, typically that of finding the *least violated constraint*. Despite having an exponential number of solutions, the problem may be "easy": for instance, for (11) it reduces to a *maximum flow/minimum cut* problem on a properly defined digraph (depending on $\bar{x}_i$), which admits efficient solution algorithms. In other cases, these *separation problems* may be "hard" in principle, but still be solvable efficiently enough for the required size to make the approach viable. The advantage of having a much stronger formulation may counterbalance the cost of solving many LPs and many (possibly, "hard") separation problems, making these "large but strong" formulations powerful tools for the solution of problems such as the BRP.

Iteratively adding violated constraints is called the *cutting-plane method*, since we add constraints (i.e., hyperplanes), a.k.a. *cuts* or *valid inequalities*, that cut away the current (fractional) solution $\bar{x}_i$. Interestingly, it is in theory possible to *entirely solve any* MILP by only relying on a cutting-plane approach, without any branching ever occurring. This can be done e.g. by relying on the *Chvàtal-Gomory* (CG) procedure, that is a systematic way of deriving valid inequalities for the convex hull of the integer solutions. Indeed, a "classic" result in polyhedral analysis shows that every valid inequality for the convex hull can be obtained by applying the CG procedure a finite number of times. In other words, a cutting-plane approach based on CG can be shown to be an exact method to solve MILP. Despite the theoretical interest, in practice pure cutting-plane algorithms are not successful on their own, because: (i) a huge number of cutting planes is typically needed, (ii) cuts tend to get weaker and weaker as the algorithm proceeds (a.k.a. "tailing off"), (iii) no feasible integer solution is obtained until the very end. Yet, these ideas are useful from a practical point of view. Unlike SEC, the CG cuts do not need any specific structure, and therefore can be applied to any MILP. Similarly, other classes of "general purpose cuts" have been developed (such as clique, cover, mixed-integer rounding, implied bounds, flow path, disjunctive cuts, ...) that are now available in most MILP solvers, where they can be automatically generated for any MILP. Even though these cutting planes alone cannot usually solve a MILP, they can be very useful to considerably strengthen the given formulation. Indeed, all current most successful solution algorithms for MILP are based on a *combination* of cutting-plane techniques, typically using several different classes of cuts, with the B&B approach. This is known as *Branch-and-Cut* (B&C), which is in essence just a B&B where cutting planes are generated throughout the tree T to improve the LP bounds. A somewhat delicate balance has to be attained at each iteration, since the approach has to decide which of the two basic operations (branching or cutting) to apply at a node that is not fathomed; significant research has gone in finding rules that work efficiently in many cases, and algorithmic parameters can be set to change this behaviour for a given instance. Finally, most current solvers allow the user to add "specific" cuts for her own MILP, say SEC for BRP, which can also considerably improve the performances of the B&C.

Adding (in principle, exponentially many) cuts is not the only way to construct "tight" formulations. Another (in some sense "dual", but we cannot delve in the precise mathematical description of this concept) way is to work in a completely different space of variables, possibly having an exponential size in n. We now briefly illustrate the idea for our BRP case study.

In BRP, a solution consists of a subset of all the *feasible vehicle routes*; a feasible route $r \in \mathcal{R}$ is a directed (simple) cycle that starts from the depot, visits a subset of the nodes exactly once and returns to the depot, such that the vehicle load never exceeds the capacity Q and never becomes negative. With this definition, BRP can be recast as the problem of selecting a *feasible* subset $X \subset \mathcal{R}$ with minimum cost. Note that there are two "levels" of feasibility: that of the *routes*, and that of the *subsets of routes*. In other words, the crucial step is distinguishing the constraints that define the feasibility of the individual routes, from those that define the feasibility of the subset of routes as a whole; the former will be encapsulated in the concept of route, and therefore will not appear explicitly in the formulation. A subset $X \subset \mathcal{R}$ is feasible if each station node is visited by exactly one route, and the number of routes does not exceed the number of available vehicles m. Hence, we can define a (*set partitioning*) formulation of BRP using $|\mathcal{R}|$ binary variables (columns) $x_r \in \{0, 1\}$, one for each $r \in \mathcal{R}$:

$$\min \sum_{r \in \mathcal{R}} c_r x_r \qquad (12)$$
$$\sum_{r \in \mathcal{R} : i \in r} x_r = 1 \qquad i \in N' \qquad (13)$$
$$\sum_{r \in \mathcal{R}} x_r \leq m \qquad (14)$$
$$x_r \in \{0, 1\} \qquad r \in \mathcal{R} \qquad (15)$$

In the objective function (12), c_r is the cost of a feasible route $r \in \mathcal{R}$, i.e., just the sum of the costs of the arcs in route r. The *set partitioning* constraints (13) guarantee that each station is served by exactly one route; they can be written in terms of the subset $\mathcal{R}(i) \subset \mathcal{R}$ of the feasible routes that visit node i. Finally, (14) are the *cardinality constraints* on the fleet of vehicles. We remark, again, that the constraints describing a feasible route do not appear in the formulation, but are "implicit" in the definition of $\mathcal{R}$; this makes (12)–(15) remarkably simpler than (1)–(10), but at the cost of using a number of variables that is typically *exponential* in the size of the graph. Simplicity is, however, not the most relevant advantage of (12)–(15): it can be seen that the formulation is usually quite "tight", i.e., it produces (much) better lower bounds than (1)–(10). Thus, (12)–(15) would be in principle a better starting point to apply a B&B approach, it it were possible to efficiently solve its continuous relaxation, which is an LP with an exponential number of variables (columns).

Fortunately, this is, again, possible. Indeed, we have presented the approach already for LPs with an exponential number of constraints (rows), and it turns out that the exactly the same idea works here just by looking at the *dual* of the LP. Again we cannot go into the mathematical details; suffices here to say that, exactly as we can define *separation sub-problems* for generating new rows (constraints), we can define *pricing sub-problem* for generating new variables (columns). Starting with a RMP with a "small" number of initial columns, the pricing problem identifies columns with *negative reduced cost* that, if added to the RMP, may decrease its optimal value; if there is none, the current optimal RMP solution is also optimal for the LP with all the columns. In the BRP application, columns correspond to routes, and the pricing problem is a *Constrained Shortest Path* problem, i.e., the problem of finding the path (route) on the graph with minimal *reduced cost* (a modified cost taking into account the current *dual* optimal solution of the RMP). This is a "hard" problem in general, but typically solvable for the size of realistic instances; besides, different variants can be used (e.g., allowing or not the path to pass multiple times through a single node) that offer a trade-off between the complexity of the pricing problem and the quality of the corresponding LP bound.

Solving LPs with a very large (say, exponential) number of variables is known as *Column Generation*; a B&B where such LPs are solved at each node is also called a *Branch-and-Price* (B&P). Often, formulations amenable to CG are quite similar to in shape to (12)–(15), and are referred to as *set partitioning* formulations. Over the last twenty years, set partitioning formulations have become very popular for many combinatorial optimization problems, and the reason for this success is twofold. First, for some problems (e.g., crew pairing), there are basically no alternative formulations. Second, in many cases, these formulations provide quite strong lower bounds and therefore can be the basis of efficient solution procedures. Unfortunately, while B&C, comprised the possibility of adding problem-specific cuts, is implemented in basically all modern MILP solvers, B&P is much less supported; among the main MILP solvers, only SCIP [29] does B&P natively. However, B&P and B&C are not alternative; there is nothing conceptually preventing using valid inequalities in a formulation with a large number of variables. The all-singing, all-dancing algorithmic framework is therefore the so called Branch-and-Price-and-Cut (B&P&C), which can be considered the ultimate way of developing "strong" formulation. Such an approach cannot typically be implemented by inexperienced users, whereas B&C using only "general purpose" cuts only requires familiarity with an algebraic modelling language and use of a general-purpose solver as a "black box"; hence, development of "large", sophisticated formulations is typically not the first step, and it is only justified if the run-of-the-mill approach is not effective. Yet, it is important to realise that relatively simple tools are available that allow to construct a "simple" MILP formulation of one's application and quickly test it, and that these same tools support increasingly sophisticated approaches that can be used when strictly required, possibly with the help of OR experts.

5 Conclusions and Further Reading

Optimization problems arise in basically all facets of human activity, and in particular whenever the system to be managed is complex. This is almost by necessity the case in sharing economy applications, as only when the set of users is large, allowing to averaging out individuals' needs, effectively sharing resources is viable; in other words, sharing economy systems are necessarily complex. Unfortunately, most optimization problems are "difficult": the cost of their solution grows exponentially fast with the size of the problem (system). However, conceptual and software tools are nowadays available that may make it possible to solve optimization problems of the size required by applications in reasonably short times. The usual steps for using them are developing an appropriate system model, and then formulating the problem as a Mixed-Integer Linear one (although formulations using convex, maybe quadratic or conic, may also be viable). If a "strong" formulation is used, chances are that the problem is solved efficiently enough. If this does not happen, possible resorts are implementing ad-hoc heuristics, or seek assistance by OR experts to develop "larger" but "stronger" formulations. However, the first attempt can be done by relatively inexperienced users, and still have some chances of yielding satisfactory results.

Of course, optimization is a vast subject, and this primer has barely scratched the surface of the huge amount of theoretical and practical tools that have been developed for the solution of optimization problems, usually combining techniques from different areas of mathematics (linear algebra, geometry, discrete mathematics, graph theory, ...), and computer science. Several textbooks are available that present modern optimization methods in details, such as the classical [4] for B&C and [6] for CG/B&P approaches. Two other valuable recent reference works, among the many others, are [7] and [8]. While MILP is usually the go-to class for most applications, MINLPs can also be (carefully) considered; a recent useful reference is [5].

Acknowledgements The first author acknowledge the financial support of the University of Pisa under the grant PRA_2017_33 "Distretti urbani a zero impatto energetico ed ambientale". The authors acknowledge the financial support of the Italian Ministry for Education, Research and University (MIUR) under the project PRIN 2015B5F27W "Nonlinear and Combinatorial Aspects of Complex Networks" and of the Europeans Union's EU Framework Programme for Research and Innovation Horizon 2020 under the Marie Skłodowska-Curie Actions Grant Agreement No 764759 "MINOA – Mixed-Integer Non Linear Optimisation: Algorithms and Applications".

References

1. Miller C, Tucker A, Zemlin R (1960) Integer programming formulations and traveling salesman problems. J Assoc Comput Mach 7:32–329
2. Toth P, Vigo D (eds) (2002) The vehicle routing problem. SIAM
3. Toth P, Vigo D (eds) (2014) Vehicle routing: problems, methods, and applications. SIAM
4. Wolsey LA, Nemhauser GL (1999) Integer and combinatorial optimization. Wiley
5. Lee J, Leyffer S (Eds) (2012) Mixed integer nonlinear programming. The IMA volumes in mathematics and its applications. Springer
6. Desaulniers G, Desrosiers J, Solomon MM (eds) (2005) Column generation. Springer
7. Conforti M, Cornuejols G, Zambelli G (2014) Integer programming. Springer
8. Korte B, Vygen J (2018) Combinatorial optimization—theory and algorithms. Springer
9. Fliege J, Kaparis K, Khosravi B (2012) Operations research in the space industry. Eur J Oper Res 217(2):233–240
10. Galli L (2011) Combinatorial and robust optimisation models and algorithms for railway applications. 4OR 9(2):215–218
11. Boyd S, Vandenberghe L (1997) Semidefinite programming relaxations of non-convex problems in control and combinatorial optimization. In: Paulraj A, Roychowdhuri V, Schaper C (eds) Communications, computation, control and signal processing. Kluwer, pp 279–288
12. van Ackooij W, Lopez ID, Frangioni A, Lacalandra F, Tahanan M (2018) Large-scale unit commitment under uncertainty: an updated literature survey. Ann OR 271(1):11–85
13. Frangioni A, Galli L, Stea G (2014) Optimal joint path computation and rate allocation for real-time traffic. Comput J 58(6):1416–1430
14. AMPL. https://ampl.com/
15. GAMS. https://www.gams.com/
16. Coliop. http://www.coliop.org/
17. ZIMPL. http://zimpl.zib.de/
18. FlopC++. https://projects.coin-or.org/FlopC++
19. PuLP. https://pythonhosted.org/PuLP/

20. Pyomo. http://www.pyomo.org/
21. JuMP. https://github.com/JuliaOpt/JuMP.jl/
22. YALMIP. https://yalmip.github.io/
23. MPS. http://lpsolve.sourceforge.net/5.0/mps-format.htm
24. LP. http://lpsolve.sourceforge.net/5.1/lp-format.htm
25. Cplex. https://www.ibm.com/analytics/cplex-optimizer
26. GuRoBi. http://www.gurobi.com/
27. MOSEK. https://www.mosek.com/
28. Cbc. https://projects.coin-or.org/Cbc
29. SCIP. http://scip.zib.de/
30. LocalSolver. http://www.localsolver.com/
31. COIN-Reharse. https://github.com/coin-or/Rehearse

Networked Systems Theory: Distributed Algorithms for Optimal Cooperation of Dynamical Systems

Lucia Pallottino

Abstract Interconnection of a large number of systems has become a reality from a technological point of view although intelligent connection among smart systems still presents a challenge in several application scenarios. More specifically, when many smart devices must accomplish an overall goal based on information exchanges with other devices, several factors render this task a real challenge, such as, knowing what specific information to exchange, with whom, how to go about it, and when this exchange will occur. Moreover, interconnected devices may require access to or the use of common resources, making the management of the overall system even more complex. The management of shared resources usually sets the goal of optimizing usage while guaranteeing achievement of the overall goals. Algorithms or protocols that grant the device a correct, safe and optimized access to resources play a fundamental role. In this chapter we will provide those algorithms that are fundamental in several different application scenarios that allow for the development of more complex algorithms in order to manage interconnected systems in the management of shared resources.

Keywords Networked dynamic systems · Consensus algorithms · Distributed Optimization

1 Introduction

Nowadays concepts such as "Internet of Things" have gained much attention from both the research and the industrial communities. For example, the Industrial Internet of Things (IIoT) has expanded the concept of different objects connected through a network to a much larger scale in the Industry 4.0 framework. Industrial machines equipped with sensors are interconnected with the employees across the supply and

---

L. Pallottino (✉)
Research Center "E. Piaggio", Dipartimento di Ingegneria dell'Informazione, University of Pisa, Pisa, Italy
e-mail: lucia.pallottino@unipi.it

© Springer Nature Switzerland AG 2020
E. Crisostomi et al. (eds.), *Analytics for the Sharing Economy:
Mathematics, Engineering and Business Perspectives*,
https://doi.org/10.1007/978-3-030-35032-1_3

delivery chains in order to monitor production and quickly respond to or predict possible malfunctions. Goods production may also be monitored to ensure timely consumer delivery by preventing potential breakdowns and subsequent need for additional supplies. Hence the overall technology will improve efficiency by minimizing unnecessary expenses and by maximizing quality. However, interconnected devices require access to and the use of common resources creating many challenges, as identified in the sharing economy concept. Nevertheless, the full potential and range of applications of this technology have not yet been exploited. So the more this technology will be made available to large numbers of people, the more ideas and applications will be discovered that will change the way we work and live in the near future. Indeed, the interconnection of smart devices, with possibly shared resources, are already used in many other scenarios beyond those of industrial applications such as online trading, cryptocurrency exchange, social networking, sensor networks, energy networks, traffic control systems, electricity distribution management, or biological systems.

Such connectivity of smart devices must be designed and controlled to make it as intelligent as possible and to manage the common resources by organizing and optimizing their use. Modelling this kind of networks is fundamental to understanding how the overall system will operate or to understand how they can be designed and developed. In this chapter we will focus on the case of networked agents (or multi-agent systems) with common resources to be properly shared in order to reach a goal for the overall system. We consider networks of devices that exchange information in order to organize themselves, to plan, optimize or learn. Such devices can therefore be processors, sensors, actuators, PCU's, Robots, or electronic supply. In order to abstract from the particular device, a more formal definition of "agent" is necessary. Several definitions of agents can be found in the literature, such as the one proposed by Wooldridge in [1]: "An agent is a computer system that is situated in some environment and is capable of autonomous action in this environment in order to meet its design objective", or a similar one proposed by Franklin and Graesser in [2]:"An autonomous agent is a system situated within and a part of an environment that senses that environment and acts on it, over time, in pursuit of its own agenda and so as to effect what it senses in the future". In this context, agents are (1) autonomous, in that they can make their own decisions, (2) reactive and proactive in that they act based on available information to achieve a particular objective, and (3) interactive, in that they exchange information with other agents to collaborate towards common goals. In the presence of shared resources, the coordination and optimization of their use is fundamental, such as, for example, autonomous vehicles that reach an intersection that must coordinate in order to avoid collisions and therefore must have exclusive access to the resource, or access to a shared database from different CPUs. The motivations behind using interconnected agents as opposed to a single one are many. First, complex tasks may be accomplished more quickly by using parallel computation (or generally, parallel executions) and by reducing complexity. Second, a single point of decision may be prone to faults and hence agent interconnection increases system robustness. Finally, agents with different characteristics and abilities may be interconnected allowing the use of simpler, or smaller, lighter, and cheaper agents while maintaining the complexity of the overall system. This agent interconnection poses new challenges such as managing the shared resources, the network, the communication, the possible high number of agents, and the agent's heterogeneity, to name a few. The aim of this chapter is to provide the fundamentals for two of the many problems that emerge in networked systems where there is no centralized decision making unit. The first is the consensus problem where, based on limited information exchange, the interconnected agents must agree on a variable. Basic definitions are introduced together with a simple application to formation control even though the range of applications of such algorithms is very wide. The second problem we consider in this chapter is the solving of an optimization problem using a distributed approach, thereby decoupling the problem into subproblems solved by agents through information exchange in order to reach the overall optimal solution.

2 Background

In this chapter we introduce the notation, definitions and fundamental concepts for the modeling and control of interconnected agent systems. Graphs are a good tool to model agents interactions and shared resources since they offer a sufficient level of abstraction while keeping essential characteristic of the networked system. Indeed, agents are generic entities (such as sensors, CPUs, robots, things in the Internet of Things) that can, in general, interact with others and with the environment (by collecting information or physically modifying the environment) and possibly move in it. More importantly, graphs can capture the agents interactions through the arcs concept. In other words, we associate an arc between two nodes if they represent two agents that are able to exchange information, to interact (such as in social networks) or they share a common resource. An arc can even model the fact that an agent decision directly influence the decision of the other agent (such as in social and economic systems).

In case of a sensor network, the abstraction allows to neglect the geometry of the sensors whose footprints can be very complex and usually can not be formally written with closed form equations, see e.g. Fig. 1. In this case, we may represent with a graph the fact that the agent R equipped with a radar is able to detect the relative position of another agent S, that in turn is able to detect the relative position of another agent C. The agent C equipped with a camera does not detect the presence of any other agent. We decode this information structure in a graph where the arc

Fig. 1 Example of sensors' footprints

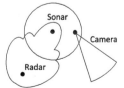

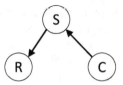

Fig. 2 Abstraction of sensors with graph

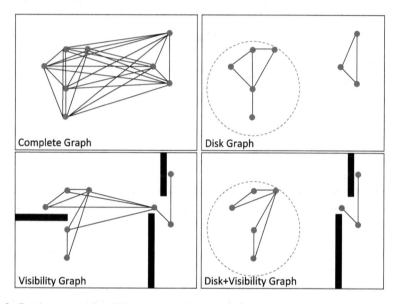

Fig. 3 Graphs representing different communication network typologies

from node S to node R represents the fact that the information of the position of S is available to R and so on, see e.g. Fig. 2.

With such abstraction some characteristics of the system are lost in the model, however several properties of interconnected agents behaviours may be derived from the topology of the network as shown next. The graph models the interconnections and interdependencies between agents and its physical interpretation strongly depends on the particular application. For example, referring to Fig. 3, agents may be interconnected with any other agent leading to a complete graph while, in case of limited range sensors or communication devices, a disk graph could be used where arcs are considered between nodes if the corresponding agents are closer than a given distance. In case of line of sight communication visibility graphs are appropriate, where arcs are considered only if the line of sight between nodes does not intersect any obstacle. More complex graphs can be considered through combination of others such as the combined visibility and disk graph reported in Fig. 3. It is worth noting that the interconnection between agent may also depend on the agent distances leading to a time varying topology (i.e. arcs can change in time) in case of moving agents.

Since interconnection is not necessarily symmetric, oriented graphs are used to represent the possible asymmetry of the network structure. In other words, a directed arc from node i toward node j is in the graph only if agent j has access to information regarding i or it is influenced by j. As mention we now, abstract form the fact that this is due to the sensing capability of j or if a message has been sent from i to j. In several scenarios, such as the one of networked mobile robots, different kind of networks can be sued from static (the sets of node and arcs are fixed), to dynamic (the sets of nodes and arcs are time-varying) or even random (network in case of probabilistic communication).

2.1 Fundamentals of Graph Theory

We now introduce several definitions on graphs in order to provide basic results that will allow the formal characterization of algorithms performance.

A **graph** $G = (V, E)$ is characterized by a finite set of nodes V ad a set of arcs $E \subset V \times V$ that connect pairs of nodes, such interconnected nodes are **adjacent**. Given $e \in E$, there exist v_i and $v_j \in V$ such that $e = (v_i, v_j)$. A **path** in G is a sequence of adjacent nodes.

A graph G is **connected** if for any pair of nodes there exists a path between the nodes. For example, the disk graph represented in Fig. 3 is not connected.

In case we are interested in modeling some costs for the information exchange (e.g. a delay) or the shared resource, a **weighted graphs** $G = (V, E, w)$ can be used where $w : E \to \mathbb{R}$ is a function that associates a value (or a cost) to any arc in E. The **cost of a path** is given by the sum of the costs of the arcs in the path.

In case of an asymmetric structure a **directed graph** (or digraph) can be used. For those graphs E is composed by ordered pairs of nodes. A **directed path** is a path formed by a sequence of ordered pairs of the form $\{\ldots, (v_i, v_j), (v_j, v_k), (v_k, v_m), \ldots\}$. A digraph is **strongly connected** if for any pair of nodes there exists a directed path between them.

In order to evaluate and quantify the properties of the networked system, other mathematical tools that capture the graph topology are necessary. For this purpose matrices can be associated to graphs by encoding the nodes interconnections. For a directed or undirected graph $G = (V, E)$ let $N_i = \{v_j \in V | (v_i, v_j) \in E\}$ be the set of nodes adjacent to node v_i. In case of undirected graph $j \in N_i \leftrightarrow i \in N_j$, that is not necessarily true in directed graphs. For a weighted directed or undirected graph the **degree** $d(v_i) = \sum_{v_j \in N_i} w((v_i, v_j))$ is the sum of the costs of all the arcs between v_i and its adjacent nodes.

Definition 1 The **Degree Matrix** of a graph $G = (V, E)$ is a $n \times n$ diagonal matrix, where n is the number of nodes, with diagonal element (i, i) equal to $d(v_i)$

Definition 2 The **Adjacency Matrix** of a graph $G = (V, E)$ is a $n \times n$ matrix A with elements

$$[A(G)]_{i,j} = a_{i,j} = \begin{cases} w((v_i, v_j)) & \text{if } (v_i, v_j) \in E \\ 0 & \text{otherwise.} \end{cases} \quad (1)$$

It is worth noting that the adjacency matrix encodes the existence of (possibly asymmetric) interconnections of the agents and their costs and hence can be not symmetric while in case of undirected graph the adjacency matrix is always symmetric. Another fundamental matrix is the Laplacian matrix:

Definition 3 The **Laplacian matrix** of a graph $G = (V, E)$ is $L(G) = \Delta(G) - A(G)$ where Δ and A are the degree and the adjacency matrices respectively.

It follows from the definition that for any Laplacian matrix (hence for any graph) the sum of the elements of each row is zero and hence the vector **1** of dimension n with elements equal to 1 is a right eigenvector associated to the eigenvalue $\lambda = 0$. In case of undirected graph the Laplacian is a symmetric and positive semi-definite matrix with real eigenvalues: $\lambda_1(G) \leq \lambda_2(G) \leq \cdots \leq \lambda_n(G)$ where $\lambda_1(G) = 0$. In case of directed graph eigenvalues symmetry of the Laplacian does not hold in general and eigenvalues can be complex numbers.

Theorems for the localization of eigenvalues such as the one by Gershgorin [3] can help in better characterizing the Laplacian eigenvalues:

Theorem 1 First Gershgorin's Theorem
Given $A \in \mathbb{C}^{n \times n}$, the eigenvalues lays in the union of Gershgorin disks centered in a_{ii} with radius $r_i = \sum_{j=1 j \neq i}^{n} |a_{ij}|$.

For the particular case of the Laplacian this implies that eigenvalues lay in the circle centered in $(-d_{max}(G), 0) \in \mathbb{C}$ with radius $d_{max}(G)$ where $d_{max}(G)$ is the maximum degree of the nodes. For the undirected case it hence holds: $\lambda_1(G) \leq \lambda_2(G) \leq \cdots \leq \lambda_n(G) \leq 2d_{max}(G)$. It is possible to prove that an undirected graph G is connected if and only if $\lambda_2(G) > 0$. This is a first result that shows how the Laplacian eigenvalues play a crucial role in the properties of the graph and hence this is a powerful tool in the network design phase. Another result is that for an undirected graph the algebraic multiplicity of the 0 eigenvalue is equal to the number of connected components in G.

Unfortunately, for an undirected graph it is only possible to prove that if the graph G is strongly connected there is only one 0 eigenvalue. The viceversa is not true. However, based on those results it is possible to prove the convergence properties of distributed algorithms among agents in the network as show in next Section.

3 Consensus Protocol

Distributed consensus algorithms are based on local information exchange between agents in case a specific action is performed when a consensus (or an agreement) in the network is achieved. Such algorithms can be applied to a wide range of application domains such as Computer Science, Robotics, Economics, Social Networks, Smart Grids, Biology etc.. Examples of applications of distributed consensus algorithms are the coordination of autonomous vehicles (often referred to as Rendezvous protocols, see e.g. [4] for an overview of the consensus protocols in multi-agent coordination), multi-retailer inventory control (see e.g. [5]), control of communication networks congestions, blockchains (see e.g. "proof-of-work"), management of shared database systems, real-time online strategy game, genome wide epigenetic modifications studies, modeling of plant developments etc. Several consensus-like algorithms have been developed for the different applications taking also into account delays, nodes failures, communication failures etc..

For the sake of brevity, we focus on the consensus protocol for a set of n interconnected agents that want to reach a consensus on a real variable. In this case a state x_i is associated to each agent i and a consensus is reached when the vector $x = (x_1, \ldots, x_n)^T$ belongs to the span of vector $\mathbf{1}$ (i.e., all components are equal). In order to reach a consensus each node may update its state based on the information received from adjacent (or neighbouring) agents. Let consider the time continuous case in which each agent update its state as follow:

$$\dot{x}_i(t) = \sum_{j \in N(i)} w_{ij} \left(x_j(t) - x_i(t) \right), \quad i = 1, \ldots, n, \tag{2}$$

where $N(i)$ are the agents adjacent to agent i in a graph $G = (V, E, w)$ and $w_{ij} = w(v_i, v_j)$. Considering the overall system with state $x \in \mathbb{R}^n$ the consensus dynamics is

$$\dot{x}(t) = -L(G)x(t). \tag{3}$$

A detailed description of the consensus protocols, proofs of theorems and applications can be found e.g. in [6] and [7]. Let $x(0) = (x_1(0), \ldots, x_n(0))^T$ be the initial state, it is possible to prove that:

Theorem 2 *For a connected graph G, for any initial condition $x(0)$ the consensus protocol (3) converges toward a vector of the form $\alpha \mathbf{1}$, with convergence rate equal to $\lambda_2(G)$. Moreover, consensus is obtained on the average of initial conditions value, i.e. $\alpha = \frac{1}{n} \sum_i x_i(0)$*

This theorem states that connectivity is fundamental for consensus convergence but also that the second Laplacian eigenvalue plays a fundamental role in the velocity of convergence.

Theorem 3 *For a strongly connected graph G, for any initial condition $x(0)$, the consensus protocol (3) converges toward a vector of the form $\beta \mathbf{1}$. Moreover, consensus is obtained on $\beta = w_r w_l^T x(0)$ with $w_l^T w_r = 1$ where w_r and w_l are left and right Laplacian eigenvectors associated to the zero eigenvalue, respectively.*

It is worth mentioning that average consensus, i.e. $\beta = \frac{1}{n} \sum_i x_i(0)$, is possible in case of balanced strongly connected graphs.

It is also possible to define a discrete time consensus of the form $x_i(k+1) = x_i(k) + \epsilon \sum_{j \in N_i} a_{ij}(x_j(k) - x_i(k))$, with overall system dynamic $x(k+1) = Px(k)$ where $P = I - \epsilon L$ is the **Perron matrix** associated to the graph with Laplacian L and parameter ϵ. The convergence properties depend on the properties of P such as being primitive. Refer to [6] for more details.

In case the status x_i is the position of the robot i on the plane, the consensus protocol can be used to reach a consensus and hence to move the robots, characterised by a simple single integrator dynamics on both coordinates, toward a common point. This behaviour is known as *rendezvous* and can be used to let the robot maintain connectivity of the network avoiding them to loose connection with the rest of the team. Similar strategies can be used to deploy a team of mobile robot in an environment to be explored, to manage energy resources in microgrids, to manage agreement on the state of a blockchain, to name few.

3.1 Consensus for Mobile Robots Formation

To show an application of the consensus algorithm, consider a team of robots that must move maintaining a precise formation (e.g. a particular shape) expressed in terms of given relative distances they have to maintain pairwise. Let $x_i(t) \in \mathbb{R}^2$ be the 2D coordinates of the robot i at time t and let $G_f = (V, E_f)$ be the formation graph where V are the robots indexes and $e = (v_i, v_j) \in E_f$ if we want to reach a given distance $d_{i,j}$ between i and j, i.e. $\|x_i(t) - x_j(t)\|^2 \to d_{i,j}^2$ for $t \to \infty$. In case each robot can decide how to vary its position $x_i(t)$ based on local information we may use the consensus protocol described in previous section. Indeed, consider a set of 2D positions $z_i, i = 1, \ldots, n$, such that for each $e = (v_i, v_j) \in E_f$ it holds $\|z_i - z_j\|^2 = d_{i,j}^2$, i.e. if the robots have coordinates z_i the desired distances have been achieved. Let $\epsilon_i(t) = x_i(t) - z_i, i = 1, \ldots, n$, by applying the consensus protocol to the variables ϵ_i it holds $\dot{\epsilon}_i = -\sum_{j \in N_{f,i}} (\epsilon_i - \epsilon_j)$ where $N_{f,i} = \{j \in \{1, \ldots, n\} | (v_i, v_j) \in E_f\}$. In case of a connected graph G_f a consensus is obtained on the variables ϵ_i. In other words $\epsilon_i(t) \to \epsilon$ for $t \to \infty$. Concluding, considering the dynamics of the robot we have $\dot{x}_i(t) = \dot{\epsilon}_i(t)$ and $\epsilon_i - \epsilon_j = x_i - x_j - (z_i - z_j)$ leading to the system

$$\dot{x}_i = - \sum_{j \in N_{f,i}} (x_i - x_j - (z_i - z_j))$$

that steers the variables $x_i(t)$ toward $z_i + \epsilon$ hence the robots will move toward 2D positions that verify the distances specified by the formation graph G_f. Indeed, $\|z_i + \epsilon - (z_j + \epsilon)\|^2 = \|z_i - z_j\|^2 = d_{i,j}^2$. It is worth noting that, in case of mobile robots the communication network can be time variant possibly jeopardising the network connectivity. To cope with this problem strategies to maintain connectivities can be used, see e.g. [7, 8]. Another application of the consensus protocol is shown in next section where agents share resources that must be optimally used.

4 Distributed Optimization

Distributed optimization is a widespread research area that has faced a growing interest in several applications of networked systems such as management of shared resources, distributed estimation in sensor networks, distributed control of networked robots. The literature of both theory and applications is very large and is transversal with respect to several research communities. The reader may refer to [10–13] and literature there in to gain a wider knowledge of the approaches and algorithms available.

Consider the optimization problem $\min_{x_i \in C_i} \sum_{i=1}^{N} f_i(x_i)$, given the separable nature of the problem it is possible to let each agent i solving the problem $\min_{x_i \in C_i} f_i(x_i)$ in parallel and obtain the overall solution by summing up all the optimal values. Note that this is possible toward a simple consensus algorithm as the one reported in previous Section.

In this framework we are hence able to model a networked system where each agent has a private cost function that must be minimized with respect to some constrained variables and the goal of the overall system is to minimize the overall cost. Moreover, this problem can be straightforwardly distributed among agents. In case of resources shared among agents this problem become more complex since other than some **local** variables x_i each **cost function** f_i may depend also on the shared resources. In this case the problem $\min_{y, x_i \in C_i} \sum_{i=1}^{N} f_i(x_i, y)$ is not trivially separable anymore due to the presence of **public** (common or shared) variables y. In order to model the shared resources we can rely again upon graphs where each node represents an agent and arcs represent that two agents share a resource. However, with graphs it is not possible to model the fact that more than two agents share a common resource. Such cases can be modeled with the concept of **hypergraphs** where arcs can connect more than two nodes.

An example of an hypergraph is reported in Fig. 4. Referring to the figure, we have that agents 1 and 2 share a resource so as agents 2 and 4 and agents 2 and 3. Moreover, there exists also a shared resource among agents 1, 2 and 3. Let $G_H = (V, E_H)$ be the hypergraph where V has N nodes and each hyperarc $e_k \in E_H$ connects nodes in V representing a shared resource characterized by a variable $y_k \in \mathbb{R}$. Let n_j the number of arcs, in E_H, node j belongs to, or equivalently the number of resources agent j shares with others. Let r be the number of hyperarcs in E_H, i.e., the number

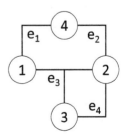

Fig. 4 Hypergraph for shared resources

of shared resources. The vector $y^T = (y_1, \ldots, y_r)^T \in \mathbb{R}^r$ is the **resource vector**. Let $E_k \in \mathbb{R}^{n_k \times r}$ be the matrix with rows equal to the p-th vectors of the canonical base where node k belongs to the p-th arc. Referring to the hypergraph in Fig. 4 we have $N = 4$ agents, $r = 4$ shared resources, $n_1 = n_3 = n_4 = 2$ nodes 1, 3, 4 share a common resource with other agents while $n_2 = 3$ hence agent 2 shares three resources. Finally, we have $E_1 = \begin{pmatrix} 1 & 0 & 0 & 0 \\ 0 & 0 & 1 & 0 \end{pmatrix}$, $E_2 = \begin{pmatrix} 0 & 1 & 0 & 0 \\ 0 & 0 & 1 & 0 \\ 0 & 0 & 0 & 1 \end{pmatrix}$, $E_3 = \begin{pmatrix} 0 & 0 & 1 & 0 \\ 0 & 0 & 0 & 1 \end{pmatrix}$, $E_4 = \begin{pmatrix} 1 & 0 & 0 & 0 \\ 0 & 1 & 0 & 0 \end{pmatrix}$. Let $z_k \in \mathbb{R}^{n_k}$ be a sort of local copies of the resource variables in y it holds $z_k = E_k y$. The optimization problem with shared resources can be written as

$$\begin{cases} \min_{y, x_i} \sum_{i=1}^{N} f_i(x_i, z_i) \\ \text{s. t. } x_i \in C_i \\ z_i = E_i y \end{cases} \qquad (4)$$

There are two typical approaches to solve the problem through a subdivision into subproblems each solved by an agent. The first is named the **primal approach** and is based on the fact that once a resource assignment y is provided to all agents, each agent must solve the following subproblem:

$$\begin{cases} \varphi_i(y) = \min_{x_i} f_i(x_i, z_i) \\ \text{s. t. } x_i \in C_i \\ z_i = E_i y \end{cases} \qquad (5)$$

The **master** problem to be solved is $\min_y \sum_{i=1}^{N} \varphi_i(y)$, i.e., the master problem is the one that decides the amount of resources assigned to agents. In case of convex functions, this problem can be solved with the iterative approach named **subgradient method** that is an extension of the classical gradient method in case of non differential functions, see e.g., [14, 15] for details. Consider a subgradient w_i of function φ_i in y, i.e. $w_i \in \partial \varphi_i(y)$ the iterative update of the resource vector is $y(j+1) = y(j) - \alpha_j \sum_{i=1}^{N} E_i^T w_i$ and, under some conditions on the update steps, the iterated variable converges toward the solution of the problem in (4). It is worth noting that also the master problem can be solved by each agent fully distributing the optimization. Indeed, once the initial resource assignement $y(0)$ and the update steps are available to the agents, the algorithm implemented by each agent is:

Different convergence conditions may be used to terminate the iterative algorithm in step 2, as well as different distributed algorithms can be used to share vectors $E_i^T w_i$ and compute w in step 6. It is worth noting that this approach allows different agents to solve an optimization subproblem (of smaller dimension) sharing few information with other agents. Indeed, each agent has no knowledge of the cost function f_i or the optimal values of local variables x_i of others. Only one of the subgradient must be shared with others without providing any further information that can hence be protected for security, privacy or confidentiality reasons. Theupdate of vector y is an

Algorithm 1: Primal Decomposition

Input: $y(0)$, α_j
Output: *Optimal resource assignement* y

1 Initialize $j = 0$
2 **while** *Convergence condition on* $y(j)$ *does not hold* **do**
3   Solve subproblem (5) obtaining $\varphi_i(y(j))$
4   Compute $w_i \in \partial \varphi_i(y(j))$
5   Communicate $E_i^T w_i$ to interconnected agents
6   Apply a consensus protocol to compute $w = \sum_{i=1}^N E_i^T w_i$
7     $y(j+1) = y(j) - \alpha_j w$
8     $j = j + 1$
9 **end**

iterative modification of the resources allocated to agents that converges toward to optimal resource assignment.

Another approach, often used in combination with the primal decomposition one, is the **dual decomposition method**. The dual problem is obtained considering the consistency constraint of the local copies of y among agents, i.e., given a node k we have n_k local copies of assigned resources with as many equality constraints. Hence, a Lagrangian multiplier $\nu_k \in \mathbb{R}^{n_k}$ can be used to embed the constraints in the cost function. Considering the vector ν of the Lagrangian multipliers, i.e., $\nu^T = (\nu_1, \ldots, \nu_N)^T$, and the matrix E with blocks of rows E_i with $i = 1, \ldots, N$, the dual problem is

$$\begin{cases} \max_\nu \inf_{x_i, y} \sum_i = 1^N f_i(x_i, z_i) + \nu_i^T z_i \\ \text{s. t. } x_i \in C_i \\ E^T \nu = 0 \end{cases} \quad (6)$$

The dual problem can be decomposed into subproblems each solved by an agent once a vector ν (with $E^T \nu = 0$) is provided:

$$\begin{cases} g_i(\nu_i) = \min_{x_i} f_i(x_i, z_i) + \nu_i^T z_i \\ \text{s. t. } x_i \in C_i \end{cases} \quad (7)$$

while the **master dual problem** is to maximize $g(\nu) = \sum_{i=1}^N g_i(\nu_i)$ with $E^T \nu = 0$, or equivalently to minimize $-g(\nu)$. Also in this case, the master dual problem can be solved with a subgradient method with the additional fact that the constraint $E^T \nu = 0$ must be ensured. For this reason we use the **projected subgradient method** by updating the Lagrangian multiplier so that the constraint is verified: given a subgradient w_j of $g(\nu(j))$ the iteration is $\nu(j+1) = P(\nu(j) + \alpha_j w_j)$ and P is the projection on the constraint $E^T \nu = 0$. It is possible to show that for the dual problem the projected subgradient method is $\nu(j+1) = \nu(j) + \alpha_j (z - E\tilde{y})$ where $\tilde{y} = (E^T E)^{-1} E^T z$. Note that in this case if $E^T \nu(j) = 0$ also $E^T \nu(j+1) = 0$ and constraint is verified at each iteration step. Under some conditions on the update steps, the iterated variable converges toward the solution of the dual problem in

(6). Also in the dual decomposition approach, the master problem can be distributed among agents. Once the initial resource price $\nu(0)$, with $E^T\nu(0) = 0$ (e.g. $\nu(0) = 0$) and the update steps are available to the agents the algorithm implemented by each agent is reported in Algorithm 2.

Algorithm 2: Dual Decomposition

Input: $\nu(0)$, α_j
Output: *Optimal resource price ν*

1 Initialize $j = 0$
2 **while** *Convergence condition on $\nu(j)$ does not hold* **do**
3 $\quad$ Solve subproblem (7) obtaining optimal $z_i(\nu(j))$
4 $\quad$ Communicate $z_i(\nu(j))$ to interconnected agents
5 $\quad$ Compute vector $z(\nu(j))^T = (z_1(\nu(j)), \ldots, z_N(\nu(j)))^T$
6 $\quad$ $\nu(j+1) = \nu(j) + \alpha_j(I - E(E^T E)^{-1} E^T)z$
7 $\quad$ $j = j+1$
8 **end**

The Lagrangian vector ν can be seen as the price of the resources that are updated based on their use. In these terms the subproblems are solved taking into account the costs or incomes of the used resource. Prices are iteratively modified (without becoming negative) as follows: price are increased in case of over-used resources and decreased otherwise toward an equilibrium among the agents sharing the resource.

Further reading for different distributed optimization algorithms can be found in [16–19] While potential applications of distributed optimization algorithms can be found in [20] where the foraging behavior of bacteria is simulated, in [21] in case of formation control, in [22] in case of optimal task assignment to robots taking into account the robots' dynamics and in [23] for applications to large scale machine learning.

5 Conclusions

This chapter faces the problem of coordinating the overall behaviour of interconnected dynamical systems also in presence of shared resources. The topic is broad, it has a wide range of application domains and of proposed approaches based on different methodologies. In the particular framework of distributed algorithms, two main classes of methodology are considered in this chapter: the Consensus protocol and the distributed optimization. Even though those topics do not cover the whole complexity and extent of the considered problem, the goal of this chapter is to describe the main challenges that must be taken into account and some typical approaches used in literature. Fundamentals and examples are reported as a starting point for detailed studies. If interested in the topics, the reader may refer to the cited literature and the works cited therein.

References

1. Wooldridge MJ (2002) An introduction to multi-agent systems. John Wiley, New York, USA
2. Franklin S, Graesser A (1996) It is, an agent, or just a program?: A taxonomy for autonomous agents. In: Müller JP, Wooldridge MJ, Jennings NR (eds) Intelligent agents III agent theories, architectures, and languages. ATAL 1996. Lecture notes in computer science (Lecture notes in artificial intelligence), vol 1193. Springer, Berlin
3. Varga RS (2004) Geršgorin and his circles. In: Springer series in computational mathematics, vol 36. Springer, Berlin
4. Ren W, Beard RW, Atkins EM (2005) A survey of consensus problems in multi-agent coordination. In: Proceedings of the 2005, American control conference, vol. 3, pp 1859–1864
5. Bauso D, Giarré L, Pesenti R (2004) Neuro-dynamic programming for cooperative inventory control. In: Proceedings of the IEEE 2004 American control conference, vol. 6, pp 5527–5532
6. Olfati-Saber R, Murray RM (2004) Consensus problems in networks of agents with switching topology and time-delays. IEEE Trans Autom Control 49(9):1520–1533
7. Mesbahi M, Egerstedt M (2010) Graph theoretic methods in multiagent networks, vol 33. Princeton University Press
8. Sabattini L, Chopra N, Secchi C (2013) Decentralized connectivity maintenance for cooperative control of mobile robotic systems. Int J Robot Res 32(12):1411–1423
9. Ji M, Egerstedt M (2006) Distributed formation control while preserving connectedness. In: Proceedings of the 45th IEEE conference on decision and control. San Diego, pp 5962–5967
10. Yang B, Johansson M (2011) Distributed optimization and games: a tutorial overview, Networked control systems. pp 109-148
11. Nedic A, Ozdaglar A, Parrilo PA (2010) Constrained consensus and optimization in multi-agent networks. IEEE Trans Autom Control 55(4):922–938
12. Notarstefano G, Bullo F (2011) Distributed abstract optimization via constraints consensus: theory and applications. IEEE Trans Autom Control 56(10):2247–2261
13. Bertsekas DP, Tsitsiklis JN (1997) Parallel and distributed computations: numerical methods. Athena Scientific, Belmont, Massachusetts
14. Bertsekas DP (2015) Convex optimization algorithms, 2nd edn. Athena Scientific, Belmont, Massachusetts
15. Boyd S, Vandenberghe L (2004) Convex optimization. Cambridge University Press, New York, USA
16. Bürger M, Notarstefano G, Allgöwer F (2014) A polyhedral approximation framework for convex and robust distributed optimization. IEEE Trans Autom Control 59(2):384–395
17. Zhu M, Martínez MS (2012) On distributed convex optimization under inequality and equality constraints. IEEE Trans Autom Control 57(1):151-164
18. Bürger M, Notarstefano G, Bullo F, Allgöwer F (2012) A distributed simplex algorithm for degenerate linear programs and multi-agent assignments. Automatica 48(9):2298–2304
19. Zhu M, Martínez S (2015) Distributed optimization-based control of multi-agent networks in complex environments. In: SpringerBriefs in control, automation and robotics. Springer International Publishing
20. Passino KM (2002) Biomimicry of bacterial foraging for distributed optimization and control. IEEE Control Syst Mag 22(3):52–67
21. Raffard RL, Tomlin CJ, Boyd SP (2004) Distributed optimization for cooperative agents: application to formation flight. In: 43rd IEEE conference on decision and control, vol. 3, pp 2453–2459
22. Settimi A, Pallottino L (2013) A subgradient based algorithm for distributed task assignment for heterogeneous mobile robots. In: 52nd IEEE conference on decision and control, pp 3665–3670
23. Tsianos KI, Lawlor S, Rabbat MG (2012) Consensus-based distributed optimization: practical issues and applications in large-scale machine learning. In: 50th annual allerton conference on communication, control, and computing (Allerton), Monticello, IL, pp 1543–1550

On Distributed Generalized Nash Equilibrium Seeking

Sergio Grammatico

Abstract Generalized Nash equilibrium problems describe multi-agent systems where each decision maker, or agent, aims at minimizing its individual cost function, yet all are subject to shared, coupling constraints. Distributed algorithms represent viable solution methods for solving generalized Nash equilibrium problems, since they require the agents to optimize and communicate locally, besides agree on the shared resources with selected other agents. The design of efficient solution algorithms is however extremely challenging from a theoretical perspective. In this chapter, we show that operator theory offers the appropriate mathematical tools to design distributed solution methods for generalized Nash equilibrium problems.

Keywords Generalized nash equilibrium problem · Monotone operator theory · Operator splitting

1 Introduction

1.1 Sharing in Multi-Agent Cooperative and Noncooperative Systems

The road traffic of the future is perhaps the most relevant example of multi-agent system with shared infrastructure. Conventional, autonomous and semi-autonomous vehicles will in fact drive towards their destinations while sharing the road space-time. From a mathematical perspective, those vehicles can be modeled as intelligent,

---

This work was partially supported by the Netherlands Organization for Scientific Research (NWO) under research projects OMEGA (grant n. 613.001.702) and P2P-TALES (grant n. 647.003.003).

S. Grammatico (✉)
Delft Center for Systems and Control (DCSC), Faculty of Mechanical, Maritime and Materials Engineering (3mE), Delft University of Technology, Delft, Netherlands
e-mail: s.grammatico@tudelft.nl
URL: https://sites.google.com/site/grammaticosergio

© Springer Nature Switzerland AG 2020
E. Crisostomi et al. (eds.), *Analytics for the Sharing Economy:
Mathematics, Engineering and Business Perspectives*,
https://doi.org/10.1007/978-3-030-35032-1_4

cooperative or noncooperative, decision makers, each with an individual cost function, e.g. a trade-off between minimum travel time and fuel consumption, subject to local constraints, e.g. physical limits on velocity and acceleration, and most importantly, subject to shared constraints, e.g. safety distance constraints, due to the interaction with other intelligent vehicles [6]. Another relevant example of sharing in multi-agent systems is the power grid of the future, where distributed generators, storage units, consumers and prosumers will act seeking their individual interest while sharing the capacity and the flexibility of the grid and possibly competing on ad-hoc energy markets [23].

A collection of inter-dependent optimization problems associated with (noncooperative) decision makers, or agents, each with an individual aim, i.e., cost function and constraints, is—mathematically speaking—a game. The inter-dependence lies in the fact that the cost function and the constraints of an agent may be affected not only by local decision variables, but also by decision variables of other agents. Game theoretic problems are motivated by several societal challenges, such as demand side management in the smart grid [23], e.g. for charging/discharging electric vehicles [8, 17, 18, 21], demand-response regulation in competitive markets [15], congestion control in traffic and communication networks [1, 9]. The common denominator is the presence of a large number of agents, possibly interacting via a communication network, whose actions may disrupt the shared infrastructure, e.g. the power grid or the transportation network, if proper coordination and control mechanisms are not in place.

1.2 Computational Game Theory and Monotone Operator Theory

Designing solution methods for multi-agent equilibrium problems has been gaining high research interest in the last years. In fact, several researchers have been developing distributed equilibrium seeking algorithms for games without coupling constraints [11, 12, 14, 24], even in the stochastic case [13], and more recently for games with shared coupling constraints [3, 9].

With focus on the generalized Nash equilibrium problem (GNEP) [7], [20, Chap. 12], the formulations in [3, 25] have introduced an elegant approach based on monotone operator theory [2] to characterize the equilibrium solutions as the zeros of the sum of two monotone operators.

1.3 Why Operator Theory?

Operator theory represents perhaps the most general framework to model and solve equilibrium problems. For instance, unlike variational inequalities, smoothness of the problem data, such as the cost functions, is not required. Most importantly, operator theory supports the design of solution methodologies that are computationally viable, e.g. operator splitting for monotone inclusion problems [2, Sect. 26].

Essentially, convex optimization and monotone equilibrium seeking problems are special instances of monotone inclusion problems [22]. It follows that most known centralized and distributed solution algorithms are the outcome of suitable operator splitting methods. For example, the asymmetric projection algorithm (APA) is nothing but a preconditioned forward-backward (p-FB) splitting algorithm [4] and the celebrated alternating direction method of multipliers (ADMM) is nothing but a Douglas–Rachford (DR) splitting algorithm.

Basic Notation

$\mathbb{R}$ denotes the set of real numbers, and $\overline{\mathbb{R}} := \mathbb{R} \cup \{\infty\}$ the set of extended real numbers. **0** (**1**) denotes a matrix/vector with all elements equal to 0 (1); to improve clarity, we may add the dimension of these matrices/vectors as subscript. $A \otimes B$ denotes the Kronecker product between matrices A and B; $\|A\|$ denotes the maximum singular value of A. Given N vectors $x_1, \ldots, x_N \in \mathbb{R}^n$, $\boldsymbol{x} := \operatorname{col}(x_1, \ldots, x_N) = \left[x_1^\top, \ldots, x_N^\top\right]^\top$.

Operator Theoretic Background

Id$(\cdot)$ denotes the identity operator. The mapping $\iota_S : \mathbb{R}^n \to \{0, \infty\}$ denotes the indicator function for the set $S \subseteq \mathbb{R}^n$, i.e., $\iota_S(x) = 0$ if $x \in S$, ∞ otherwise. For a closed set $S \subseteq \mathbb{R}^n$, the mapping $\operatorname{proj}_S : \mathbb{R}^n \to S$ denotes the projection onto S, i.e., $\operatorname{proj}_S(x) = \operatorname{argmin}_{y \in S} \|y - x\|$. The set-valued mapping $\mathrm{N}_S : \mathbb{R}^n \rightrightarrows \mathbb{R}^n$ denotes the normal cone operator for the the set $S \subseteq \mathbb{R}^n$, i.e., $\mathrm{N}_S(x) = \varnothing$ if $x \notin S$, $\{v \in \mathbb{R}^n \mid \sup_{z \in S} v^\top(z - x) \leq 0\}$ otherwise. For a function $\psi : \mathbb{R}^n \to \overline{\mathbb{R}}$, $\operatorname{dom}(\psi) := \{x \in \mathbb{R}^n \mid \psi(x) < \infty\}$; $\partial \psi : \operatorname{dom}(\psi) \rightrightarrows \mathbb{R}^n$ denotes its subdifferential set-valued mapping, defined as $\partial \psi(x) := \{v \in \mathbb{R}^n \mid \psi(z) \geq \psi(x) + v^\top(z - x)$ for all $z \in \operatorname{dom}(\psi)\}$; A set-valued mapping $\mathcal{F} : \mathbb{R}^n \rightrightarrows \mathbb{R}^n$ is ℓ-Lipschitz continuous, with $\ell > 0$, if $\|u - v\| \leq \ell \|x - y\|$ for all $x, y \in \mathbb{R}^n$, $u \in \mathcal{F}(x)$, $v \in \mathcal{F}(y)$; $\mathcal{F}$ is (strictly) monotone if $(u - v)^\top(x - y) \geq (>) 0$ for all $x \neq y \in \mathbb{R}^n$, $u \in \mathcal{F}(x)$, $v \in \mathcal{F}(y)$; $\mathcal{F}$ is η-strongly monotone, with $\eta > 0$, if $(u - v)^\top(x - y) \geq \eta \|x - y\|^2$ for all $x \neq y \in \mathbb{R}^n$, $u \in \mathcal{F}(x)$, $v \in \mathcal{F}(y)$; $\mathcal{F}$ is η-averaged, with $\eta \in (0, 1)$, if $\|\mathcal{F}(x) - \mathcal{F}(y)\|^2 \leq \|x - y\|^2 - \frac{1-\eta}{\eta} \|(\operatorname{Id} - \mathcal{F})(x) - (\operatorname{Id} - \mathcal{F})(y)\|^2$, for all $x, y \in \mathbb{R}^n$; $\mathcal{F}$ is β-cocoercive, with $\beta > 0$, if $\beta \mathcal{F}$ is $\frac{1}{2}$-averaged. With $\mathrm{J}_\mathcal{F} := (\operatorname{Id} + \mathcal{F})^{-1}$, we denote the resolvent operator of $\mathcal{F}$, which is $\frac{1}{2}$-averaged if and only if $\mathcal{F}$ is monotone; $\operatorname{fix}(\mathcal{F}) := \{x \in \mathbb{R}^n \mid x \in \mathcal{F}(x)\}$ and $\operatorname{zer}(\mathcal{F}) := \{x \in \mathbb{R}^n \mid 0 \in \mathcal{A}(x)\}$ denote the set of fixed points and of zeros, respectively.

2 Generalized Nash Equilibrium Problem Setup

2.1 Mathematical Background

We consider a set of N autonomous agents, where each agent $i \in \mathcal{N} := \{1, 2, \ldots, N\}$ shall choose its decision variable (i.e., strategy) x_i from the local decision set $\Omega_i \subseteq \mathbb{R}^n$ with the aim of minimizing its local cost function $(x_i, \boldsymbol{x}_{-i}) \mapsto J_i(x_i, \boldsymbol{x}_{-i}) : \mathbb{R}^n \times \mathbb{R}^{n(N-1)} \to \overline{\mathbb{R}}$, which depends on both the local variable x_i (first argument) and on the decision variables of the other agents, $\boldsymbol{x}_{-i} = \operatorname{col}\left(\{x_j\}_{j \neq i}\right)$ (second argument).

Furthermore, we consider *generalized* games, where the coupling among the agents arises not only via the cost functions, $\{J_i\}_{i \in \mathcal{N}}$, but also via their feasible decision sets. In particular, we describe the coupling constraints via an affine function, $x \mapsto Ax - b = \sum_{i \in \mathcal{N}} \{A_i x_i\} - b$, where $A \in \mathbb{R}^{m \times nN}, b \in \mathbb{R}^m, A_i \in \mathbb{R}^{m \times n}$ and $A = [A_1, \ldots, A_N]$. Thus, the collective feasible set, $\mathcal{X} \subseteq \mathbb{R}^{nN}$, reads as

$$\mathcal{X} := \boldsymbol{\Omega} \cap \left\{ y \in \mathbb{R}^{nN} \mid Ay - b \leq \mathbf{0}_m \right\}. \tag{1}$$

It follows that the feasible decision set of each agent $i \in \mathcal{N}$ is characterized by the set-valued mapping $\mathcal{X}_i$, defined as $\mathcal{X}_i(\boldsymbol{x}_{-i}) := \left\{ y_i \in \Omega_i \mid A_i y_i \leq b - \sum_{j \neq i}^N A_j x_j \right\}$.

The set Ω_i represents the local decision set for agent i, while the matrix A_i defines how the decision variable of each agent i is involved in the coupling constraints. For instance, the shared constraints in (1) may contain a sparsity pattern that can be defined via a graph, where each agent has a set of neighbors with whom to share some constraints.

Remark 1 (Affine coupling constraint) Affine coupling constraints as in (1) are common in generalized Nash games, see [9, 16, 19, 25]. Note that the more general case with separable convex coupling constraints can be reformulated as game with affine coupling constraints under mild regularity assumptions [10, Remark 2]. □

Then, let us postulate standard convexity and compactness assumptions for the constraint sets, and convexity of the cost functions with respect to their local decision variable.

Standing Assumption 1 (**Compact convex constraints**) *For each $i \in \mathcal{N}$, the set Ω_i is nonempty, compact and convex. The set $\mathcal{X}$ satisfies Slater's constraint qualification.* □

Standing Assumption 2 (**Convex functions**) *For each $i \in \mathcal{N}$ and $y \in \prod_{j \neq i} \Omega_j$, the function $J_i(\,\cdot\,, y)$ is convex.* □

2.2 Generalized Nash Equilibrium Problem

The aim of each agent $i \in \mathcal{N}$, given the decision variables of the other agents, $\boldsymbol{x}_{-i}$, is to choose a strategy, x_i, that solves its local optimization problem according to the game setup previously described, i.e.,

$$\begin{cases} \min_{x_i \in \Omega_i} & J_i(x_i, \boldsymbol{x}_{-i}) \\ \text{s.t.} & A_i x_i \leq b - \sum_{j \neq i}^{N} A_j x_j \end{cases} \quad \forall i \in \mathcal{N}. \tag{2}$$

From a game-theoretic perspective, this is the problem to compute a generalized Nash equilibrium, as formalized in the following definition.

Definition 1 (*Generalized Nash equilibrium*) A collective strategy $\boldsymbol{x}^*$ is a generalized Nash equilibrium (GNE) of the game in (2) if $\boldsymbol{x}^* \in \mathcal{X}$ and, for all $i \in \mathcal{N}$,

$$J_i(x_i^*, \boldsymbol{x}_{-i}^*) \leq \inf \{ J_i(y, \boldsymbol{x}_{-i}^*) \mid y \in \mathcal{X}_i(\boldsymbol{x}_{-i}^*) \}.$$

□

For each $i \in \mathcal{N}$, the set of optimal solutions to (2) is the so-called best-response mapping.

Definition 2 (*Best response mapping*) For each $i \in \mathcal{N}$, the set-valued mapping $\mathcal{B}_i : \mathbb{R}^{n(N-1)} \rightrightarrows \Omega_i$ defined as

$$\mathcal{B}_i(\boldsymbol{x}_{-i}) := \begin{cases} \operatorname*{argmin}_{x_i \in \Omega_i} & J_i(x_i, \boldsymbol{x}_{-i}) \\ \text{s.t.} & A_i x_i \leq b - \sum_{j \neq i}^{N} A_j x_j \end{cases} \tag{3}$$

is the best-response (BR) mapping of agent i. □

In plain words, a set of strategies is a generalized Nash equilibrium if no agent can improve its objective function by unilaterally changing its strategy to another feasible one. The problem to find such a set of strategies is known as generalized Nash equilibrium problem (GNEP).

Based on Definition 2, we note that a GNE, $\boldsymbol{x}^*$, is a fixed point of the Cartesian product of the BR mappings, i.e.,

$$\boldsymbol{x}^* \in \Pi_{i \in \mathcal{N}} \mathcal{B}_i(\boldsymbol{x}_{-i}^*),$$

or, in short, $\boldsymbol{x}^* \in \text{fix}\,(\Pi_{i \in \mathcal{N}} \mathcal{B}_i)$.

Under Assumptions 1–2, the existence of a GNE of the game in (2) follows from Brouwer's fixed-point theorem [20, Prop. 12.7]. Instead, uniqueness does not hold in general.

3 Operator Theoretic Characterization

In this section, we use optimization theory and operator theory to formulate the GNEP in (2) as a special operations research problem.

3.1 Karush–Kuhn–Tucker System

With the aim to decouple the affine coupling constraints in (2), let us first introduce the Lagrangian functions of the game, $(L_i)_{i \in \mathcal{N}}$, one for each agent $i \in \mathcal{N}$:

$$L_i(x_i, \boldsymbol{x}_{-i}, \lambda_i) := J_i(x_i, \boldsymbol{x}_{-i}) + \iota_{\Omega_i}(x_i) + \lambda_i^\top (A\boldsymbol{x} - b), \qquad (4)$$

where $\lambda_i \in \mathbb{R}^m$ is the dual variable of agent i.

Since each optimization problem in (2) is convex and Slater's constraint qualification holds, the Karush–Kuhn–Tucker (KKT) conditions are necessary and sufficient for optimality. The following result summarizes the equivalence between the set of solutions to the coupled KKT system and the set of GNE.

Lemma 1 ([20, Sect. 12.2.3]) *A strategy set $\boldsymbol{x}^*$ is a GNE of the game in (2) if and only if the following KKT conditions are satisfied for all $i \in \mathbb{N}[1, N]$:*

$$\mathrm{KKT}_i : \quad \begin{cases} 0 \in \partial_{x_i} J_i(x_i^*, \boldsymbol{x}_{-i}^*) + \partial_{x_i} \iota_{\Omega_i}(\boldsymbol{x}_{-i}^*) \\ 0 \leq \lambda_i \perp -(A\boldsymbol{x}^* - b) \geq 0. \end{cases} \qquad (5)$$

□

Note that the constraint qualification in Standing Assumption 1 ensures that the dual variables $\{\lambda_i\}_{i \in \mathcal{N}}$ are bounded [5, Sect. 5.2.3].

Now, to characterize the solutions to the KKT system in (5), we can introduce the mapping $\mathcal{T} : \mathbb{R}^{nN+mN} \rightrightarrows \mathbb{R}^{nN+mN}$, defined as

$$\mathcal{T}(\boldsymbol{x}, \boldsymbol{\lambda}) := \left(\Pi_{i \in \mathcal{N}} \partial_{x_i} L^i(\boldsymbol{x}, \lambda_i) \right) \times \left(\Pi_{i \in \mathcal{N}} N_{\mathbb{R}^m_{\geq 0}}(\lambda_i) - (A\boldsymbol{x} - b) \right), \qquad (6)$$

also called KKT operator [22], where $\boldsymbol{\lambda} = \mathrm{col}\,((\lambda_i)_{i \in \mathcal{N}}) \in \mathbb{R}^{mN}$ is the set of dual variables.

Lemma 2 *A set of strategies $\boldsymbol{x}^*$ is a GNE of the game in (2) if and only if $(\boldsymbol{x}^*, \boldsymbol{\lambda}^*) \in \mathrm{zer}(\mathcal{T})$, for some $\boldsymbol{\lambda}^* \in \mathbb{R}^{mN}_{\geq 0}$.* □

3.2 Monotone Games

In the reminder of the paper, we consider monotone games, i.e., we assume that the pseudo sub-differential mapping, so-called game mapping, is a monotone operator. Monotonicity in game theory, and more generally in equilibrium problems, is in fact the natural counterpart of convexity in optimization.

Definition 3 (*Game mapping*) The set-valued mapping $F : \mathbb{R}^{nN} \rightrightarrows \mathbb{R}^{nN}$ defined as

$$F(x) := \Pi_{i \in \mathcal{N}} \partial_{x_i} J_i(x_i, x_{-i}) \tag{7}$$

is the so-called *game mapping*, i.e., the *pseudo sub-differential mapping*, of the game in (2). □

Standing Assumption 3 *The game mapping F in (7) is monotone.* □

Next, we show that a GNE of the game in (2) is a zero of a set-valued mapping which is monotone if the game mapping in (7) is monotone.

Definition 4 (*Extended game mapping*) The set valued mapping $\widetilde{F} : \mathbb{R}^{nN+m} \rightrightarrows \mathbb{R}^{nN+m}$ defined as

$$\widetilde{F}(x, \lambda) := \begin{bmatrix} F(x) + A^\top \lambda \\ -(Ax - b) \end{bmatrix}, \tag{8}$$

with F as in (7), is the *extended game mapping* of the game in (2). □

Lemma 3 *The mapping $\widetilde{F}$ in (8) is monotone.* □

Proposition 1 *Define the mapping $T : \mathbb{R}^{nN+m} \rightrightarrows \mathbb{R}^{nN+m}$ as*

$$T(x, \lambda) := N_{\mathcal{X} \times \mathbb{R}^m_{\geq 0}}(x, \lambda) + \widetilde{F}(x, \lambda), \tag{9}$$

with $\widetilde{F}$ as in (8). The mapping T is monotone and $(x^*, \lambda^*) \in \mathrm{zer}(T)$, for some $\lambda^* \in \mathbb{R}^m_{\geq 0}$, if and only if $(x^*, (\lambda^*)_{i \in \mathcal{N}}) \in \mathrm{zer}(\mathcal{T})$. □

Clearly, a solution to the KKT system in (5) with the same dual variable is a GNE. The equivalence holds for so-called *variational equilibria*, see [7, Def. 3, Th. 3.1].

Theorem 1 ([3, Th. 1]) *A strategy set $x^* = (x_i^*)_{i \in \mathcal{N}}$ is a so-called variational GNE of the game in (2) if and only if there exists $\lambda^* \in \mathbb{R}^m_{\geq 0}$ such that (x_i^*, λ^*) satisfies the KKT system in (5), for all $i \in \mathcal{N}$.* □

In view of Lemma 1 and Theorem 1, the problem to find a GNE can be recast into the problem to find a zero of the monotone operator T in (9). The latter belongs to a class of long-studied problems in monotone operator theory [2].

4 Distributed Computation via Operator Splitting

4.1 Operator Splitting

Operator splitting methods are based on the equivalent characterization between the set of zeros of an operator, e.g. the KKT operator, and the set of fixed points of some other operator, typically, the operator that defines the equilibrium seeking algorithm. For example, we have that $\mathrm{zer}(\mathcal{T}) = \mathrm{fix}(\mathrm{Id} - \mathcal{T})$ and that $\mathrm{zer}(N_S + F) = \mathrm{fix}(\mathrm{proj}_S \circ (\mathrm{Id} - F))$, where N stands for the normal cone operator of the set S. The latter is called forward backward (FB) splitting [2, Sect. 26.5] and is known in smooth convex optimization, $\min_{y \in S} f(y)$, for generating the so-called projected-gradient algorithm, $y(k+1) = \mathrm{proj}_S (y(k) - \epsilon \nabla f(y(k)))$.

The starting point is to split the operator $\mathcal{T}$ in (6), or $\mathcal{T}$ in (9), into the sum of two monotone operators. The most natural choice is to consider $\mathcal{T} = \mathcal{A} + \mathcal{B}$, where $\mathcal{A} = N_{\mathcal{X} \times \mathbb{R}_{\geq 0}^{mN}}$ and $\mathcal{B}$ is the rest. Alternatively, one can move all the non-differentiable terms to $\mathcal{A}$ and the differentiable ones to $\mathcal{B}$, if these are separable.

Several operator splitting choices have been proposed in the literature besides the FB, e.g. the Douglas–Rachford (DR) [2, Sect. 26.7] splitting and the forward-backward-forward (FBF) [2, Sect. 26.6]. In the following subsections, we show the outcome of two variants of preconditioned FB (pFB) operator splitting.

4.2 Distributed Computation

Let us apply the pFB splitting to general network games with coupling constraints. We assume that the agents can communicate though a strongly connected graph with adjacency matrix $[a_{i,j}] \in \mathbb{R}^{N \times N}$. The outcome is the following algorithm [25, Alg. 1], where $z = [(z_i)_{i \in \mathcal{N}}]$ are auxiliary variables used for reaching consensus on the optimal dual variables λ.

Algorithm 1: Distributed pFB

$$\forall i \in \mathcal{N}: \quad x_i(k+1) = \mathrm{proj}_{\Omega_i}\left(x_i(k) - \alpha(\nabla_{x_i} J_i(x_i(k), \boldsymbol{x}_{-i}(k)) - A_i^\top \lambda_i(k))\right)$$

$$\forall i \in \mathcal{N}: \quad z_i(k+1) = z_i(k) + \alpha \sum_{j=1}^{N} a_{i,j} \left(\lambda_i(k) - \lambda_j(k)\right)$$

$$\forall i \in \mathcal{N}: \quad \lambda_i(k+1) = \mathrm{proj}_{\mathbb{R}_{\geq 0}^m} (\lambda_i(k) + \epsilon(2A_i x_i(k+1) - A_i x_i(k) - \tfrac{1}{N} b$$
$$+ \sum_{j=1}^{N} a_{i,j}(\lambda_i(k) - \lambda_j(k)) \quad (10)$$
$$+ 2(z_i(k+1) - z_j(k+1)) - (z_i(k) - z_j(k))))$$

The convergence of Algorithm 1 to a GNE is supported by the following technical statement.

Theorem 2 ([25, Th. 2]) *Assume that the game mapping F in (7) is strongly monotone and Lipschitz continuous, and that $[a_{i,j}]$ is the adjacency matrix of a strongly connected graph. The sequence $((x(k), \lambda(k)))_{k=0}^{\infty}$ defined by Algorithm 1, with step size $\alpha > 0$ small enough, globally converges to some $(x^*, \lambda^*) \in \mathrm{zer}(\mathcal{T})$, with $\mathcal{T}$ as in (6).* □

4.3 Semi-Decentralized Computation

In Algorithm 1, the agents shall exchange both primal variables, in order to compute $\partial_{x_i} J_i(x_i(k), x_{-i}(k))$ in $F(x(k))$, and dual variables, $\lambda(k), z(k)$. The exchange of the latter can be avoided if a supervisor agent is present that collects aggregate information on the coupling constraints, $Ax(k)$, and broadcasts the same dual variable $\lambda(k) \in \mathbb{R}^m$ to all agents. The following algorithm, the outcome of a tailored pFB splitting [4, Alg. 1], realizes that semi-decentralized computational scheme.

Algorithm 2: Semi-decentralized pFB

$$\forall i \in \mathcal{N}: \quad x_i(k+1) = \mathrm{proj}_{\Omega_i}\left(x_i(k) - \alpha(F(x(k)) + A^\top \lambda(k))\right)$$
$$\lambda(k+1) = \mathrm{proj}_{\mathbb{R}_{\geq 0}^m}\left(\lambda(k) + \alpha(2Ax(k+1) - Ax(k) - b)\right)$$

The convergence of Algorithm 2 to a GNE is supported by the following technical statement.

Theorem 3 ([4, Th. 1]) *Assume that the game mapping F in (7) is strongly monotone and Lipschitz continuous. The sequence $((x(k), \lambda(k)))_{k=0}^{\infty}$ defined by Algorithm 2, with step size α small enough, globally converges to some $(x^*, \lambda^*) \in \mathrm{zer}(\mathcal{T})$, with $\mathcal{T}$ as in (9).* □

5 Conclusion

Operator theory is a powerful framework to support and develop computational game theory via elegant and effective mathematical tools. In this chapter, we have adopted an operator theoretic perspective and briefly sketched the basic steps to solve a generalized Nash equilibrium problem via distributed algorithms.

References

1. Barrera J, Garcia A (2015) Dynamic incentives for congestion control. IEEE Trans Autom Control 60(2):299–310
2. Bauschke HH, Combettes PL (2010) Convex analysis and monotone operator theory in Hilbert spaces. Springer
3. Belgioioso G, Grammatico S (2017) Semi-decentralized Nash equilibrium seeking in aggregative games with coupling constraints and non-differentiable cost functions. IEEE Control Syst Lett 1(2):400–405
4. Belgioioso G, Grammatico S (2018) Projected-gradient algorithms for generalized equilibrium seeking in aggregative games are preconditioned forward-backward methods. In: Proceedings of the European control conference
5. Boyd S, Vandenberghe L (2004) Convex optimization. Cambridge University Press
6. Fabiani F, Grammatico S (2018) A mixed-logical-dynamical model for automated driving on highways. In: Proceedings of the IEEE conference on decision and control. Miami, USA
7. Facchinei F, Kanzow C (2007) Generalized Nash equilibrium problems. 4OR 5(3):173–210
8. Grammatico S (2016) Exponentially convergent decentralized charging control for large populations of plug-in electric vehicles. In: Proceedings of the IEEE conference on decision and control. Las Vegas, USA
9. Grammatico S (2017) Dynamic control of agents playing aggregative games with coupling constraints. IEEE Trans. on Automatic Control 62(9):4537–4548
10. Grammatico S Proximal dynamics in multi-agent network games. IEEE Trans Control Netw Syst (in press)
11. Grammatico S, Parise F, Colombino M, Lygeros J (2016) Decentralized convergence to Nash equilibria in constrained deterministic mean field control. IEEE Trans Autom Control 61(11):3315–3329
12. Kannan A, Shanbhag U (2012) Distributed computation of equilibria in monotone Nash games via iterative regularization techniques. SIAM J Optim 22(4):1177–1205
13. Koshal J, Nedic A, Shanbhag U (2013) Regularized iterative stochastic approximation methods for stochastic variational inequality problems. IEEE Trans Autom Control 58(3):594–609
14. Koshal J, Nedić A, Shanbhag U (2016) Distributed algorithms for aggregative games on graphs. Oper Res 64(3):680–704
15. Li N, Chen L, Dahleh M (2015) Demand response using linear supply function bidding. IEEE Trans Smart Grid 6(4):1827–1838
16. Liang S, Yi P, Hong Y (2017) Distributed Nash equilibrium seeking for aggregative games with coupled constraints. Automatica 85:179–185
17. Ma Z, Callaway D, Hiskens I (2013) Decentralized charging control of large populations of plug-in electric vehicles. IEEE Trans Control Syst Technol 21(1):67–78
18. Ma Z, Zou S, Ran L, Shi X, Hiskens I (2016) Efficient decentralized coordination of large-scale plug-in electric vehicle charging. Automatica 69:35–47
19. Paccagnan D, Gentile B, Parise F, Kamgarpour M, Lygeros J (2016) Distributed computation of generalized Nash equilibria in quadratic aggregative games with affine coupling constraints. In: Proceedings of the IEEE conference on decision and control. Las Vegas, USA
20. Palomar D, Eldar Y (2010) Convex optimization in signal processing and communications. Cambridge University Press
21. Parise F, Colombino M, Grammatico S, Lygeros J (2014) Mean field constrained charging policy for large populations of plug-in electric vehicles. In: Proceedings of the IEEE conference on decision and control. Los Angeles, USA, pp 5101–5106
22. Ryu E, Boyd S (2016) Primer on monotone operator methods. Appl Comput Math 15(1):3–43
23. Saad W, Han Z, Poor H, Başar T (2012) Game theoretic methods for the smart grid. IEEE signal processing magazine pp 86–105

24. Salehisadaghiani F, Pavel L (2018) Distributed nash equilibrium seeking in networked graphical games. Automatica 87:17–24
25. Yi P, Pavel L (2017) A distributed primal-dual algorithm for computation of generalized nash equilibria via operator splitting methods. In: Proceedings of the IEEE conference on decision and control. Melbourne, Australia, pp 3841–3846

Queueing Theory in the Context of Shared Resources

Christopher King

Abstract In many sharing economy applications, it may happen that the request for a given resource exceeds the available supply. One way to address this problem is by regulating access to the resource through a queueing system. This chapter offers a primer on queueing theory with focus on sharing economy systems. The chapter has been organized so that the mathematical development is linked to concrete examples.

Keywords Queueing theory · Charging of electric vehicles · Car sharing

1 Introduction

It may happen that the demand for a shared resource exceeds the available supply. There are some obvious ways to address this problem, for example by controlling access through a lottery system or by using a first-come, first-served regimen. However if the resource is renewable, and if clients are willing to delay their access to the resource, then another option is to employ some kind of queueing system which will provide access for everyone who needs it, albeit with some added delays. In order to design such a system it is necessary to build a mathematical model using tools from queueing theory. Queueing theory is a well-developed field that can be used to design and analyze models for controlling access to a shared resource (for example see [3] where a queueing model is used to analyze the optimal design of a shared fleet of electric vehicles), and the presentation in this chapter is intended to provide the background and context for this kind of application.

C. King (✉)
Northeastern University, Boston, MA, USA
e-mail: c.king@northeastern.edu

© Springer Nature Switzerland AG 2020
E. Crisostomi et al. (eds.), *Analytics for the Sharing Economy:
Mathematics, Engineering and Business Perspectives*,
https://doi.org/10.1007/978-3-030-35032-1_5

1.1 Basic Notions of Queueing Theory

The basic ingredients of a queueing model are (a) a sequence of arrival events, where each arrival signals a new customer needing access to the service, (b) a server or group of servers which can satisfy the customer needs, and (c) a protocol for assigning service to waiting customers. Queueing models may evolve in either continuous time or discrete time, depending on the application. This chapter will give a brief overview of some results in queueing theory, with emphasis on the underlying mathematical techniques (exact solutions and inequalities). For more details the interested reader is referred to the many excellent sources available, for example [1, 2, 4, 7].

1.2 Concrete Models

The mathematical description of queueing theory can be quite technical and this may be an obstacle for readers new to the subject. So in order to increase the readability of this chapter we will build the development around some concrete applications, and we use these examples as motivation for the results.

Our concrete example for continuous time models is a charging station for electric vehicles. We picture one or more charging nodes (the servers), and suppose that a queue forms when all nodes are busy. The quality of service for customers is measured by the time spent waiting for a free node (more time spent waiting is lower quality), while for the operator the cost of service is the amount of time when the nodes are idle (more idle nodes means less revenue).

For discrete time models, our concrete example is a car sharing program, where a fleet of cars is available for loan each day. Cars are assigned only at the start of the day, and there is a fixed number available for loan. If a customer is not able to obtain a car then they can join a queue and try to get service the following day. The quality of service for customers is the number of days spent waiting for a car, while for the operator the cost of service is the number of idle cars.

1.3 General Principles

We will describe two useful results which apply in broad generality to queueing systems. The first concerns stability of the queue, and the second is Little's Law. Stability means that the queue length does not grow to infinity, and the condition for stability is that the average arrival rate must be less than the service rate when the queue is non-empty. Little's Law [6] is a relation between average quantities in a stable queueing system. Let λ be the average rate at which customers enter the system, let $\overline{N}$ be the average number of customers in the system, and let $\overline{T}$ be the average time spent by a customer in the system. Then we have the relation

$$\lambda \overline{T} = \overline{N} \tag{1}$$

The quantities $\lambda, \overline{T}, \overline{N}$ can be computed as long-run averages over the entire history of the system. Alternatively, if the system is in equilibrium so that its stochastic behavior is described by a time-invariant distribution, then these quantities can be computed as expected values using the equilibrium distribution. Note that Little's Law is also valid for a subsystem, for example the queue or the servers, as long as the parameters are suitably defined.

1.4 Queueing Protocol

The simplest and most widely studied queueing protocol is FIFO (first-in, first-out), meaning that customers are served in the order in which they arrived. So the customer at the head of the queue is assigned to the next free server. We will only consider queueing models with the FIFO protocol. A common variation is priority queueing, where there are several classes of customers, and service is prioritized according to the class of the customer. The interested reader is referred to [1] for more information on this topic.

2 Continuous Time Models

In general we will denote by c the number of servers, and we let $N(t)$ denote the number of customers in the system at time t. The arrival rate is denoted by λ. We also let $N_q(t)$ denote the number of customers in the queue at time t, so that

$$N_q(t) = \begin{cases} 0 & \text{for } N(t) \leq c \\ N(t) - c & \text{for } N(t) > c \end{cases} \tag{2}$$

Customers arrive sequentially (in continuous time models the probability of a simultaneous arrival is zero), so customers can be labeled $\{1, 2, \ldots, n, \ldots\}$ according to the order in which they arrived. The time spent waiting in the queue by the nth customer is W_n, and the time in service is S_n. Thus the total time in the system is $T_n = W_n + S_n$. In single server systems the customers depart in the same order as they arrived; in multi-server systems this need not be the case.

Little's law can be applied to the entire system, and also to the queue. For example, when applied to the queue it yields

$$\lambda w = n_q \tag{3}$$

where w and n_q are respectively the average time spent in queue and the average number in the queue. These can be computed as the long-run averages

$$w = \lim_{n \to \infty} \frac{1}{n} \sum_{k=1}^{n} W_k, \quad n_q = \lim_{T \to \infty} \frac{1}{T} \int_0^T N_q(t)\, dt \qquad (4)$$

2.1 The $M/M/1$ Queue

The most completely analyzed queueing model is the continuous time $M/M/1$ queue. In this model (a) new customers arrive according to a Poisson process with a constant arrival rate λ, (b) a single server provides service to one customer at a time, and the service times are independent exponential random variables with constant mean μ^{-1}, and (c) if a customer arrives while the server is busy then the customer joins the end of a FIFO queue and must wait until the current service completes and all customers ahead in the queue are served (when service ends for a customer, service immediately starts for the next customer at the head of the queue).

2.2 Markov Chain Exact Results

The stochastic process $\{N(t)\}$ for the $M/M/1$ system is a continuous time Markov chain (CTMC). This special property holds thanks to its two key assumptions (a) and (b); namely a Poisson arrival process and IID (independent, identically distributed) exponential service times. The general theory of CTMC's shows that the chain is stable if $\lambda < \mu$, and in this case there is a probability distribution π on the non-negative integers $\{0, 1, 2, \ldots\}$ such that the distribution of $N(t)$ converges to π as $t \to \infty$, so that for all $n \geq 0$

$$\mathbb{P}(N(t) = n) \to \pi(n) \quad \text{as } t \to \infty \qquad (5)$$

We will call π the equilibrium or stationary distribution of the chain. It can be easily computed using standard methods. In particular, it can be shown that

$$\pi(n) = (1 - \rho)\, \rho^n, \quad \rho = \frac{\lambda}{\mu} \qquad (6)$$

where ρ is called the traffic density of the model. The equilibrium distribution describes the long-term behavior of the chain. Note that stability is the condition $\rho < 1$.

Using the exact values (6) we can compute many quantities of interest; for example the mean number in the system and in the queue are respectively

$$\mathbb{E}[N] = \sum_{n=0}^{\infty} n\,\pi(n) = \frac{\lambda}{\mu - \lambda}, \quad \mathbb{E}[N_q] = \frac{\lambda^2}{\mu(\mu - \lambda)} \tag{7}$$

In the limit $t \to \infty$ the time spent by a customer in the system also converges to an equilibrium distribution, and this distribution turns out to be exponential. Denoting this equilibrium random variable by T we have

$$\mathbb{P}(T > t) = e^{-(\mu - \lambda)t}, \quad t \geq 0 \tag{8}$$

It follows that T has an exponential distribution with mean $(\mu - \lambda)^{-1}$. The mean waiting time in the queue can also be computed using the distribution (6), or more simply by using (7) in Little's Law (3) for the queue:

$$\mathbb{E}[W] = \frac{\lambda}{\mu(\mu - \lambda)} \tag{9}$$

2.3 The $M/M/c$ Queue

In order to make a more realistic model for the example of the charging station, we want to allow more than one server, corresponding to having more than one charging node available for use. There is a natural extension of the $M/M/1$ system called the $M/M/c$ system which provides this model. We now have c servers which can work in parallel, so the queue starts to grow only when there are more than c customers in the system. The arrivals process is still Poisson, and each server works independently; service times are all exponential, with the same mean. This system is again a CTMC and the equilibrium distribution can be computed exactly. The expressions are a little more complicated but still tractable.

The condition for stability, i.e. to have an equilibrium distribution, is

$$\lambda < c\mu \tag{10}$$

and the traffic density now is

$$\rho = \frac{\lambda}{c\mu} \tag{11}$$

The probability that there are $j \leq c$ customers in the system, meaning that there is no queue, is

$$\pi(j) = \frac{(c\rho)^j}{j!}\pi(0), \quad 0 \leq j \leq c \tag{12}$$

where

$$\pi(0) = \left(\sum_{j=0}^{c-1} \frac{(c\rho)^j}{j!} + \frac{(c\rho)^c}{c!(1-\rho)} \right)^{-1} \qquad (13)$$

and the probability for more than c customers is

$$\pi(c+j) = \rho^j \pi(c), \quad j \geq 0 \qquad (14)$$

The time spent waiting in queue is the weighted sum of an atom at 0 and an exponential random variable with rate $\mu c - \lambda$. The probability of the atom at 0 is $\mathbb{P}(W=0) = \pi(0) + \cdots + \pi(c-1)$.

2.4 Application to Charging Station Model

Here λ is the rate of arrival of customers to the charging station, and μ^{-1} is the mean charging time needed by each customer. The $M/M/c$ queue can provide an appropriate model in this case, where c is the number of charging nodes. The quality of service to customers is measured by the time spent waiting until a node is free. The mean time spent in queue $\mathbb{E}[W]$ provides an average measure of this delay, however a customer may be more interested in the likelihood that they will have to wait longer than 20 minutes (say). The corresponding measure of quality of service is

$$Q_{user} = \mathbb{P}(W \leq t_c) = 1 - \mathbb{P}(W > t_c) \qquad (15)$$

where t_c is the threshold time of interest. Using the results above we find

$$\mathbb{P}(W > t_c) = \mathbb{P}(W > 0) e^{-(\mu c - \lambda) t_c} = [1 - \mathbb{P}(W = 0)] e^{-(\mu c - \lambda) t_c} \qquad (16)$$

The cost of delays for the operator of the charging station is measured by the amount of 'downtime', or the number of charging nodes that are not being used. In this case it is appropriate to look at the long-run average quantity. Let I be the number of idle charging nodes, then the cost to the station is

$$C_{st} = \mathbb{E}[I] = \sum_{j=0}^{c-1} (c-j) \pi(j) \qquad (17)$$

There is a tension between these quantities; in order to keep (17) small it is necessary to let the delay for customers grow, which will decrease (15). So there may be a tradeoff between the goals of the customers and the charging station.

We consider a specific example: suppose that there are $c = 5$ charging nodes, and the average charging time per vehicle is $\mu^{-1} = 4$ min. Figure 1 shows the value of

Fig. 1 Quality of service for customer

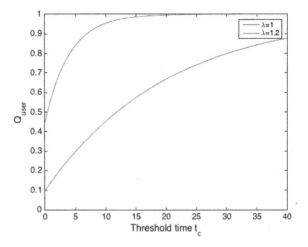

Fig. 2 Cost for charging station

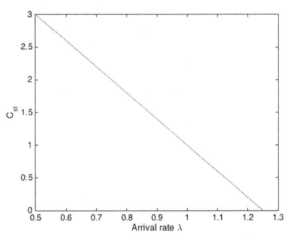

Q_{user} given by (15) as a function of the threshold time t_c for the arrival rates $\lambda = 1$, and $\lambda = 1.2$.

The cost to the station C_{st} given by (17) can be computed easily using Little's Law, when applied to the 'system' consisting of just the charging nodes. The average charging time spent by a customer is $\mu^{-1} = 4$ minutes, and the arrival rate is λ, since in equilibrium the rate of arrivals into the queue is the same as the rate of departures out of the queue. Thus by Little's Law the average number of customers being serviced at any time is $\lambda \mu^{-1}$. But this is the same as the average number of busy nodes, hence the average number of idle nodes is

$$C_{st} = \mathbb{E}[I] = c - \lambda \mu^{-1} \tag{18}$$

Figure 2 shows a plot of this cost as a function of the arrival rate λ for the scenario described above, where $c = 5$ and $\mu^{-1} = 4$.

3 Discrete Time Models

We will build the analysis of the discrete model around the concrete example of the car sharing program introduced before. Let X_n denote the number of new requests at the start of day n, and let Y_n denote the number of outstanding requests which cannot be satisfied on day n. These outstanding requests constitute the queue, and they are carried over to the following day. Thus the total number of requests on day $n+1$ is $R_{n+1} = X_{n+1} + Y_n$, where X_{n+1} is the number of new requests arriving on day $n+1$. If $Y_n > 0$ then the cars in queue from day n will get first priority for service on day $n+1$, and thereafter any remaining cars will be released to subsequent requests. Note that requests may build up over several days, so a client may have to wait several days until their request is satisfied.

We will use N to denote the number of customers, and M to denote the number of cars. In order to allow the possibility of demand exceeding supply we assume that $M < N$. The client demand is the number of cars requested, and as noted we will let X_n denote the demand on day n. Clearly $0 \leq X_n \leq N$, but it may happen that $X_n > M$, in which case it would not be possible to accommodate all requests on day n. A similar example was analyzed in the paper [3], and we will make use of some of those results here.

If $R_{n+1} = X_{n+1} + Y_n > M$ then there will be unfulfilled requests on day $n+1$, and these requests will be put in queue. Thus

$$Y_n + X_{n+1} > M \Rightarrow Y_{n+1} = Y_n + X_{n+1} - M \qquad (19)$$

If $Y_n + X_{n+1} \leq M$ then supply will meet demand and the queue will be empty at the end of the day, so

$$Y_n + X_{n+1} \leq M \Rightarrow Y_{n+1} = 0 \qquad (20)$$

So the update rule is

$$Y_{n+1} = \max\{0, Y_n + X_{n+1} - M\} = (Y_n + X_{n+1} - M)^+ \qquad (21)$$

Note that this update equation is a special case of the Lindley equation for queues [5]. Also note that if $Y_n + X_{n+1} < M$ then some cars will not be used on day $n+1$, and so we define the number of idle cars on day $n+1$ to be

$$I_{n+1} = \min\{0, Y_n + X_{n+1} - M\} = (Y_n + X_{n+1} - M)^- \qquad (22)$$

Define

$$B_n = X_n - M \qquad (23)$$

then we have

$$Y_{n+1} = \max\{0, Y_n + B_{n+1}\} \qquad (24)$$

Writing out the first few terms this gives

$$Y_0 = 0$$
$$Y_1 = \max\{0, B_1\}$$
$$Y_2 = \max\{0, Y_1 + B_2\} = \max\{0, B_2, B_1 + B_2\}$$

so in general we get

$$Y_n = \max\{0, B_n, B_n + B_{n-1}, B_n + B_{n-1} + B_{n-2}, \ldots\} \qquad (25)$$

Since the arrival numbers $\{X_n\}$ are IID, it follows that $\{B_n\}$ are also IID, and so there is no significance to the labels of these random variables. Therefore Y_n has the same distribution as

$$Y'_n = \max\{0, B_1, B_1 + B_2, \ldots, B_1 + \cdots B_n\} \qquad (26)$$

We define

$$\Sigma_0 = 0, \quad \Sigma_j = \sum_{i=1}^{j} B_i, \quad j = 1, \ldots, n \qquad (27)$$

then

$$Y'_n = \max\{\Sigma_0, \Sigma_1, \ldots, \Sigma_n\} \qquad (28)$$

and letting $n \to \infty$ we get

$$Y_\infty = \sup_{n \geq 0} \Sigma_n \qquad (29)$$

The random variable Y_∞ describes the stationary or equilibrium distribution of the queue. Unlike the case of the $M/M/1$ queue there are no simple formulas for its distribution, however we will show it does provide some useful results about the model. In particular we can regard $\{\Sigma_n\}$ as a random walk on the integers (both positive and negative), which implies that Y_∞ is the maximum of the random walk. Clearly this maximum value diverges unless the mean is negative at each step, that is unless $\mathbb{E}[B] = \mathbb{E}X - M < 0$, so for stability it must be a random walk with negative drift.

3.1 Quality of Service

The cost of delays can be assessed from the viewpoint of the user or from the viewpoint of the system. From the user's viewpoint we consider the probability that any user needs to wait k extra days or more, which is the probability that the queue length exceeds kM. So we define

$$Q_{user} = 1 - \mathbb{P}(Y > kM) \tag{30}$$

where Y is the equilibrium distribution of the $\{Y_n\}$. From the system's viewpoint the relevant metric is the average usage of cars, and whether there are idle cars in the fleet, so we define the cost for the system to be

$$C_{system} = \mathbb{E}[I] \tag{31}$$

where I is the equilibrium distribution of $\{I_n\}$, the number of idle cars.

3.2 Bounds for Stationary Distribution

By assuming that the number of customers in the queue reaches a stationary distribution Y, it is possible to derive some bounds for the expected waiting time, and hence for the customer's quality of service. These bounds are similar to the well-known Pollaczek–Khinchin bound for the $M/G/1$ queue, and hold quite generally without additional assumptions on the probability distributions of the random variables. Note that the bound depends on M, the size of the car fleet, but does not depend on N, the number of customers.

Lemma 1 *Let μ_X and σ_X^2 denote the mean and variance of the demand X. Assume that the model is stable, so that $\mu_X < M$, and is in its stationary distribution. Then the expected value of Y (the number in the queue) satisfies*

$$\mathbb{E}[Y] \leq \frac{\sigma_X^2}{2(M - \mu_X)} \tag{32}$$

Proof Recall that

$$Y_{n+1} = \max\{0, Y_n + B_{n+1}\} = (Y_n + B_{n+1})^+ \tag{33}$$

We also have from (22)

$$Y_n + B_{n+1} = (Y_n + B_{n+1})^+ - (Y_n + B_{n+1})^- = Y_{n+1} - I_{n+1} \tag{34}$$

where I_{n+1} is the number of idle cars on day $n+1$. We have

$$\text{VAR}[Y_n + B_{n+1}] = \text{VAR}[Y_{n+1}] + \text{VAR}[I_{n+1}] - 2\mathbb{E}[Y_{n+1} I_{n+1}] + 2\mathbb{E}[Y_{n+1}]\mathbb{E}[I_{n+1}] \quad (35)$$
$$= \text{VAR}[Y_{n+1}] + \text{VAR}[I_{n+1}] + 2\mathbb{E}[Y_{n+1}]\mathbb{E}[I_{n+1}] \quad (36)$$

where we used the fact that

$$Y_{n+1} I_{n+1} = (Y_n + B_{n+1})^+ (Y_n + B_{n+1})^- = 0 \quad (37)$$

We now note that Y_n and B_{n+1} are independent and thus

$$\text{VAR}[Y_n + B_{n+1}] = \text{VAR}[Y_n] + \text{VAR}[B_{n+1}] = \text{VAR}[Y_n] + \text{VAR}[X_{n+1}] \quad (38)$$

Furthermore the assumption that the system is in its stationary distribution implies that

$$\text{VAR}[Y_n] = \text{VAR}[Y_{n+1}], \quad \text{VAR}[X_n] = \sigma_X^2, \quad \mathbb{E}[Y_{n+1}]\mathbb{E}[I_{n+1}] = \mathbb{E}[Y]\mathbb{E}[I] \quad (39)$$

Hence we deduce (in the limit $n \to \infty$)

$$\sigma_X^2 = \text{VAR}[I] + 2\mathbb{E}[Y]\mathbb{E}[I] \quad (40)$$

Finally we note that the queue length Y_n at the end of day n is the difference between the total number of requests received up to that time and the total number of completed trips up to that time, that is

$$Y_n = \sum_{k=1}^{n} X_k - \sum_{k=1}^{n} (M - I_k) \quad (41)$$

The stability assumption implies that

$$\lim_{n \to \infty} \frac{1}{n} Y_n = 0 \quad (42)$$

and therefore

$$\lim_{n \to \infty} \frac{1}{n} \sum_{k=1}^{n} X_k = \lim_{n \to \infty} \frac{1}{n} \sum_{k=1}^{n} (M - I_k) \quad (43)$$

which gives

$$\mu_X = M - \mathbb{E}[I] \quad (44)$$

Hence returning to our formula we deduce that

$$\mathbb{E}[Y] = \frac{\sigma_X^2 - \text{VAR}[I]}{2(M - \mu_X)} \leq \frac{\sigma_X^2}{2(M - \mu_X)} \qquad (45)$$

which completes the proof.

3.3 The Operator's Cost of Service

The cost of service for the operator is determined by the number of idle vehicles, which was labeled I_n. The mean or expected value is the appropriate metric in this case, and this was computed in (44) (assuming the stationary distribution). Therefore we have

$$C_{system} = \mathbb{E}[I] = M - \mu_X \qquad (46)$$

3.4 Bound for the Customer's Quality of Service

Using the result of Lemma 1 we can easily deduce a lower bound for the customer's quality of service. Applying Markov's inequality and (32) gives

$$\mathbb{P}(Y > kM) \leq \frac{\mathbb{E}[Y]}{kM} \leq \frac{\sigma_X^2}{2kM(M - \mu_X)} \qquad (47)$$

and therefore

$$Q_{user} \geq 1 - \frac{\sigma_X^2}{2kM(M - \mu_X)} \qquad (48)$$

Figure 3 shows two plots of the right side of (48) as a function of k, with parameters $\mu_X = 100$ and $\sigma_X^2 = 90$, for the values $M = 102$, $M = 112$.

3.5 Binomial Distribution

In order to obtain more precise bounds we now specialize to a particular (and familiar) case where the number of arrivals is modeled by a binomial random variable. So we assume that each client has the same probability p of requesting a car each day, and that all clients choose independently each day whether or not to request a car. In this case the values X_n follow a binomial probability distribution $p(x)$, where

Fig. 3 Lower bound for quality of service

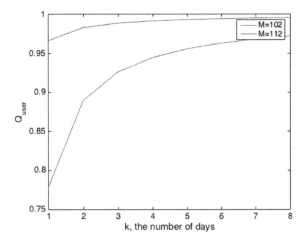

$$p(x) = \mathbb{P}(X = x) = \binom{N}{x} p^x (1-p)^{N-x}, \quad x = 0, 1, \ldots, N \tag{49}$$

The mean and variance of X are

$$\mu_X = Np, \quad \sigma_X^2 = Np(1-p) \tag{50}$$

Note: in the limit $N \to \infty$, $p \to 0$ the distribution of X becomes Poisson, which would be closer to the continuous time model with Poisson arrival process.

3.6 Bound on Queue Length

Again we consider the probability that any user needs to wait k extra days or more, which is the probability that the queue length exceeds kM. Define

$$b = M - \mu_X = M - Np, \quad \alpha = \frac{b}{\sigma_X^2} = \frac{M - Np}{Np(1-p)}$$

Lemma 2 *For all* $k \geq 1, n \geq 1$

$$P(Y_n > kM) \leq e^{-kb/2} \left(e^{b^2/2M} - 1 \right)^{-1}. \tag{51}$$

Proof Recall that Y_n has the same distribution as

$$Y_n' = \max \{\Sigma_0, \Sigma_1, \ldots, \Sigma_n\}.$$

where

$$\Sigma_j = \sum_{i=1}^{j}(X_i - M) \qquad (52)$$

Then we have for $k \geq 1$

$$P(Y_n > kM) = P(\max\{0, \Sigma_1, \Sigma_2, \ldots, \Sigma_n\} > kM)$$

$$= P\left(\bigcup_{j=1}^{n}\{\Sigma_j > kM\}\right)$$

$$\leq \sum_{j=1}^{n} P(\Sigma_j > kM).$$

There are several ways to estimate the right side of this inequality. One way uses the normal approximation for the binomial distribution, and this approach was explored in [3]. Here we will use Chernoff's bound, which is related to the large deviation bound for the binomial distribution. Recall that the arrival numbers $\{X_n\}$ are independent and binomial, and therefore $\sum_{i=1}^{j} X_i$ is also binomial, with mean jNp. The Chernoff bound states that for any $\delta > 0$,

$$\mathbb{P}\left(\sum_{i=1}^{j} X_i \geq (1+\delta)jNp\right) \leq e^{-\delta^2 jNp/(2+\delta)} \qquad (53)$$

The bound (53) will be derived in Sect. 3.8. Noting that $\Sigma_j = \sum_{i=1}^{j} X_i - jM$, we can use this to get the desired bound by taking

$$(1+\delta)jNp = (j+k)M \qquad (54)$$

Substituting this in (53) and noting that $2 + \delta < 2(1+\delta)$ we deduce that

$$P(\Sigma_j > kM) \leq e^{-(\delta jNp)^2/2(j+k)M} = e^{-(kM+jb)^2/2(j+k)M} \qquad (55)$$

Some simple algebra shows that

$$\frac{(kM+bj)^2}{2(k+j)M} \geq \frac{kb}{2} + \frac{b^2}{2M} j \qquad (56)$$

and therefore

$$P(\Sigma_j > kM) \leq e^{-kb/2 - jb^2/2M} \qquad (57)$$

This gives

$$P(Y_n > kM) \leq e^{-kb/2} \sum_{j=1}^{n} e^{-jb^2/2M}$$

The sum is geometric, and is bounded by

$$\sum_{j=1}^{n} e^{-jb^2/2M} < \sum_{j=1}^{\infty} e^{-jb^2/2M} = \frac{e^{-b^2/2M}}{1-e^{-b^2/2M}} = \left(e^{b^2/2M} - 1\right)^{-1}$$

as required.

3.7 Bound on Customer Quality of Service

From Lemma 2 we deduce the following bound for quality of service:

$$Q_{user} = 1 - \mathbb{P}(Y > kM) \geq 1 - e^{-kb/2} \left(e^{b^2/2M} - 1\right)^{-1} \qquad (58)$$

Fig. 4 compares the bounds (48) and (58) as a function of M for the binomial model with $N = 1000$, $p = 0.1$, $\mu_X = Np = 100$, and $k = 1$. As the graph shows, the bound (58) is weaker than (48) for low values of M, and is tighter for larger values of M.

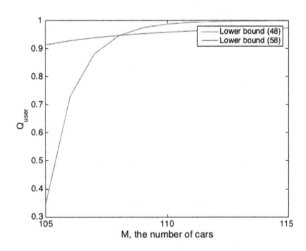

Fig. 4 Comparing lower bounds for quality of service

3.8 Derivation of the Chernoff Bound

Consider a collection of IID random variables $\{V_i\}$ with mean μ, and let $V = \sum_{i=1}^{m} V_i$. Then for any $\theta > 0$, $a > 0$ we may apply the Markov inequality to deduce that

$$\mathbb{P}(V \geq a) = \mathbb{P}(e^{\theta V} \geq e^{\theta a}) \leq e^{-\theta a}\, \mathbb{E}[e^{\theta V}] \tag{59}$$

Furthermore the IID property of $\{V_i\}$ implies that

$$\mathbb{E}[e^{\theta V}] = \left(\mathbb{E}[e^{\theta V_1}]\right)^m = e^{m\, \ln \mathbb{E}[e^{\theta V_i}]} \tag{60}$$

Now we apply this to the situation of interest, where V_i is a Bernoulli random variable with mean p, $V = \sum_{i=1}^{j} X_i$, $m = Nj$, and $a = (1+\delta)Njp$. We then have

$$E[e^{\theta V_i}] = (1-p) + pe^{\theta} \leq e^{p(e^{\theta}-1)} \tag{61}$$

and therefore from (59) and (60) we get

$$\mathbb{P}\left(\sum_{i=1}^{j} X_i \geq (1+\delta)Njp\right) = \mathbb{P}(V \geq a) \leq e^{-\theta a}\, e^{Nj\, p(e^{\theta}-1)} \tag{62}$$

We choose $\theta = \ln(1+\delta)$, leading to

$$\mathbb{P}\left(\sum_{i=1}^{j} X_i \geq (1+\delta)Njp\right) \leq e^{-Njp[(1+\delta)\ln(1+\delta)-\delta]} \tag{63}$$

Finally the bound (53) follows from (63) by noting the inequality

$$(1+\delta)\ln(1+\delta) - \delta \geq \frac{\delta^2}{2+\delta} \tag{64}$$

which holds for all $\delta \geq 0$.

4 Summary and Further Directions

This chapter has discussed some selected topics from queueing theory, and has presented the results in the context of applications which are relevant to the sharing economy theme. There are many other interesting and important results in queueing theory, and the interested reader is referred to the vast literature on queueing theory for details (see the references as a starting point).

References

1. Bertsekas D, Gallager R (1992) Data networks, 2nd edn. Prentice Hall
2. Borovkov AA (1976) Stochastic processes in queueing theory. Stochastic modelling and applied probability series, vol 4. Springer-Verlag, New York
3. King C, Griggs W, Wirth F, Quinn K, Shorten R (2015) Alleviating a form of electric vehicle range anxiety through On-Demand vehicle access. Int J Control 88(4):717–728
4. Kleinrock L (1976) Queueing systems: theory, vol 1. Wiley
5. Lindley DV (1952) The theory of queues with a single server. Proceedings of the Cambridge Philosophical Society, vol. 48, pp 277–289
6. Little JDC (2011) Little's law as viewed on its 50th anniversary. Oper Res 59(3):536549
7. Ross S (2014) Introduction to probability models, 11th edn. Academic Press

Enablers for Collaborative Consumption

Emanuele Crisostomi, Bissan Ghaddar, Florian Häusler,
Joe Naoum-Sawaya, Giovanni Russo and Robert Shorten

1 Introduction

Globally, a multitude of peer-to-peer businesses are disrupting traditional economic models and threatening established markets including hospitality, transportation, and banking among many others. While certainly, the appetite of consumers to the sharing economy is the main driver to the fast evolving markets, the factors that have accelerated the adoption of collaborative consumption have been founded on the development of leading technologies that enable a better user experience while creating a strong feeling of trust and convenience. It is not a coincidence, that the rapid growth of the market leaders in the sharing economy is aligned with a sense of disbelief in classic companies [1] and has strengthened the sense of belongingness

---

E. Crisostomi
University of Pisa, Pisa, Italy
e-mail: emanuele.crisostomi@unipi.it

B. Ghaddar · J. Naoum-Sawaya
Ivery Business School, Canada
e-mail: bghaddar@ivey.ca
e-mail: jnaoum-sawaya@ivey.ca

F. Häusler
Moovel, Germany
e-mail: florian.haeusler@gmail.com

G. Russo
Department of Information & Electrical Engineering and Applied Mathematics,
University of Salerno, Fisciano, Salerno, Italy
e-mail: giovarusso@unisa.it

R. Shorten
Dyson School of Design Engineering, Imperial College London, London, UK
e-mail: r.shorten@imperial.ac.uk

of people and their desire of being part of a community [2], and on the other side with the widespread of ubiquitous computing most notably the adoption of the smart phone in our every day life.

This section thus outlines the various Key (?) enablers of the sharing economy focusing on the recent research in technological and economic developments to present a holistic view of the challenges and opportunities both to businesses and to consumers. In this regard, Chap. 6 outlines the recent developments in Information and Communication Technologies and the Internet of Things which enable the pervasive access to shared resources in real-time. Secure, trustworthy, privacy-preserving, peer-to-peer networks are also fundamental for the large scale adoption of sharing systems and the emerging paradigms as discussed in Chap. 7 which include Blockchain and distributed ledger technologies. In this collaborative consumption framework, new business models need to be adopted to sustain the paradigm shift where classic customers evolve to prosumers that share ownership of products. Chapter 8 presents such business models together with a critical discussion of the main expected sources of revenues. Centralized decision making then becomes infeasible and distributed control algorithms are becoming increasingly important as described in Chaps. 9 and 10. Finally, the sharing economy is based on a network of individuals that collaborate to share resources. As discussed in Chap. 11, the active participation in exchanging resources requires a behavioral change among the participating individuals which can then be leveraged to support societal developments.

References

1. L Gansky (2010) The mesh: why the future of business is sharing, 251 p. Portfolio/Penguin, USA
2. R Botsman, R Rogers (2010) What's mine is yours: the rise of collaborative consumption. 279 p. HarperCollins Publishers, USA
3. E Crisostomi, R Shorten, F Wirth (2016) Smart Cities: A Golden Age for Control Theory? IEEE Technology and Society Magazine, pp 23–24

Advances in Cloud Computing, Wireless Communications and the Internet of Things

Gopika Premsankar and Mario Di Francesco

Abstract There is a growing amount of data generated by a variety of devices in the Internet of Things (IoT). Sharing economy applications can leverage such data to provide solutions of high societal impact. Several technologies together enable the collaborative use of data through software services. This chapter describes the key developments in these technological areas. In particular, it describes advances in cloud computing that have resulted in new software architectures and deployment practices. Such improvements enable the rapid creation and deployment of new services on the cloud. Next, it highlights recent developments in wireless networks that allow heterogeneous devices to connect and share information. Furthermore, this chapter describes how IoT platforms are becoming interoperable, thus fostering collaborative access to data from diverse devices. Finally, it elaborates on how the described technologies jointly enable new sharing economy solutions through a case study on car sharing.

Keywords Internet of Things · IoT · Cloud computing · Mobile networks · LPWAN · LoRa

1 Introduction

Several advances in the field of Information and Communications Technology (ICT) have influenced how we are able to share our time, material and skills [46]. Indeed, there are many examples of online sharing platforms such as Airbnb, Uber, marketplaces and food sharing applications. The success of these applications depend on several factors. First, the ubiquitous availability of *Internet connectivity* enables us to access such services everywhere and at any time. Furthermore, the advent of new *cloud computing* service models enables application developers to quickly deploy

---

G. Premsankar (✉) · M. Di Francesco
Department of Computer Science, School of Science, Aalto University, Espoo, Finland
e-mail: gopika.premsankar@aalto.fi
e-mail: mario.di.francesco@aalto.fi

new services and functionality. The economic barrier to entry has been lowered with the availability of pay-per-use, on-demand computing and data storage resources. Now, we are witnessing an increasing amount of machine-generated data from *Internet of Things* (IoT) devices such as sensors, household appliances, wearable devices, vehicles and much more. Indeed, sharing economy applications can make use of the wide variety of data and enable sharing in a more immersive and collaborative manner. The IoT enables new services that rely on seamless inter-operation between home networks, neighborhood networks and global suppliers of goods and services [35]. This chapter describes recent developments in the fields of cloud computing (Sect. 2), wireless connectivity (Sect. 3) and IoT (Sect. 4) that enable such new applications and services. Section 5 describes how these technologies together enable a car sharing application. Finally, Sect. 6 provides some concluding remarks.

2 Cloud Computing

Traditionally, software applications were deployed on bare metal servers. Developers typically had to invest in the infrastructure on which services were deployed, and manage the hardware in addition to the application itself [36]. Furthermore, they had to overprovision their deployments, i.e., make more computing resources available than required so as to meet the peak demand. However, this resulted in under-utilizing the hardware when the demand is low. The advent of *virtualization* enabled multiple virtualized servers (or instances) to run on a single physical machine. The virtualized instances, also known as Virtual Machines (VMs), run in isolation from each other and run their own operating system. This resulted in the initial wave of *cloud computing*, wherein cloud providers manage large data centers while developers (or service providers) deploy applications as VMs [31]. The main benefits of cloud computing are the availability of an infinite amount of resources (computing, network and storage) that can be used on-demand and released when no longer required [36]. This allowed software developers and organizations the flexibility to start with a lower level of resources and scale them as required. This also meant that computing resources could be treated as utility, which lowered the barrier for innovative services as a large initial commitment of resources was no longer required [48].

Cloud computing services are made available under different service models, described next.

- *Infrastructure-as-a-Service* (IaaS) allows the developers to deploy virtualized instances, typically VMs, within which custom software can be run. In this paradigm, the end users manage the deployed instances including the OS, storage and networking [50].
- *Platform-as-a-Service* (PaaS) allows the end user to deploy applications on the cloud while the cloud provider manages the programming tools (software libraries, language runtimes), networking, storage and OS. Typically, the user controls the

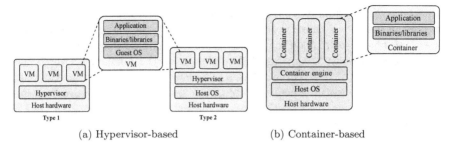

Fig. 1 Architecture of virtualized instances (**a** VMs and **b** containers) and host hardware

application settings and relies on the cloud provider for maintaining the underlying infrastructure [50].
- *Function-as-a-Service* (FaaS) [23] is a new paradigm that allows end users to deploy stateless functions on the cloud platform. The functions are executed only when explicitly invoked or triggered by an event (such as user input or database changes). The platform itself takes care of executing the functions and scaling them based on the demand. FaaS again relies on the cloud provider for managing the infrastructure; however, it differs from PaaS in that the functions are not billed when they are idle.
- *Software-as-a-Service* (SaaS) allows the end users to use applications provided by the cloud platform. The applications themselves are accessible through a thin client interface such as a web browser or an application programming interface (API). All aspects of the application and infrastructure are managed by the cloud provider [50].

2.1 Virtualization

Virtualization is the key enabler of cloud computing and its associated service models. The two most popular virtualization technologies are *hypervisor-based* and *operating system-based* (or *container*-based) virtualization.

Hypervisor-Based Virtualization

In hypervisor-based virtualization (Fig. 1a), a software abstraction layer called a virtual machine monitor (VMM) or hypervisor lies between the VMs and the underlying physical hardware. The hypervisor manages the virtual machines and has full control of system resources. It provides an environment for execution that is identical to the underlying server [59]. The hypervisor provides complete isolation between VMs and also from the underlying hardware, thereby allowing multiple OSes to run at the same time. For instance, it is possible to run a VM with Windows OS on top of a machine with a Linux-based OS. There are two types of hypervisors: a *type 1* or *bare-metal* hypervisor runs directly on top of the host hardware, whereas a *type 2* or *hosted* hypervisor runs on top of the host OS. Examples of type 1 hypervisors include Xen [24] and VMWare ESX [51], and those of type 2 include KVM [43] and Oracle VirtualBox [75]. Hypervisor-based virtualization supports multi-tenancy and provides excellent isolation between VMs; however, it introduces an overhead that affects performance of the VMs [34, 56]. Moreover, the time taken to start a VM can be in the order of minutes as a complete OS needs to be started.

Container-Based Virtualization

Container-based virtualization (Fig. 1b) is a lightweight form of virtualization that is becoming more prevalent on the cloud today [70]. This form of virtualization does not rely on a hypervisor and uses the host OS kernel-level features to provide isolation. Thus, virtualized instances (known as containers) do not need to run a separate OS [56]. Containers start faster than VMs and generally achieve a better performance [34]. Linux containers is one of the most popular implementations of this form of virtualization [34]. We refer to Linux containers as containers from now on. Containers use the following kernel features: *namespaces* and *cgroups* [34]. Namespaces provide isolation between containers so that one container has no visibility of objects outside it. Linux cgroups are used to limit the CPU and memory consumption of containers.

For completeness, we also discuss Docker [13], a popular open source platform for building, deploying and managing containers. Docker containers are generally used for deploying applications[1] instead of complete machines as described earlier. Docker provides a set of tools that allows the simplified use of container technology [55]. For instance, Docker packages the software application and its dependencies into a standalone package called an *image* [12]. Containers can then be started or created from Docker images. Docker also provides *registries* through which Docker images can be easily shared [55]. Finally, Docker simplifies application development by providing portability between different machines, i.e., the same container can be run on different machines while still exposing the same execution environment to the application. Docker is being widely adopted by the software development community and is being actively developed with contributions from the key IT players such as Amazon, Microsoft and Google [55].

2.2 Application Development on the Cloud

The emergence of Docker containers has changed the way applications and services are developed. Traditionally, software applications were written as single standalone modules that contain all the logic required to run the application [38]. However, such

---

[1] https://docs.docker.com/engine/faq/.

a *monolithic architecture* is difficult to manage as the application size and complexity grows [54]. For instance, problems arise when software bugs have to be traced in a large codebase. Furthermore, even small changes in the application requires the complete application to be rebooted. Finally, such an architecture cannot fully utilize the scaling benefits of the cloud as the whole copy of the application instance needs to be created even if only a single component of the application requires more CPU or memory.

Microservices

With the increasing adoption of Docker containers, a more cloud-native form of architecture known as the *microservice* architecture emerged. In this approach, applications are partitioned into smaller independent components (or microservices), each performing some business logic [38]. Although Docker containers were not created specifically for microservices, they present an ideal way for deploying independent modules easily with low cost [52]. The microservices themselves should be easy to understand and able to scale independently [38]. Communication between the microservices occur through network calls, typically using lightweight Representational State Transfer (REST) application programming interfaces (APIs). There are several advantages with this approach as compared to a monolithic application [54]. First, the microservices can be developed with different technology stacks if required. Second, the individual microservices can be scaled as required. This is especially useful when different services have different requirements; for instance, some services may be more CPU-intensive than others and thus, scaled out faster. Next, such an architecture reduces the time required to deploy new services or functionalities. For instance, a microservice can be deployed with few changes and even rolled back if required without affecting other services. Finally, the functionality provided by a microservice can be reused by other components for different purposes. The many benefits listed above meant that several organizations such as Netflix, SoundCloud and Amazon have successfully used this architecture to build large-scale fault-tolerant systems. However, some of the disadvantages of the microservice architecture arise from its distributed nature and reliance on the network for service calls [54]. This requires careful design of the microservices and a way to manage consistency between distributed components.

Function-as-a-Service

Function-as-a-Service (*FaaS*) is a relatively new extension of the microservice architecture. As briefly discussed earlier, in such a paradigm, applications are decomposed into multiple independent *stateless functions*. FaaS is also known as *serverless computing* as all operational concerns of the underlying infrastructure are abstracted away from the developers [65]. Again, the common approach for deploying functions is to run them in containers. The serverless platform itself takes care of executing the function and scaling it when required. Thus, software developers can concentrate on business logic and do not have to manage scaling, function runtimes and lifecycles of virtualized instances. Such an approach significantly reduces the time taken to develop and deploy applications, which will result in more innovative services [65]. However, there are a few disadvantages with this approach. First, there can be performance issues as functions (containers) are not running all the time. This implies that a container may need to first start and then install application dependencies before being able to respond to a request. Again, the distributed nature of FaaS-based applications requires careful consideration of consistency and network calls. Furthermore, as the FaaS approach is in a nascent stage, carrying out end-to-end tests is difficult as tooling for development, management and deployment are limited [65]. Finally, using a public cloud offering for serverless functions implies that there are certain limitations on the duration of function execution and supported language runtimes. However, there are several open source serverless frameworks (Fission,[2] Kubeless[3] and OpenFaas[4]) that provide more flexibility in implementation.

Container Orchestrators

Finally, we discuss the role of *container orchestrators* in application development. The overall software application consisting of multiple Docker-based microservices (or stateless functions) needs to be reliable, highly available and scalable. Container orchestrators represent an automated mechanism to create, manage and deploy distributed applications. Kubernetes [30] is one of the most widely-used container orchestrators. It has radically helped to increase the speed at which applications can be deployed due its features of *immutability*, *declarative configuration* and *online self-healing* [40]. First, Kubernetes relies on the immutable nature of Docker images. As described earlier, Docker packages both the application and its dependencies into an image that can be used to deploy the application. This greatly simplifies deployment; for instance, if an error occurs, it is simply possible to roll back to the older image. Second, Kubernetes uses a declarative configuration object: the developer only has to specify the desired state of the system and Kubernetes ensures that the desired state is fulfilled. An example of such a state is that the application module needs three running replicas (or identical running instances). Such a declarative approach is easy to understand as there is no ambiguity in the specification of the desired state. Finally, Kubernetes takes all the necessary steps to maintain the desired state through online self-healing. This implies that it continuously monitors the state of the application and takes necessary corrective action. For instance, if a running instance of the application goes down, Kubernetes detects the change and brings up a new instance to maintain the desired state. Thus, Kubernetes abstracts away the operations tasks from software developers and is designed to give developers velocity, efficiency and agility [40].

---

[2]https://fission.io.
[3]https://kubeless.io.
[4]https://www.openfaas.com.

2.3 Edge and Fog Computing

Edge [66] and *fog* [27] computing[5] involve bringing cloud computing capabilities closer to the end devices. The main objective is to efficiently process the growing amount of data generated by Internet of Things (IoT) devices (described in Sect. 4) such as smart meters, connected cars, home and building automation systems. The number of such connections is expected to grow to 3.3 billion by 2021 [1]. Currently, the data from such devices are transferred to the cloud to obtain a meaningful analysis. Indeed, the elastic and on-demand nature of the cloud computing resources make it ideally suited for this purpose. However, cloud computing resources are usually available in large data centers located far away from the end devices generating the data. Thus, the large volume of data sent places immense stress on the backhaul links to the cloud. Moreover, there are several applications that require low-latency processing of the data, such as vehicular safety applications [62], virtual/augmented reality applications [61] and real-time data analytics [67]. To address these issues, computing resources are made available at the edge of the network, i.e., closer to the end devices generating the data (Fig. 2). This allows data to be processed with very low latency and removes the need for data to be sent to the distant cloud data centers. Moreover, such an approach addresses privacy concerns by processing the data at the edge and removing private information before sending to the cloud [67]. It is important to note that edge and fog computing are expected to co-exist with cloud computing. Some long-term forecasting and analytics can still be carried out on the cloud without having to send all data to it.

Edge computing and fog computing differ in the approach to bring computing resources closer to the user. Edge computing relies on co-locating resource-rich servers along with access points, i.e., one hop away from end devices [62, 66]. Multi-Access Edge Computing (previously known as Mobile-Edge Computing) (MEC) follows this approach wherein software applications and cloud computing capabilities are made available in the radio access network [4]. MEC servers are typically deployed at wireless base station sites. This approach is standardized by the European Telecommunications Standards Institute (ETSI). Similarly, our previous work [62] considers the deployment of edge devices for vehicular applications. This scenario assumes that access points (called roadside units) deployed along the road are augmented with computing resources. The cars send data (such as location, direction, speed) every second to the roadside units; this data is processed immediately at the edge and a response sent back to the car. Such processing includes predicting whether collisions occur or deciding when autonomous vehicles can change lanes. Thus, the latency of processing and sending the response is critical for such applications. We demonstrate that even with a small amount of computing resources at the edge, such an approach is able to meet the computational demands of vehicular applications in a city without having to send data to the cloud.

On the other hand, fog computing utilizes the computing resources of heterogeneous devices, including the end devices themselves. For instance, the authors in [26,

---

[5]Image adapted from http://www.ntt.co.jp/news2014/1401e/140123a.html.

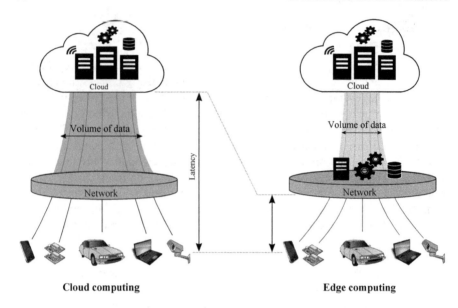

Fig. 2 The core concept of edge computing is to reduce latency of processing and volume of data sent to the cloud

41] describe a fog platform comprising of a diverse set of devices, including edge routers, access points, set-top boxes and smartphones. The OpenFog Consortium, an alliance of industrial companies and universities, has recently standardized the fog computing architecture [14]. In this approach, the fog consists of multiple layers or tiers, each with different levels of compute and storage resources. For instance, in a fog-based vehicular application scenario (Fig. 3), the cars themselves are fog computing nodes which can process data on-board. The cars (fog nodes) can connect to other cars within the same tier as well as to fog nodes in other tiers. This allows the devices within a layer to carry out processing in case connectivity to the higher tier goes down. Each tier provides additional processing, networking and storage capabilities than the tier below it. Thus, data from each tier is aggregated and sent up to the next layer. The next layer (above the cars) consists of roadside devices such as roadside access points or traffic cameras. The layers above consist of neighborhood and regional fog tiers, each representing devices with increased capability. The key feature of this architecture is that interactions are possible between tiers as well as within the tier itself. Thus, the services should be available even if connectivity between the tiers is temporarily not available.

The approaches described above rely on virtualization to enable software applications to run seamlessly across different devices [47, 66]. Furthermore, the cloud is still expected to play a role in managing the applications at scale. Together they can enable applications that require very low latency while still relying on the more resource-rich cloud for large-scale batch analytics.

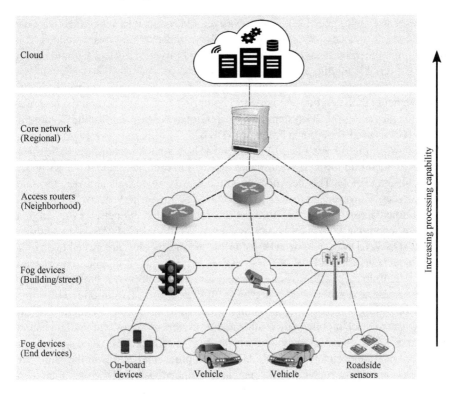

Fig. 3 The multi-tier architecture of the fog [14]

3 Wireless Connectivity

Today's wireless networks provide ubiquitous Internet connectivity accessible anywhere and at all times. The amount of Internet Protocol (IP) data handled by wireless networks has increased by a factor of 100 between 2010 to 2018 [20]. This section focuses on long-range wireless communications, i.e., cellular (or mobile) networks as well as wireless connectivity solutions specific to IoT devices.

3.1 Mobile Networks

The demand for higher data rates and lower latencies are the driving forces behind the evolution of mobile (or cellular) networks. There has been a tremendous increase in data rates from second generation (2G) networks (the first to allow mobile data access) to the current fourth generation (4G) networks, with up to 100 Mbps supported today [74]. The adoption of mobile data was also driven by attractive flat-rate pricing schemes and the availability of new smartphone devices with high-resolution screens and better user interfaces [49]. The pervasive, always-on Internet connectivity has enabled users to access diverse mobile services whenever they want. Indeed, it is hard to imagine the success of taxi-sharing applications such as Uber without having a fast and reliable mobile Internet connection.

The advancement in mobile Internet connectivity was made possible due to improvements in the end-to-end network, comprising of the radio and the core network. On the radio side, 2G systems known as Global Systems for Mobile Communication (GSM) used a combined Frequency-Division Multiple Access (FDMA)/Time-Division Multiple Access (TDMA) system [44]. In such a system, the radio spectrum is divided into frequency sub-bands and within each sub-band, time is divided into frames and slots. This network supported mostly voice calls and very low data rate Internet. However, there was a growing demand for high data rates required by multimedia communication. To meet this requirement, 3G networks relied on a different technique known as Direct Sequence Wideband Code Division Multiple Access (DS-WCDMA) within TDMA [44], wherein time slots are available on multiple frequencies. This allowed for a growth in the capacity of mobile networks and a data rate of up to 14 Mbps. As the demand for data services over mobile networks increased, the next generation of networks, 4G Long Term Evolution (LTE), introduced the orthogonal frequency division multiplexing (OFDMA) technique [44]. In such a system, each mobile node is allocated time slots in one or more radio channel frequencies. The OFDMA technique ensures that interference between signals sent on different frequencies is minimal. Along with other innovations in the radio network, the capacity of such networks increased tremendously and the maximum data rate increased to 100 Mbps in the downlink and 50 Mbps in the uplink [44].

At the same time, the core network has evolved from circuit-switched networks to the current packet-switched Evolved Packet System (EPS). Packet switched networks were introduced in 2G networks and removed the need for dedicated end-to-end circuit-switched connections. The current 4G EPS networks are completely IP-based; this allows for fewer protocol conversions and thus a higher performance [15]. This architecture also separated the control plane and data plane elements. The control plane transports signaling messages (related to mobility and management), whereas the data plane is responsible for handling user data packets. The separation of these planes in the mobile core network allows the network operators to scale the data plane and control plane elements independently and better meet the demands of end users.

The next step in the evolution of mobile networks, i.e. the fifth generation (5G), is expected to support a growing amount of data from mobile and Internet of Things (IoT) devices with low latency communication. Specifically, 5G networks will incorporate technologies needed for low latency, energy-efficient and reliable communications from heterogeneous devices. In the radio network, modification of the radio frame structure, millimeter wave [20] and non-orthogonal multiple access (NOMA) [58] are among the key features proposed. NOMA achieves better spectral efficiency than before by allowing multiple users to share the same radio resources (frequency, timeslots and spreading code). Furthermore, the use of a different mmWave spectrum (30–300 GHz range) supports massive bandwidth and ultra-low latency applications such as virtual and augmented reality [20, 58].

Another interesting development is the softwarization of the network through *Network Function Virtualization* (NFV) [33] and *Software Defined Networking* (SDN) [57]. Currently the mobile core network consists of proprietary hardware designed to meet the high performance requirements of such networks [60]. This implies that mobile network operators need to dimension their networks and plan for peak loads as upgrading of such infrastructure is expensive and slow. NFV utilizes the virtualization techniques described in Sect. 2.1 to enable the deployment of mobile network elements as software modules on general-purpose hardware [33]. This technique brings the benefits of cloud computing, namely scalability and reduced expenses, to the core network. The core elements can be scaled out or in depending on the actual demand. Furthermore, NFV aims to bring new innovative services to the mobile network as software-based deployments have a shorter deploy cycle than hardware-based implementations. Finally, as the elements are deployed as virtualized instances, they can be dynamically moved to the location most suitable for low-latency communication [71].

SDN involves the virtualization of networking itself [42] and is complementary to NFV. The main features of SDN are: the separation of the control and data plane, centralized network intelligence at a programmable controller, and standardized application programming interfaces (APIs) [42, 57]. By separating the control plane from the data plane, the complexity of network devices such as switches are greatly simplified. The devices only need to receive instructions from the central SDN controller and forward data packets based on these instructions. The centralized controller allows network operators to have an overview of the entire network. Programs running on the SDN controller can make real-time changes to any part of the network. Finally, the standardized APIs abstract away the networking infrastructure from the applications. The APIs also enable the management of devices from multiple vendors.

SDN and NFV together enable the flexible management and programming of complex networks. These technologies make it easier for network operators to react to changes in the network. Furthermore, the time to deploy services is considerably reduced from hardware-based implementations.

5G networks are expected to support several new applications and services [18]. For instance, multiple person video communication is expected to become pervasive. This will enable collaboration at a scale not seen before. This will be further enhanced by the support for augmented and virtual reality. Sensor- and user-generated data can be used to replicate the movement and gestures of people from a physical setting to a virtual one [61]. Furthermore, applications that rely on tactile signals and haptic feedback will become reality with the low latency communication offered by 5G networks. In the core network, edge computing (described in Sect. 2.3) has been proposed to move computing closer to the end user and thereby meet the low latency requirements of such applications. This will allow remote control of machinery and robots, thereby enabling applications in remote health care as well.

3.2 Low Power Wide Area Networks

Low power wide area networks (LPWANs) are a class of networks specifically targeted for resource constrained and battery-powered IoT devices. Such networks offer low power, long range connectivity (in the range of kilometers) and support only low data rates. Thus, LPWANs are highly suited for smart city and machine-to-machine applications, including smart metering, smart grid and agricultural monitoring [22, 64]. They support a class of applications and devices that cannot be otherwise served by existing wireless technologies. For instance, mobile networks (described in Sect. 3.1) are not energy efficient as they require much more complex processing on the end devices [64]. Moreover, this increased complexity would increase the cost of end devices. On the other hand, LPWANs promise a battery lifetime of ten years and a communication range of several kilometers [64]. It is important to note that these technologies are designed for a specific class of applications that require only low data rates (in the range of kilobits per second) and can tolerate some latency in communication [64]. This section describes the following LPWAN technologies: LoRa [45, 68], narrowband-IoT [63] and Sigfox [9].

LoRa networks consist of two main components, LoRa and LoRaWAN. LoRa refers to the proprietary physical layer developed by Semtech [68], whereas LoRaWAN [45] corresponds to the medium access control (MAC) and network layers of the protocol stack. The physical layer achieves long distance communication by using the *chirp spread spectrum modulation* technique, wherein the signal is encoded into *chirp* pulses spread over a wide spectrum [64]. Chirp pulses can go from low to high frequencies (up-chirp) or vice-versa (down-chirp) over time. This modulation technique makes the signal robust to interference from other transmissions. This is very useful as LoRa operates in the unlicensed sub-GHz band. Each LoRa transmission can be configured with the following parameters: carrier frequency, bandwidth, coding rate, spreading factor and transmission power [32]. These parameters affect the communication range, the data rate and the occurrence of collisions. Thus, it is possible to assign the parameters in such a way to minimize collisions between transmissions. For instance, LoRaWAN specifies an Adaptive Data Rate (ADR) algorithm that dynamically manages the communication parameters to increase the capacity of the network and maximize the battery life of the end devices [69]. LoRaWAN also specifies the architecture of LoRa networks. The end devices or LoRa nodes communicate with *gateways* over the LoRa physical layer. Gateways simply relay the messages received from the nodes to a central *network server*. Nodes are not associated with a single gateway; this implies that gateways can receive and process messages from all nodes within its communication range. The network server manages the network and further sends the messages to the required *application server*. Such an architecture allows to increase the capacity of the network by increasing the number of gateways [64]. For instance, The Things Network [73] (TTN) is an open, community-driven LoRa network which allows the general public to place gateways of their own and thereby expand coverage. TTN itself provides the network server and the means to integrate applications to the network. Such a community-driven network allows end users to develop applications that can use the coverage provided by deployed gateways. Moreover, several mobile network operators (such as KPN, Orange, Swisscom, Softbank and others) have started deploying LoRaWAN networks to meet the growing demand for services that rely on such networks [53].

Narrowband-IoT (NB-IoT) is standardized by the 3GPP and is based on the LTE technology (described in Sect. 3.1). It operates in the licensed radio spectrum and thus will be available through telecom service providers. Furthermore, there are no duty cycle restrictions as the spectrum is licensed. In contrast to LoRa, NB-IoT uses the *narrowband modulation* technique, wherein the signal is encoded in a low bandwidth which also minimizes the noise level [64]. This technique also ensures that the spectrum is efficiently utilized by all the links. NB-IoT reuses several concepts from LTE including the frequency-division multiple access (FDMA). This implies that the end device needs to synchronize to the carrier frequency. Thus, the complexity of devices increases as compared to LoRa. However, this also implies that NB-IoT can ensure a higher quality of service (QoS) and lower communication latency than LoRa. At the time of writing there are 58 commercial networks[6] that use NB-IoT.

Sigfox is an LPWAN solution provider that operates its eponymous network based on a proprietary technology [9]. Sigfox networks use the *ultra narrowband modulation* technique, wherein the signal is encoded into a very narrow bandwidth (of less than 100 Hz) [64]. This reduces the amount of noise resulting in higher receiver sensitivity and thus, long distance communication. However, this is achieved at the expense of the data rate which is limited to 100 bps [64]. The devices communicate with proprietary Sigfox base stations using a random access MAC protocol and thus devices are not complex. Sigfox networks are deployed in partnership with other service providers (including telecommunication service providers) and 62 countries[7] are expected to be covered by the end of 2018.

Each LPWAN solution has its own modulation scheme and relies on a separate network architecture. However, the goals of each network is similar: to provide low-cost, long-range, low-power network connectivity to resource constrained devices. The value of connecting these devices to the network will be more apparent from the discussion in the next section.

4 Internet of Things

The Internet of Things (IoT) comprises of billions of Internet-connected devices equipped with sensors and actuators that are able to interact and cooperate with each other to achieve a common goal [21]. Indeed, we already see examples of the IoT today: RFID sensor-based tracking of shipments, routing of vehicles based on GPS data and real-time control of home devices through home assistants [76]. The rapid increase in the number of deployed devices is due to the availability of more efficient

---

[6]https://www.gsma.com/iot/mobile-iot-commercial-launches/.

[7]https://www.sigfox.com/en/coverage/become-so.

and low cost IoT devices, improved wireless connectivity (described in Sect. 3) as well as advances in the cloud (described in Sect. 2).

4.1 IoT Devices

The devices forming the IoT can range from low-cost, low-complexity devices such as temperature/humidity sensors to full-fledged connected cars equipped with sophisticated sensors. This section highlights the main features of the IoT devices [17, 19, 39]. One of the most important features is that the device can *sense* its environment and collect measurements. Examples of these measurements include temperature, humidity, pressure, location, motion, light and sound. An on-board sensor can measure the data and convert it to a machine-friendly, digital representation. Certain devices are also equipped with *actuators*, i.e., components that produce a physical effect (such as motion or an electromechanical signal) in response to an input. A *processing unit* (such as a microcontroller, microprocessor, CPU, FPGA, etc.) handles several tasks, including processing sensed data and managing the remaining systems on the device. The capability of the processing unit depends on the requirements of the specific IoT application. Next, the device should be able to *communicate* to other devices or a network gateway or controller. Thus, the device typically contains a wireless transceiver for the chosen connectivity option. Again, here, the choice of connectivity standard depends on the requirements of the application. Finally, the device needs a *power source* to be able to function. Typically, the devices are battery-powered; but they can sometimes scavenge energy from other sources such as solar cells [19].

The growing availability of low-cost, low-power devices is leading to the pervasive deployment of IoT devices [16, 39]. For instance, micro-electro-mechanical systems (MEMS) are an attractive method to package sensors and actuators as they can integrate these elements on a very small scale at low costs. Thus, sensors can be easily added to everyday objects. Moreover, the on-board processors are getting more powerful while simultaneously becoming smaller. Finally, the power consumption of such devices are further reduced when using IoT-specific connectivity options such as LoRa.

4.2 Connecting IoT Devices

Next, we discuss how IoT devices communicate with each other or to the network. We have already discussed the available physical layer protocols in Sect. 3. Above this, there are several options for networking and transport protocols designed for IoT devices. At the network layer, IoT devices typically rely on the Internet Protocol (IP) and specifically the IPv6 protocol for networking [39]. However, there are several practical concerns to this approach especially for resource-constrained devices and unreliable wireless networks. For instance, some devices may not have sufficient energy resources (being battery-operated) to run the whole IP networking stack. Furthermore, some wireless networks such as LPWANs can have high latency and packet losses. In such scenarios, an *IoT gateway* acts as an intermediary on the communication path between the device and the application [39, 72]. These gateways translate non-IP packets from the devices to IP-based packets that can be sent to the application server. To this end, the Internet Engineering Task Force (IETF), an Internet standards organization, has developed standardized protocols to incorporate the resource-constrained non-IP devices into an IP-based network. The interested reader can refer to the IETF working groups[8] (*6lo, 6tisch, lpwan, ipwave*) for the specifics of the protocols for IoT networks.

At the transport layer, IoT networks may use either the Transmission Control Protocol (TCP) or User Datagram Protocol (UDP) [39]. The choice of the protocol depends on both the upper and lower layers of the networking stack as well as the capability of the devices. TCP ensures reliable communication through a connection-oriented scheme (a session is established between the sender and receiver) and several error correction mechanisms including retransmission of packets. On the other hand, UDP is a connectionless protocol wherein data packets are sent between sender and receiver without error control. Next, at the application layer, there are several options, including generic web-based HTTP protocols, existing messaging protocols (XMPP) and more lightweight IoT-specific protocols such as CoAP and MQTT. CoAP and MQTT aim to support messaging for resource-constrained devices that may operate in networks with packet loss and low bandwidth [17]. A detailed description of application protocols is available in [17, 39]. Again, the choice of protocol relies on the requirements of the specific application and capability of the devices.

4.3 Inter-operating Networks

The growing availability of smart, connected IoT devices has resulted in their rapid adoption in several areas, such as smart homes, industrial settings and transportation. The discussion above highlights the variety of communication protocols, which are decided based on the device capability and application requirements. Thus, the IoT solutions usually operate in their own vertical silos with specific communication protocols for the domain they operate in [25, 29]. Although such IoT solutions bring tremendous benefits within their particular area of operation, they do not meet the original vision of smart devices being able to cooperate with each other to meet a common goal. For instance, consider a scenario wherein IoT devices are deployed at bus stops across a city. These devices can detect the number of people currently waiting for a bus. This data can be used by the public transport provider to send more buses when there is a surge in demand. A truly IoT solution would also allow this data to be accessed by different services, such as a registered car pooling ser-

---

[8]https://datatracker.ietf.org/wg/.

vice that can also serve this demand. Such a scenario requires that devices can be seamlessly *discovered* and their sensed data *accessed* across different domains and architectures [72].

There are several approaches for the discovery of IoT devices. They can categorized into three main categories [10, 28]. One class relies on small- or medium-range wireless communication for discovering devices nearby. For instance, Google and Apple both use Bluetooth Low Energy for their UriBeacon (now part of the Eddystone project) and iBeacon projects that enable devices to discover and interact with each other. A second category of discovery relies on searching for device endpoints in a network. For instance, *multicast DNS*[9] (mDNS) is a distributed version of DNS service discovery (DNS-SD) that translates the IoT service/host name to an IP address within small networks. Finally, a third category relies on querying of centralized directories and thus scales to larger networks. The CoRE Resource Directory[10] is an example of such an approach. This approach allows services from other domains and networks to discover resources based on attributes; for instance, all devices matching a certain criteria (type or interface) can be listed by querying this directory [72].

Another important aspect of interoperability is that the data should be sent in a standardized format so that different applications and services can extract useful information from it [17]. To this end, the IPSO Alliance [2], a global body of multiple IoT companies, aims to enable IoT device interoperability by specifying open standards for semantics, security, device identity and other protocols. They have specified a data model known as Smart Objects [3] to describe IoT device resources. For instance, a temperature sensor could be represented as 3300/0/5700 where 3300 represents that it's a temperature sensor, 0 represents the 0th instance of the sensor and 5700 refers to the most recent reading. This abstraction allows the software application to use simple APIs to access and read the IoT device resources. The model is designed to work on top of any REST-based protocol. The Web of Things Thing Description, a standard developed by the World Wide Web Consortium (W3C), also describes a formal model for representing IoT devices [11]. In this model, device resources are typically represented in JSON-LD format and different application layer protocols such as MQTT, CoAP and HTTP are supported. A listing of other data models is available at [10]. Another approach relies on using a separate entity, i.e., a *data broker*, between the devices and the application server [39]. The broker converts the data from multiple devices to a common format that can be accessed by authorized applications. Such an approach is suitable for non-IP based devices as well and where an application protocol is not used.

The approaches highlighted above are a step towards achieving the goal of interoperable IoT networks that will further enable new and innovative services.

---

[9]https://www.ietf.org/rfc/rfc6762.txt.
[10]https://core-wg.github.io/rd-dns-sd/#resource-directories.

5 Towards a Sharing Economy

The increasing maturity of cloud-based services, wireless connectivity solutions and IoT devices creates a growing opportunity for data-driven solutions for the sharing economy. Such solutions can have a societal impact through the collaborative use of data from multiple providers, including individuals and public or private organizations [37]. To this end, a cloud-based IoT platform represents an ideal way to aggregate data from multiple sources and expose this information to different service providers. There are several commercial and open source IoT platforms available today [17]. OpenMTC [7] is a prominent example of an open source platform built to enable shared data access across several application domains. We briefly describe the key features of this platform and how it enables sharing economy solutions.

5.1 OpenMTC

OpenMTC is an open source implementation of the standardized IoT platform architecture by oneM2M.[11] The goal of such a platform is to provide a horizontal layer for IoT devices from different domains (such as healthcare, transport and utilities) to communicate with the application layer. The platform was released in 2012 and made open source in 2017. The architecture follows a hierarchy of three layers, described next.

- The *application layer* consists of *application entities* that implement the service logic. Examples of such entities include applications to monitor vehicle fleets or to track power consumption.
- The *common services layer* provides the core functionality of the platform including data and device management, service subscription management, discovery and others (illustrated in Fig. 4). The functionality of each entity is detailed in [5].
- The *network services layer* provides transport and connectivity services to the devices connected to the platform.

The interfaces between the different layers (Mca, Mcn and Mcc) are also standardized. Thus, the overall architecture is designed to be protocol-agnostic and thereby support devices from different domains. A node in the OpenMTC platform can implement one or more of the entities from the different layers (described above) and expose standardized interfaces to other nodes in the network. The *gateway* and *backend* are the main components of the OpenMTC platform. The gateway interconnects devices from different domains. *Protocol adapters* are used to provide inter-connectivity with other IoT platforms. Furthermore, the platform provides a *software development kit* (SDK) to allow application developers to write applications compliant with

---

[11]oneM2M (http://www.onem2m.org) is a global standards initiative comprising of eight regional ICT standards organizations and over 200 companies.

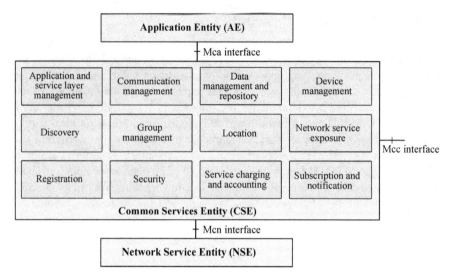

Fig. 4 oneM2M entities

oneM2M [8]. The source code of all components are available at [8] along with the respective Docker images.

5.2 Case Study: Car Sharing

A car sharing service is ideally suited for urban users who prefer not to own a car, but still have the convenience of using a car when required. When the user needs a car, he/she simply goes to the nearest available shared car, unlocks it and drives it to the preferred destination. The user then parks the car in a designated spot and has to pay only for the duration of the trip. The success of such a service depends on the seamless use of the cars with minimal user intervention. This can be achieved by leveraging data from IoT devices and other service providers (such as insurance companies). The following discussion is based on the oneM2M specification [6], which describes how IoT platforms (such as OpenMTC) can enable car sharing applications.

Cars today are equipped with a variety of sensors, including door control sensors, tire pressure sensors, fuel level sensors and GPS. The rich set of data from these sensors can be used to automate and simplify the process of utilizing shared cars for the end user. Sensor data is communicated to a cloud-based *IoT platform* via a *smartphone* which acts as the gateway. Other devices such as access points deployed along the road [62] can also behave as gateways. However, it is feasible to offer the car sharing service as a smartphone application and thus, the phone can behave as a gateway. Furthermore, the smartphone itself has additional sensors that can provide useful data, for instance, for navigation. The IoT platform collects the status

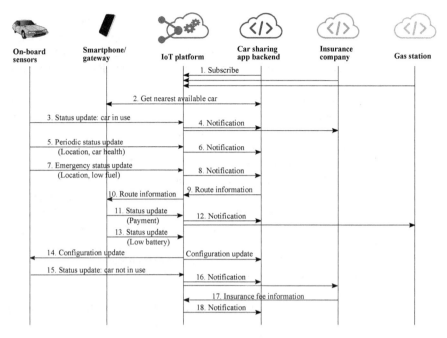

Fig. 5 Information flow for a car sharing application

and configuration information from the vehicles as well as the *service providers*. Service providers include the car sharing provider itself, insurance companies and gas stations. Providers are assumed to have an agreement to provide a unified service.

Figure 5 illustrates how the data from multiple sources can be exchanged for a car sharing service takes place.

1. First, the service provider applications register to the IoT platform. Each application subscribes to the specific information it requires. Such a subscription allows the service provider application to monitor updates or changes to the subscribed information. Examples of such information include location, health and fuel status of the car. This step also requires the IoT platform to ensure that only authorized applications are granted access to the information.
2. When a user intends to use a shared car, he/she obtains the location of the nearest available car from the smartphone application.
3. Next, he/she proceeds to the nearest car (pointed to by the application), opens the car door and starts the car through the application. The smartphone can interact with the car through Bluetooth or NFC. The application ensures that the user has a valid subscription and is authorized to use the car. The car's on-board sensors report that the car is in use to the IoT platform via the smartphone (or gateway).

4. The IoT platform communicates the status of the car as occupied to the backend of the car sharing application. This status update can be used to trigger an update to the front-end smartphone application or website to mark the car as unavailable to other users. The insurance provider is also informed of the status of the car.

5–6. The car periodically reports its status—including location, fuel status and health—to the IoT platform. The IoT platform reports the information to the applications subscribed to the particular data.

7–8. In case of an emergency, for instance, if the fuel is low, the car sends an update with an "urgent" status. This ensures that the IoT platform immediately notifies the car sharing backend application.

9–10. The backend application computes the location of the nearest gas station (with which it has a service agreement) based on the most recent location of the car. The application then communicates the route to the nearest gas station to the IoT platform, which relays this to the smartphone.

10. At the gas station, the user can make the payment through the smartphone's NFC interface. The car sharing application also sends a status update to the IoT platform.

11–12. The IoT platform communicates the payment details to both the car sharing backend and the gas station provider. This step could also allow the car sharing provider to directly make the payment to the gas station without having the user to pay.

13. Next, the smartphone reports its low battery status to the IoT platform.

14. The IoT platform changes the configuration of the status updates to minimize battery usage. For example, the platform can increase the time interval between status updates or configure only urgent notifications to be sent. The platform then informs both the smartphone and the car sharing backend application of the new configuration.

15–16. When the user arrives at the destination, the user stops the car and turns off the ignition. All backlog data (if the configuration was changed) is reported to the IoT platform along with the notification that the car is no longer in use. The platform, in turn, notifies the backend application and the insurance provider.

17–18. The insurance provider application sends a message containing the insurance fee information to the IoT platform, which is then communicated to the car sharing application backend.

6 Conclusion

This chapter reviewed the major developments in cloud computing, wireless connectivity and IoT in the context of collaborative consumption. First, cloud computing has significantly lowered the economic barrier for deploying software applications. The software development process is further simplified by the growing popularity of Docker-based microservices and stateless functions. This in turn results in faster time to market for new services. Next, improvements in wireless connectivity have resulted in the ubiquitous availability of Internet connectivity. Low power wireless solutions specific to IoT devices have also emerged in the past few years. This allows a large number of low-cost IoT devices to connect to the network and share their sensed data. Thus, new sharing economy solutions can utilize the large volume of data from a diverse set of IoT devices to provide services in an automated and collaborative manner. Furthermore, edge and fog computing are expected to support a new class of applications that require processing of data with a very low latency.

Acknowledgements This work was partially supported by the Academy of Finland under grants number 299222 and 319710.

References

1. Cisco visual networking index: global mobile data traffic forecast update, 2016–2021 white paper. http://www.cisco.com/c/en/us/solutions/collateral/service-provider/visual-networking-index-vni/mobile-white-paper-c11-520862.pdf. Accessed 16/07/2018
2. IPSO smart objects. https://www.omaspecworks.org/develop-with-oma-specworks/ipso-smart-objects/. Accessed 23.07.2018
3. IPSO smart objects. https://github.com/IPSO-Alliance/pub. Accessed 23.07.2018
4. Mobile-edge computing (MEC); service scenarios. https://www.etsi.org/technologies-clusters/technologies/multi-access-edge-computing. Accessed 24/07/2018
5. oneM2M functional architecture, ETSI standard TS-0001-V3.11.0. http://www.onem2m.org/technical/published-drafts. Accessed 07.08.2018
6. oneM2M vehicular domain enablement, draft technical report TR-0026-V4.1.0. http://www.onem2m.org/technical/published-drafts. Accessed 07.08.2018
7. OpenMTC. http://www.open-mtc.org/index.html. Accessed 07.08.2018
8. OpenMTC. https://github.com/OpenMTC/OpenMTC. Accessed 08.08.2018
9. Sigfox. https://www.sigfox.com/en. Accessed 18.07.2018
10. Web of things – Technology landscape. http://w3c.github.io/wot/landscape.html. Accessed 23.07.2018
11. Web of things (WoT) thing description. https://www.w3.org/TR/wot-thing-description/. Accessed 23.07.2018
12. What is a container? https://www.docker.com/what-container. Accessed 16/07/2018)
13. What is docker? https://www.docker.com/what-docker. Accessed 16/07/2018
14. IEEE approved draft standard for adoption of openfog reference architecture for fog computing. IEEE P1934/D2.0, April 2018, pp 1–175 (2018)
15. 3GPP: the evolved packet core. http://www.3gpp.org/technologies/keywords-acronyms/100-the-evolved-packet-core. Accessed 04.07.2018
16. Akyildiz IF, Su W, Sankarasubramaniam Y, Cayirci E (2002) Wireless sensor networks: a survey. Comput Netw 38(4):393–422
17. Al-Fuqaha A, Guizani M, Mohammadi M, Aledhari M, Ayyash M (2015) Internet of things: a survey on enabling technologies, protocols, and applications. IEEE Commun Surv Tutor 17(4):2347–2376
18. Alliance N (2015) 5g white paper. Next generation mobile networks, white paper pp. 1–125
19. Anastasi G, Conti M, Di Francesco M, Passarella A (2009) Energy conservation in wireless sensor networks: a survey. Ad Hoc Netw 7(3):537–568

20. Andrews JG, Buzzi S, Choi W, Hanly SV, Lozano A, Soong AC, Zhang JC (2014) What will 5G be? IEEE J Sel Areas Commun 32(6):1065–1082
21. Atzori L, Iera A, Morabito G (2010) The internet of things: a survey. Comput Netw 54(15):2787–2805
22. Augustin A, Yi J, Clausen T, Townsley WM (2016) A study of LoRa: long range & low power networks for the internet of things. Sensors 16(9):1466
23. Baldini I, Castro P, Chang K, Cheng P, Fink S, Ishakian V, Mitchell N, Muthusamy V, Rabbah R, Slominski A et al (2017) Serverless computing: current trends and open problems. Research advances in cloud computing. Springer, Berlin, pp 1–20
24. Barham P, Dragovic B, Fraser K, Hand S, Harris T, Ho A, Neugebauer R, Pratt I, Warfield A (2003) Xen and the art of virtualization. ACM SIGOPS Oper Syst Rev 37:164–177. ACM
25. Bello O, Zeadally S, Badra M (2017) Network layer inter-operation of device-to-device communication technologies in internet of things (IoT). Ad Hoc Netw 57:52–62
26. Bonomi F, Milito R, Natarajan P, Zhu J (2014) Fog computing: a platform for internet of things and analytics. Big data and internet of things: a roadmap for smart environments. Springer, Berlin, pp 169–186
27. Bonomi F, Milito R, Zhu J, Addepalli S (2012) Fog computing and its role in the internet of things. In: Proceedings of the 1st edition of the MCC workshop on mobile cloud computing. ACM, pp 13–16
28. Bröring A, Datta SK, Bonnet C (2016) A categorization of discovery technologies for the internet of things. In: Proceedings of the 6th international conference on the internet of things. ACM, pp 131–139
29. Bröring A, Schmid S, Schindhelm CK, Khelil A, Kabisch S, Kramer D, Le Phuoc D, Mitic J, Anicic D, Teniente López E (2017) Enabling IoT ecosystems through platform interoperability. IEEE Softw 34(1):54–61
30. Burns B, Grant B, Oppenheimer D, Brewer E, Wilkes J (2016) Borg, omega, and kubernetes. Queue 14(1):10
31. Buyya R, Yeo CS, Venugopal S, Broberg J, Brandic I (2009) Cloud computing and emerging it platforms: vision, hype, and reality for delivering computing as the 5th utility. Future Gener Comput Syst 25(6):599–616
32. Croce D, Gucciardo M, Mangione S, Santaromita G, Tinnirello I (2018) Impact of lora imperfect orthogonality: analysis of link-level performance. IEEE Commun Lett 22(4):796–799
33. ETSI: network functions virtualisation - An introduction, benefits, enablers, challenges, call for action. Technical Report (2012)
34. Felter W, Ferreira A, Rajamony R, Rubio J (2015) An updated performance comparison of virtual machines and linux containers. In: 2015 IEEE international symposium on performance analysis of systems and software (ISPASS). IEEE, pp 171–172
35. Fischer JE, Colley JA, Luger E, Golembewski M, Costanza E, Ramchurn SD, Viller S, Oakley I, Froehlich JE (2016) New horizons for the IoT in everyday life: proactive, shared, sustainable. In: Proceedings of the 2016 ACM international joint conference on pervasive and ubiquitous computing: adjunct. ACM, pp 657–660
36. Fox A, Griffith R, Joseph A, Katz R, Konwinski A, Lee G, Patterson D, Rabkin A, Stoica I (2009) Above the clouds: a Berkeley view of cloud computing. Department of Electrical Engineering and Computer Science, University of California, Berkeley, Rep. UCB/EECS 28(13):2009
37. García JM, Fernández P, Ruiz-Cortés A, Dustdar S, Toro M (2017) Edge and cloud pricing for the sharing economy. IEEE Internet Comput 21(2):78–84. https://doi.org/10.1109/MIC.2017.24
38. Garriga M (2018) Towards a taxonomy of microservices architectures. In: Cerone A, Roveri M (eds) Software engineering and formal methods. Springer International Publishing, Cham, pp 203–218
39. Hanes D, Salgueiro G, Grossetete P, Barton R, Henry J (2017) IoT fundamentals: networking technologies, protocols, and use cases for the internet of things. Cisco Press, Indianapolis
40. Hightower K, Burns B, Beda J (2017) Kubernetes: up and running: dive into the future of infrastructure. O'Reilly Media, Sebastopol

41. Hong K, Lillethun D, Ramachandran U, Ottenwälder B, Koldehofe B (2013) Mobile fog: a programming model for large-scale applications on the internet of things. In: Proceedings of the 2nd ACM SIGCOMM workshop on mobile cloud computing. ACM, pp 15–20
42. Jain R, Paul S (2013) Network virtualization and software defined networking for cloud computing: a survey. IEEE Commun Mag 51(11):24–31
43. Kivity A, Kamay Y, Laor D, Lublin U, Liguori A (2007) KVM: the linux virtual machine monitor. In: Proceedings of the linux symposium, vol 1, Ottawa, Ontorio, Canada, pp 225–230
44. Kurose JF, Ross KW (2013) Computer networking: a top-down approach: international edition. Pearson Higher Education
45. LoRa Alliance: LoRaWAN Specification (V1.0.3). https://www.lora-alliance.org/resource-hub/lorawantm-specification-v103 (2018). Accessed 18.07.2018
46. Malmborg L, Light A, Fitzpatrick G, Bellotti V, Brereton M (2015) Designing for sharing in local communities. In: Proceedings of the 33rd annual ACM conference extended abstracts on human factors in computing systems. ACM, pp 2357–2360
47. Marín-Tordera E, Masip-Bruin X, García-Almiñana J, Jukan A, Ren GJ, Zhu J (2017) Do we all really know what a fog node is? current trends towards an open definition. Comput Commun 109:117–130
48. Marston S, Li Z, Bandyopadhyay S, Zhang J, Ghalsasi A (2011) Cloud computing-the business perspective. Decis Support Syst 51(1):176–189
49. Mcqueen D (2009) The momentum behind lte adoption [sgpp lte]. IEEE Commun Mag 47(2):44–45
50. Mell P, Grance T et al (2011) The NIST definition of cloud computing
51. Muller A, Wilson S (2005) Virtualization with VMware ESX server
52. Nadareishvili I, Mitra R, McLarty M, Amundsen M (2016) Microservice architecture: aligning principles, practices, and culture. O'Reilly Media, Sebastopol
53. Navarro-Ortiz J, Sendra S, Ameigeiras P, Lopez-Soler JM (2018) Integration of LoRaWAN and 4G/5G for the Industrial internet of things. IEEE Commun Mag 56(2):60–67
54. Newman S (2015) Building microservices: designing fine-grained systems. O'Reilly Media, Sebastopol
55. Nickoloff J (2016) Docker in action, 1st edn. Manning Publications, Greenwich
56. Nider J (2018) A comparison of virtualization technologies for use in cloud data centers. IBM research report H-0330 (HAI1801-001)
57. ONF: software-defined networking: the new norm for networks. ONF white paper (2012)
58. Parvez I, Rahmati A, Guvenc I, Sarwat AI, Dai H (2018) A survey on low latency towards 5g: ran, core network and caching solutions. IEEE Commun Surv Tutor
59. Popek GJ, Goldberg RP (1974) Formal requirements for virtualizable third generation architectures. Commun ACM 17(7):412–421
60. Premsankar G, Ahokas K, Luukkainen S (2015) Design and implementation of a distributed mobility management entity on openstack. In: 2015 IEEE 7th international conference on cloud computing technology and science (CloudCom). IEEE, pp 487–490
61. Premsankar G, Di Francesco M, Taleb T (2018) Edge computing for the Internet of Things: a case study. IEEE Internet Things J 5(2):1275–1284
62. Premsankar G, Ghaddar B, Di Francesco M, Verago R (2018) Efficient placement of edge computing devices for vehicular applications in smart cities. In: NOMS 2018-2018 IEEE/IFIP network operations and management symposium. IEEE
63. Ratasuk R, Mangalvedhe N, Zhang Y, Robert M, Koskinen JP (2016) Overview of narrowband IoT in lte rel-13. In: 2016 IEEE conference on standards for communications and networking (CSCN). IEEE, pp 1–7
64. Raza U, Kulkarni P, Sooriyabandara M (2017) Low power wide area networks: an overview. IEEE Commun Surv Tutor
65. Roberts M, Chapin J (2017) What is serverless? Understanding the latest advances in cloud and service-based architecture. O'Reilly Media, Sebastopol
66. Satyanarayanan M, Bahl P, Caceres R, Davies N (2009) The case for VM-based cloudlets in mobile computing. IEEE Pervasive Comput 8(4):14–23

67. Satyanarayanan M, Simoens P, Xiao Y, Pillai P, Chen Z, Ha K, Hu W, Amos B (2015) Edge analytics in the internet of things. IEEE Pervasive Comput 14(2):24–31
68. Semtech: What is LoRa? https://www.semtech.com/technology/lora/what-is-lora. Accessed 18.07.2018
69. Slabicki M, Premsankar G, Di Francesco M (2018) Adaptive configuration of lora networks for dense IoT deployments. In: 16th IEEE/IFIP network operations and management symposium (NOMS 2018), pp 1–9
70. Strauss D (2013) Containers-not virtual machines-are the future cloud. Linux J 228:118–123
71. Taleb T (2014) Toward carrier cloud: potential, challenges, and solutions. Wirel Commun IEEE 21(3):80–91
72. Tanganelli G, Vallati C, Mingozzi E (2018) Edge-centric distributed discovery and access in the internet of things. IEEE Internet Things J 5(1):425–438
73. The things network: the thing network mission. https://github.com/TheThingsNetwork/Manifest/blob/master/Mission.md (2015). Accessed 18.07.2018
74. Wang CX, Haider F, Gao X, You XH, Yang Y, Yuan D, Aggoune H, Haas H, Fletcher S, Hepsaydir E (2014) Cellular architecture and key technologies for 5G wireless communication networks. IEEE Commun Mag 52(2):122–130
75. Watson J (2008) Virtualbox: bits and bytes masquerading as machines. Linux J 2008(166):1
76. Woetzel J, Remes J, Boland B, Lv K, Sinha S, Strube G, Means J, Law J, Cadena A, von der Tann V (2018) Smart cities: digital solutions for a more livable future. McKinsey Global Institute, San Francisco

Distributed Ledger Technologies and the Collaborative Economy

Pietro Ferraro and Daniel Conway

Abstract Distributed Ledger Technologies (the agnostic term for Blockchain and related technologies) have recently become one of the most controversial and debated topics in industry and academy alike. They represent a way to realize databases of replicated, shared, and synchronized digital data, spread across multiple agents. Rather than having a central server, distributed ledgers make use of a peer to peer consensus system to ensure the consistency of the database, across every copy in the network. The emergency of this technology enabled, in the past ten years, the proliferation of well known cryptocurrencies (e.g., Bitcoin, Ethereum, IOTA) and promises to reshape the way we think about business and trust, in the Internet of Things era. In this chapter, after a brief introduction to Distributed Ledger Technologies in the context of the sharing economy, we present and compare the two main architectures on which distributed ledgers are built: the Blockchain and Directed Acyclic Graphs.

Keywords Distributed ledger technology · Blockchain · Directed acyclic graph

1 Introduction

In the past ten years, we have seen the emergence of Distributed Ledger Technologies (DLTs) as the new foundational technology that promises to change not only the way we conduct business but also the way of thinking human interactions. Since the publication of Nakamoto's white paper where Bitcoin, and its underlying architecture, the Blockchain, was firstly introduced [1], academia and industry alike have been

---

P. Ferraro (✉)
Dyson School of Design Engineering, Imperial College London,
Imperial College Rd, Kensington, London SW7 1AL, United Kingdom
e-mail: p.ferraro@imperial.ac.uk

D. Conway
University of South Florida, Tampa, USA
e-mail: dconway@usf.edu

probing the boundaries of this new technology beyond the domain of cryptocurrencies. Possible applications range from, storing healthcare and ID information [2], arbitrating roles and permissions in IoT [3], improvement in the insurance sector [4] and as a means to enforce compliance in a smart city environment [5]. In the specific context of the Sharing economy, the use of DLTs promises to transform the business models of today, based on the existence of a centralized entity (e.g., Uber, Airbnb), by allowing the creation of networks of peers that can interact with each other without the need for a trusted third party. Moreover, DLTs enable users to run software in a secure and decentralized manner making it possible to develop distributed applications that no longer need to be deployed on a centralized server. These DLT-based applications can be used to coordinate the activities of a large number of individuals or organizations (such as companies), who can manage their interactions without the necessity of a third party to orchestrate them [6, 7]. DLTs allow individuals to interact directly with one another in a decentralized way and, thus, it is no surprise that the sharing economy represents one of the domains in which this new technology might have the largest impact on, in the near future. Therefore, due to their potential, in what follows we provide an overview on the fundamentals of DLTs and the two main architectures namely the Blockchain and Directed Acyclic Graphs (DAGs).

2 The Blockchain

All transactions require some type of trust to succeed. Typically, that trust is provided by a government agency, a third party such as an attorney or accounting firm, or a financial institution, such as a bank or regulated exchange. These kinds of services are normally provided in exchange for a fee. Collaborative economy transactions are no different. Lift and Airbnb have created trust as part of their offerings through brand, help functionality, ratings, and they charge a fee for this, close to 25%. How could transactions occur without a formal trusted third party, thus lowering the transaction cost significantly? This is one problem DLTs solve, as the consensus mechanism plays the role of the trusted third party.

2.1 Bitcoin and the Blockchain

Bitcoin emerged in 2008, at a time when financial institutions were being heavily criticised and attacked by the public opinion. It is not the first among the virtual currencies, but it was the first one to solve the problem on double-spend, disintermediating trust, and incentivizing participation [1]. The actual underlying technologies of hash functions [8] and elliptical encryption [9] had been already well known in the scientific literature. In his white paper, Nakamoto introduces Blockchain as the architecture on which the Bitcoin was developed. Since then, following the success of Bitcoin, a large number of other currencies have been devised on the basis of the original design. Almost all these currencies, at their core, share the same functioning architecture introduced by Nakamoto. A Blockchain is a peer to peer (P2P) distributed ledger of transactions, meaning that the ledger file (i.e., the spreadsheet that holds every transaction record) is not stored at a central node, in a classical client/server architecture, but rather copies are distributed across a network. In this sense a Blockchain (and DLT in general) is a method to achieve consensus on a completely decentralized database. In order to update the ledger (i.e., exchange currency or information), nodes issue transactions among each other using a public/private key cryptography [6]. Every account-holder has a public key and a private (secret) key. The latter is used to sign/authenticate transactions, whereas the public one provides a unique address to the user. In a Blockchain there are two kinds of agents: regular ones and miners. Regular users issue transactions whereas miners are special nodes whose task is to update the ledger in exchange for a monetary reward. The exact mechanics on how miners perform this task are provided later in this chapter. One of the key features of every Blockchain is the immutability of the ledger, i.e. the ability for a Blockchain ledger to remain a permanent, indelible, and unalterable history of transactions. This is achieved making use of cryptographic hash functions (CHF) [8]. A CHF possess a number of desirable properties for the Blockchain, namely:

- It maps any input onto data of fixed size;
- It is not feasible to invert (i.e., given the output of a CHF the only way to obtain the input is through a brute force attempt);
- Small changes in the input produce large variations in the output (such that the new output appears uncorrelated with the old one).

For example, a transaction T1 might be "Dan would like to give Bob 2 bitcoins." The corresponding SHA256 hash [8] would be: "b12a0b5137e00b47ec642a96760e 424a895554313190b1b111f51a6bbb7b2f97". Refer to this transaction as T1. If someone tried to change it to 3 bitcoins, and thus the transaction to "Dan would like to give Bob 3 bitcoins", the resulting SHA256 hash would be "d016e430392b8c20ad 443a79480cbeed4ea1ac8784355fdbca0a3a1bf8775ce2". This is a totally different hash, a clear indication that the original document has been modified. In a Blockchain, we also want our hash function to guarantee that a set of transactions is not tampered with. Consider T1 above and T2, which is "Emanuele would like to give Joe 1.5 bitcoins", which has the corresponding hash of "431f992cab10d92d85359b6e316f52e 226d22b234e238a25bc3c068d93b43df1". In order to record that T1 came before T2, the transactions would be ordered, and the hash of T1 and T2 would be hashed together as well. In this case if one were to append T1 to T2, then hash that combination, one would get "cd856d896ec7671640001e4ccf4572bc57611c87375b796f91d2a4241 13b 2263". In this way it is possible to order transactions sequentially and establish a temporal ordering in the Blockchain (as any change in the transactions ordering would result in a completely different hash). Now refer to Fig. 1.

In the Blockchain transactions are collected by miners and issued to the network in the form of "blocks". Each block has a top-level hash in it. If we were to combine the previous block's hash with the current block's hash, as described above, the new hash would serve to protect against changes of the last block as well as the current block

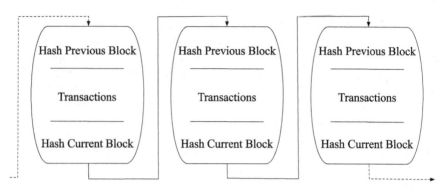

Fig. 1 Visual representation of three blocks of a Blockchain. Each block references the previous one and therefore any change in any part of the chain would result in an inconsistency

(if one of the blocks was changed, even slightly, then every other successive hash in the chain would be different, resulting in an inconsistency). This mechanism is what ensures the immutability property of the Blockchain. Finally, the first block on any Blockchain is called "the genesis block". It contains configuration information and is itself immutable.

2.2 Bitcoin Addresses

In order to exchange currency, users issue transactions among each other using public/private key cryptography. Every account-holder has a "wallet" to which are associated a public key and a private (secret) key. Private keys are created randomly by a random number generator. Using elliptical curve cryptography, a public key is created from the private key. A public key can be shared publicly, and the corresponding hash can be calculated by anyone to create the wallet address. So, how does a person prove they have the right to spend tokens? How can a user prove he is the rightful owner of the currency on a certain wallet? Consider the following example in the Bitcoin Blockchain: Dan has a wallet with 5 Bitcoin (BTC). Dan decides to send Joe 1 BTC. Dan builds a transaction, sending 1 BTC to Joe. He then hashes the transaction and encrypts the hash with his private key. He appends the encrypted hash along with a copy of the public key. The transaction is broadcast to the Bitcoin network and received by miners for consideration. How do miners verify that Dan has the right to spend the tokens in his wallet? The owner has to show that they have the private key associated with the public key. First, the public key is hashed and compared to the Bitcoin address. As public keys are known by everyone, this does not show ownership. Next, the miner hashes the transaction. As Dan has sent a hash encrypted in his private key, the miner decrypts that hash with Dan's public key. If the decrypted hash matches the hash the miner computed, then the transaction is considered valid. Only the person with the corresponding private key could have encrypted the hash so that the public key decrypted it. If the hashes do not match, or if there is an inconsistency, then the miner drops the transaction and does not include it in the next block.

2.3 Consensus and Proof of Work

As mentioned previously, miners are the nodes in the network responsible to update the ledger and to ensure consensus in the network. This task is performed in two steps. First a miner selects transactions to be added as block to the Blockchain. Secondly, assuming consistency of transactions, miners must perform a certain amount of computations to validate the next block in the Blockchain. This is generally referred to as Proof of Work (PoW). PoW involves solving a computationally-hard puzzle; more specifically, the node that performs it needs to calculate a hash in a high dimensional space that satisfies certain conditions (due to the nature of the problem, it is not feasible to perform anything different than a brute force approach [1]). The first miner who is able to compute a valid hash and hence solve the puzzle is also the miner that is eligible to add the next block. The amount of computations needed to find the solution to this puzzle is very high and as long as the total computational power of the honest miners is greater than the computational power of dishonest users, honest nodes will outpace dishonest ones and only legitimate transactions will be part of the Blockchain. In other words, any malicious attempt to tamper with the Blockchain ledger (e.g. trying to alter past transactions) can succeed only if attackers possess a sufficient amount of computational power (that depends on the total amount of computational power employed by honest miners). For more information regarding the security of Blockchain systems, the interested reader can refer to [10–13]. In the event two or more miners manage to find the correct hash at the same time, to avoid conflicts, the Blockchain protocol forces each node to build immediately on the longest chain available. In other words, if two valid blocks are issued at the same time, they will both be accepted in the Blockchain: at this stage there will be multiple legitimate chains on which miners can try to add further blocks. Due to the random nature of the PoW, it is very unlikely that two chains will continue growing at the same time. Therefore, eventually the longest chain will be accepted by the network and all the other ones will be discarded [5]. In this way, the Blockchain enforces consensus and avoids possible forks that would compromise consistency. The Blockchain has shown a very successful proof of concept for a virtual currency, but due to its slow convergence, long time between blocks, and expensive proof-of-work, it is likely to be replaced by new architectures that address particular use cases more effectively. In this perspective, in the next section, we present an alternative architecture, based on a particular graph structure, to achieve consensus on a distributed ledger.

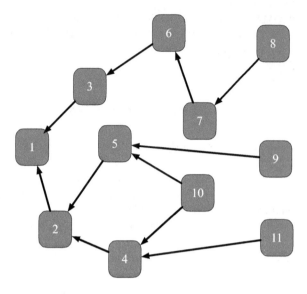

Fig. 2 Example of a DAG with 11 vertices and 10 edges. All the vertices are directed and it is impossible to find a path that connects any vertex with itself

3 DAGS, an Alternative to Blockchain?

Blockchain is not the only choice to achieve consensus among multiple users on a common ledger: another option, firstly proposed in [14], is to make use of Directed Acyclic Graphs (DAGs) as a data structure. A DAG is a graph with a finite number of vertices and directed edges, such that there is no directed path that connects any vertex of the graph with itself. An example of a DAG is depicted in Fig. 2.

One particular instance of a DAG-based DLT is The Tangle [14]. According to the original white paper, its aim is to provide a feeless and low energy consumption cryptocurrency for the IoT industry. In The Tangle each vertex, or site, of the graph represents a transaction (in the remainder of this chapter, site, transaction and vertex will be used interchangeably), edges represent approvals (described in detail later in this chapter) and the ledger information lies in the interconnections of the graph. Users can issue transactions to the ledger by taking part in the validation process: a new transaction must approve m (usually two) previously added transactions (basically adding a new vertex with m edges to the graph). The selected transactions need to be compatible with each other; while the compatibility between transactions is going to be discussed in detail in the next section, intuitively two transactions are compatible with each other if they do not introduce any inconsistency in the ledger. The recently added sites, which have not yet received approval from other vertices are called tips and the set of all unapproved transactions is called the tips set. New transactions will select sites from the tips set for approval (it is reasonable to expect that, at least the majority of them, will). Whenever a transaction is approved a new edge is added to the graph: a directed edge from site i to site j means that i directly approves j. If there is a directed path, longer than one single edge, from i to j we say that j is indirectly

Fig. 3 Transaction 8 directly approves 5 and 6. It indirectly approves 1, 2 and 3. It does not approve 4 and 7

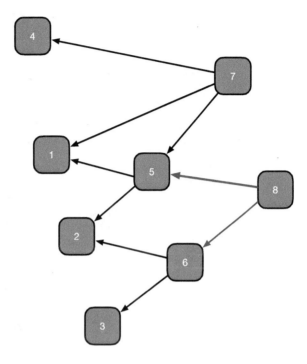

approved by i (e.g., see Fig. 3). The core metric of the Tangle, further discussed in the next section, is the Cumulative weight of a transaction: this value represents the total number of vertices that approve, directly or indirectly, a given site. Figure 4 shows an example of how the Cumulative weight changes in time.

The first transaction in the Tangle is called the genesis site (namely, the transaction where all the tokens were issued from the original account) and all transactions indirectly approve it (and therefore its Cumulative weight, at every time, is equal to the number of sites in the Tangle). Furthermore, in order to prevent malicious users from spamming the network, the approval step requires PoW. This step is less computationally intense than its Blockchain counterpart, and can be easily carried out by common IoT devices (e.g., smartphones, smart appliances, etc.). As mentioned above the PoW introduces a delay for new transactions before they are added to the tangle. Due to the time necessary to carry out the PoW, these sites may no longer be tips when the transaction is added to the tangle.

As a final note, to illustrate the time evolution of the Tangle, Fig. 5 shows an instance of the Tangle with three new incoming sites (upper panel).

The green block (the leftmost) is the genesis transaction, blue blocks are transactions that have already been approved, red blocks represent the current tips of the Tangle and grey blocks are new incoming vertices. Immediately after being issued, a new transaction tries to attach itself to m (in this instance two) of the network tips (middle panel). If any of the selected tips was inconsistent with the previous transactions, or with each other, the selection would be rejected and the process would be

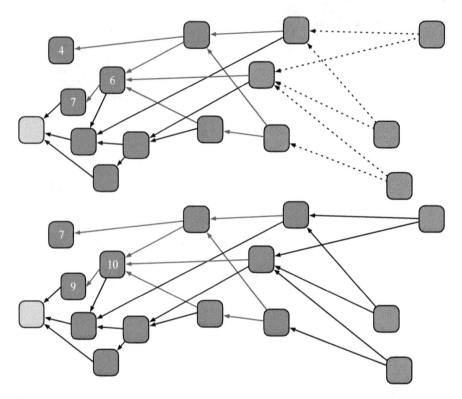

Fig. 4 Representation of the evolution of the Cumulative weight of three sites as three new transactions enter the Tangle

performed again, until two consistent sites are found. Notice that at this stage, the newly arrived transactions are carrying out the required PoW, and that the tips remain unconfirmed (dashed lines) until this process is over. Once the PoW is finished, the selected tips become confirmed sites and the grey blocks are added to the tips set (lower panel).

3.1 Double Spending Attack and Tips Selection Algorithms

In the previous section we discussed about compatibility and consistency between transactions. Checking if two transactions are consistent, during the selection process, is referred as the "verification step". We will not go into much detail to check if two transactions are consistent with each other but we can assume that the process is fast enough to be considered instantaneous and can be carried out with ease by any device. If verification fails, the selection process must be re-run until a set of consistent transactions is found. This consistency property is needed in order to pre-

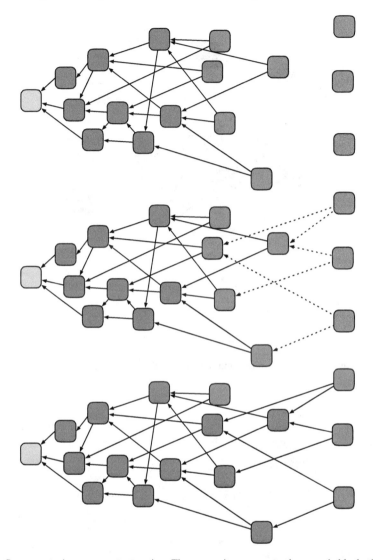

Fig. 5 Sequence to issue a new transaction. The green site represents the genesis block, the blue sites represent the approved transactions and the red ones represent the tips. The back edges represent approvals, whereas the dashed ones represent transactions that are performing the PoW in order to approve two tips

vent malicious users from tampering with the ledger by means of a double spending attack. To see how a double spending attack might be carried out on the Tangle, let us discuss a specific example to explain with more detail the process of approval. Figure 6 shows an instance of the Tangle.

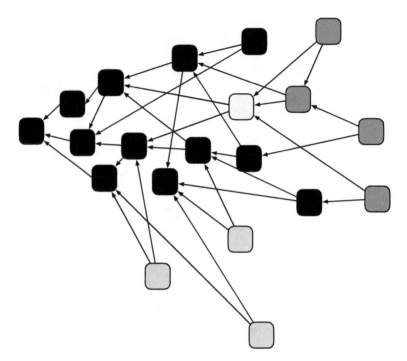

Fig. 6 The blue and the green transactions are incompatible with each other

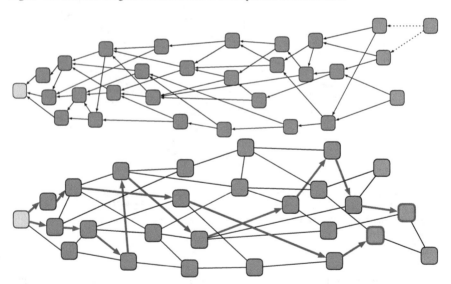

Fig. 7 Representation of the two main tips selection algorithms: the upper panel shows an instance of the Random Selection algorithm whereas the lower one presents a possible example of the Markov Chain Monte Carlo algorithm

Alice sent some currency to Bob, in exchange for some goods or for some service. The corresponding transaction is the yellow site. The same user, afterwards, produces other transactions, where she spends again the same tokens that were initially sent to Bob (e.g., she could write another transaction where she sends the same amount of currency to another address that she owns). These ones correspond to the green blocks. It is worth stressing again, at this point, that there is no mechanism to force a user to select certain sites for approval: since every user possesses a copy of the ledger they can choose any transaction for approval. Despite that, it is reasonable to assume that the vast majority of users would have little interest in targeting specific sites for approval, as their only interest is to issue their own transaction, and would instead follow the tips selection algorithm that the main protocol proposes (for a game theoretic analysis of this subject refer to [15]). In this instance, all the transactions that approve (directly or indirectly) the original yellow site (the blue blocks) are incompatible with the green ones, therefore any new transactions can either approve the green/black sites or the blue/black ones. The green/blue selection would be considered inconsistent and a new selection needs to be performed. At this point, Alice's objective would be to wait for Bob to accept their payment, receive their goods, then create one or more double spending transactions that get approved by other honest sites. It is not possible for two conflicting branches of the Tangle to both continue growing indefinitely (for a formal proof on the subject see [5]). Therefore it follows that, if an attacker succeeds in their double spending attempt, the Tangle will continue to grow from the illegitimate double spending transaction, and the legitimate branch with the original payment to the merchant will be orphaned, meaning that its sites will not be approved by new transactions. The success probability of such an attack depends on the particular selection algorithm employed by the majority of the users. So far, two algorithms have been proposed:

- Random Selection Algorithm: The Uniform Random Selection (URS) algorithm selects m (generally two) tips randomly from the pool of all possible tips. The upper panel of Fig. 7 shows an illustrative example of this procedure. This algorithm, due to its simplicity, makes the Tangle vulnerable to double spending attacks. The interested reader can refer to [14] for a detailed discussion on this topic.
- Markov Chain Monte Carlo Algorithm: In the Markov Chain Monte Carlo Selection Algorithm (MCMC) m (generally two) independent random walks are created on the tangle; the walks start at the genesis transaction and jump randomly, from one vertex to another, along the edges of the tangle. The jumping probability from transaction j to transaction k is proportional to $f(-\alpha(\mathcal{H}_j - \mathcal{H}_k))$, where $f(\cdot)$ is a monotonic increasing function (generally an exponential), α is a positive constant and $\mathcal{H}_i$ represents the Cumulative weight of transaction i. The jumping process stops when the particle jumps on a tip (an absorbing state), which is then selected for approval.

The main difference with the RS algorithm lies in the use of the graph structure: an attacker would need to create enough transactions, with Cumulative weights equal to or larger than the Cumulative weight of the main DAG, in order to make the double spending successful as, in any other scenario, the probability of jumping on one of the double spending tips would be too low. This, in turn, would require the malicious user to possess an amount of computational power comparable to the network of honest users. The interested reader can find more details on this topic in [14, 16].

4 DLTs as Database

A prudent question at this point might be, why would a database not do the same thing as a DLT? Can we not replicate a database to give us the same redundancy that DLTs offer? Would a database not store all transactions in a log, and make them much more efficient to retrieve? DLTs are not going to replace classical database technology in many application. As Blockchains continue to store transactions made, exploring them through linked blocks will probably result in poor performance. For example, at the time of this publication, Bitshares Blockchain (a cryptocurrency) has a size of approximately 600 GB. Building applications on top of a Blockchain generally will require transferring the data into a searchable optimized relational database before it can be efficiently queried. The value Blockchain offers over database is that it is better suited for inter-organizational systems where there has been a traditional lack of trust infrastructure or friction caused by multiple versions of the truth. Supply chain ecosystems have traditionally suffered from this. If different versions of the truth emerge, what are the options to have the difference resolved? Collaborative economy marketplaces have this characteristic-they are an inter-organizational system with a lack of trust infrastructure or trusted third party. Often the trust infrastructure when it does exist is fractured by different jurisdictions in the case of systems involving international parties. Blockchain is the context that offers a single version of the truth, transparency to those with permission, and immutability to avoid situations where records are either accidentally or purposely manipulated.

5 Conclusion

Clearly, we are in the very early days of DLTs, and most of the literature on this topic revolves around finding adequate use cases for this new technology and analysing their mathematical properties. It is reasonable to expect, due to the many practical advantages, that many inter-organizational systems such as those driving the collaborative economy will eventually adequate their system to a DLT-based back-end. The benefits of creating an immutable framework, with a robust consensus mechanism make DLT a very desirable technology. The Bitcoin experiment has shown, in the past ten years, that the Blockchain is a secure and reliable technology, and even though fees, energy consumption and speed of transactions, currently limit its scope of applications, it has spawned a different way of thinking and of building systems that offer new markets for types of value that were never considered before.

References

1. Nakamoto S (2008) Bitcoin: a peer-to-peer electronic cash system. https://bitcoin.org/bitcoin.pdf
2. Esposito C, De Santis A, Tortora G, Chang H, Choo KKR (2018) Blockchain: a panacea for healthcare cloud-based data security and privacy? IEEE Cloud Comput 5(1):31–37
3. Novo O (2018) Blockchain meets IoT: an architecture for scalable access management in IoT. IEEE Internet Things J 5(2):1184–1195
4. Gatteschi V, Lamberti F, Demartini C, Pranteda C, Santamaria V (2018) To blockchain or not to blockchain: that is the question. IEEE IT Prof 20(2):62–74
5. Ferraro P, King C, Shorten B (2018) Distributed ledger technology for smart cities, the sharing economy, and social compliance. IEEE Access 6:62728–62746
6. Puthal D, Malik N, Mohanty SP, Kougianos E, Das G (2018) Everything you wanted to know about the blockchain: its promise, components, processes, and problems. IEEE Consum Electron Mag 7(4):6–14
7. Conoscenti M, Vetro A, De Martin JC (2016) Blockchain for the internet of things: a systematic literature review. In: IEEE/ACS 13th international conference on computer systems and applications, pp 1–6
8. Ramakrishna MV, Fu E, Bahcekapili E (1997) Efficient hardware hashing functions for high performance computers. IEEE Trans Comput 46(12):1378–1381
9. Bhanot R, Hans R (2015) IA review and comparative analysis of various encryption algorithms. Int J Secur Appl 9(4):289–306
10. Karame GO, Androulaki E (2016) Bitcoin and blockchain security. Artech House, Norwood
11. Karame G (2016) On the security and scalability of bitcoin's blockchain. In: Proceedings of the 2016 ACM SIGSAC conference on computer and communications security, pp 1861–1862
12. Li X, Jiang P, Chen T, Luo X, Wen Q (2017) A survey on the security of blockchain systems. Future Gener Comput Syst (Elsevier). https://doi.org/10.1016/j.future.2017.08.020
13. Gervais A, Karame GO, Wüst K, Glykantzis V, Ritzdorf H, Capkun S (2016) On the security and performance of proof of work blockchains. In: Proceedings of the 2016 ACM SIGSAC conference on computer and communications security, pp 3–16
14. Popov S (2017) The tangle-version 1.4.2. https://iota.org/IOTA_Whitepaper.pdf
15. Popov S, Saa O, Finardi P (2017) Equilibria in the tangle. arXiv:1712.05385
16. Ferraro P, King C, Shorten B (2018) IOTA-based directed acyclic graphs without orphans. arXiv:1901.07302

Sharing Economy: A Business Perspective

Paolo Roma

Abstract The developments in information technology on the one hand, and the reduced purchasing power of many people due to the global crisis as well as the increasing sustainability concerns on the other hand, have contributed in recent years to the creation of a multitude of Internet-enabled peer-to-peer platforms and the consequent rise of a new paradigm of economy, namely the sharing economy. Under this paradigm, consumers on a massive scale share and make use of underutilized resources upon payment. In this chapter, we present an overview of the sharing economy phenomenon from a business perspective by identifying the major factors guiding its emergence and growth, by presenting some important business principles that characterize the sharing economy and a classification of business models utilized in this context, together with the major revenue streams associated with them. Our general overview can be useful to researchers and practitioners to gain initial understanding of the sharing economy, its causes and effects from a business perspective.

Keywords Sharing economy · Business model · Internet-enabled economy · Information technology

1 Introduction

The developments in information technology on the one hand, and the reduced purchasing power of many people due to the global crisis as well as the increasing sustainability concerns on the other hand, have contributed in recent years to the creation of a multitude of Internet-enabled peer-to-peer platforms and the consequent rise of a new paradigm of economy, namely the sharing economy. Under this paradigm, consumers on a massive scale share and make use of underutilized resources upon payment. This implies that concepts such as product/service

P. Roma (✉)
Università degli Studi di Palermo, Viale delle Scienze, 90128 Palermo, Italy
e-mail: paolo.roma@unipa.it

accessibility and utilization are emphasized and have priority over the concept of ownership. At the same time, it implies that each consumer can become a resource provider by exploiting the ownership of certain underutilized resources. Many businesses have been affected by this phenomenon, encompassing hospitality, transportation (cars, bikes, boats), babysitting, home/garden tool business, financial sector, etc. The success of Airbnb and Uber, two peer-to-peer platforms in the contexts of hospitality and car transportation, respectively, and currently two of the most valuable unicorns worldwide, has certainly contributed to the rapid growth of the sharing economy. For instance, recent industry estimates suggest that sharing economy in Europe will potentially soar up to €570 billion by 2025 compared with €28 billion in 2016 [29]. While fascinating, the idea that, by means of an online platform or an app, consumers may share certain resources with other consumers and profit from them, has opened up room for a plethora of issues that have started generating huge interest across many disciplines including economics, management, law, and sociology. Primary questions pertain to the economic implications of this disruptive economy model for players operating in traditional industries (e.g., hotels, taxi drivers, car manufacturers, tool manufacturers, financial investors), their employees, consumers in these industries, platforms themselves, and society at large. As discussed in the state of the art section, academic research has just started investigating how the sharing economy is shaping traditional industries and their consumers. However, due to the infancy of the phenomenon, there is still no clear and unanimous understanding on how this is occurring and which economic implications this impact will generate. Most of the studies do not go beyond theoretical discussions (e.g., [1, 10, 13, 23]), with only a few studies developing analytical frameworks to assess the economic impact of the sharing economy on traditional industries including their consumers (e.g., [4, 21, 36]). In addition, even fewer studies provide empirical evidence and they are mostly confined to the US context [6, 39], thus reducing the generalization of findings to other countries (e.g., Italy) due to the differences in terms of market structure and consumer attitudes. In this chapter, we present an overview of the sharing economy phenomenon from a business perspective by identifying the major factors guiding its emergence and growth, by presenting some important business principles that characterize the sharing economy and a classification of business models utilized in this context, together with the major revenue streams associated with them. Our general overview can be useful to researchers and practitioners to gain initial understanding of the sharing economy, its causes and effects from a business perspective. The chapter is organized as follows. We first provide a literature overview in Sect. 2. In Sect. 3, we discuss the drivers of the sharing economy phenomenon. In Sect. 4, we present the four business principles guiding the sharing economy. In Sect. 5, we discuss the key sectors that have been affected by the rise of the sharing economy. In Sect. 6, we provide different classifications of business models utilized in the context of the sharing economy, whereas in Sect. 7 we discuss the most common revenue streams for sharing economy platforms. Finally, in Sect. 8 we conclude by providing directions for future business research on the sharing economy.

2 Literature Overview

Academic research has just started investigating the multiple facets of the sharing economy phenomenon and thus most research questions are far from having been satisfactorily addressed. Initial studies can be divided in three categories: (a) studies analyzing the motivations behind the rapid growth of the sharing economy and the dynamics and behaviors characterizing users of sharing economy platforms, (b) studies examining the economic impact of the sharing economy on traditional industries, and (c) studies focusing on the regulation of the sharing economy. With regard to the reasons behind the fast growth of the sharing economy phenomenon, authors have indicated a number of factors guiding users to adopt this economy model. As a consequence of wealth erosion of many people due to the global crisis, economic motives naturally play the most prominent role in driving adoption of sharing economy practices [2, 19, 33]. An additional identified driver is the opportunity to derive social utility from connecting to a community or to satisfy hedonic needs by obtaining access to expensive products at lower prices [5, 33]. Other studies have shed light on some dynamics arising in sharing economy platforms, such as trust and reputation of service providers [16], the emergence of moral hazard issues [38], users' utility and satisfaction in using sharing economy services [4, 26], and even the emergence of racial discrimination [14]. Our chapter will significantly contribute to this research stream by shedding light on the factors driving consumers' behavior and their choices between services offered by incumbents and services offered by sharing economy players, this being an issue never studied before in spite of its clear relevance for the evolution of the hospitality industry. Regarding the impact of the sharing economy on traditional industries, the initial efforts have been directed to understand how peer-to-peer sharing affects physical product (e.g., cars) usage levels and the effects on manufacturers. For instance, Benjafaar et al. [4] have analytically shown that peer-to-peer sharing is more likely to lead to higher product usage levels when the cost of product ownership is relatively high, and that consumers strictly prefer an economy with peer-to-peer product sharing over an economy where such type of sharing is precluded. Fraiberger and Sundararajan [18] have provided empirical support to these insights. Using game theory, Jiang and Tian [21] have also added that consumers' sharing of products with high marginal costs is win-win for both product manufacturer and consumers, whereas sharing of products with low marginal costs can be lose-lose. Tian and Jiang [36] have pointed out that, in a distribution channel setting, peer-to-peer product sharing tends to benefit the retailer at the expenses of the manufacturer. More closely related to our chapter, other studies have examined how the growth of sharing economy players could affect incumbents in traditional industries. However, most of these studies have not gone beyond mere theoretical discussions of the characteristics of the sharing economy that may influence these industries [1, 10, 13, 23, 24, 35], with only few works providing empirical evidence of this impact. Most prominently, Zervas et al. [39] have empirically shown that hotel profitability in Texas has decreased significantly due to the growth of Airbnb. Similarly, Blal et al. [6] have found negative relationship between the growth of

Airbnb and hotels' sales in San Francisco. Fang and Ye [17] have instead provided evidence that in Idaho the growth of Airbnb has reduced the number of employees in low-end hotels. Our work will significantly contribute to this stream of research by considering the active role of incumbents and providing empirical evidence of how the strategic interactions between sharing economy players and incumbents affect market outcomes and contribute to shape the entire hospitality industry. Finally, some studies have started debating about the multiple issues associated to the regulation of the sharing economy [20, 22, 25, 30]. In line with the arguments advanced by Quattrone et al. [30], our chapter will contribute to this stream by enabling suggestions for regulation.

3 Drivers of Sharing Economy Diffusion

Although sharing economy encompasses diverse meanings and includes several business models, it should not be considered as an emergent sector, but it should be regarded as a global movement relevant in the socio-economic system. As a matter of fact, the American magazine Time included the sharing economy as one of the ten ideas able to change the world [37]. In general, the traditional business model consists of firms producing and offering a good and/or a service, through strategies pushing consumers and/or other firms to buy. Under sharing economy, the business model does not focus on product ownership. Rather, it focuses on product use. Customers share goods/services and rent them instead of buying them. Indeed, beyond the mere consumption, the traditional business entails a waste of energy and resources given that often tangible and intangible products are not utilized after purchases. In this regard, the drill is a classical example: the usage life of a drill is on average 12–13 min per owner. It is not economically sensible to own a drill and do not use it most of the time. Sharing a drill among neighbors (for instance through exchange platforms) would definitely be more suitable in this case. As Botsman and Rogers [7] suggests, consumers do not need to buy a drill, they need to make a hole in the wall. Therefore, it is the availability of the drill, not its ownership, that matters. The diffusion of the sharing economy is a recent phenomenon that has started developing since the financial crisis in 2008, which has decreased people's trust on the traditional economic system in the Western countries. However, this is not the only driving factor behind the rapid development of the sharing economy. Indeed, we need to take into account the notable accelerating impact of Internet and other digital technologies in the last ten years, as well as the increasing concerns related to the environment and resource consumption and waste. In particular, in this section, we identify and discuss in detail four main factors contributing to the global diffusion of the sharing economy, namely economic, technological, social and environmental factors.

3.1 Economic Drivers

There is a consensus on identifying the financial crisis as the main economic factor behind the rapid surge of the sharing economy [7, 35]. Indeed, the financial crisis has caused wealth erosion and unemployment, which, on the one hand, have jointly induced resource owners (e.g., landlords) to look for additional economic returns from their assets, and, on the other hand, consumers to look for less expensive product/service solutions to satisfy their needs. Indeed, the fact that the sharing economy relies on a multitude geographically distributed peer-to-peer resource providers implies that the access to these resources can occur at lower prices as compared with the case of traditional businesses, given that these providers incur low fixed costs as they use their owned underutilized small resources and incur relatively low marginal costs as they normally operate at a micro-business level. In turn, these characteristics translate into very competitive prices available in sharing economy platforms, which gives sharing economy players a competitive advantage against traditional businesses, such as hotels, taxi companies, etc. Moreover, the financial crisis has changed the behavior not only of sellers, but also and most importantly of consumers, which are now much more inclined toward sharing and borrowing than before. It is not a case that the most important sharing economy platforms were launched during the initial period of the financial crisis, i.e., 2008–2010. In those years, famous platforms, such as Airbnb, Taskrabbit, Landshare, were launched across a variety of industries. More in general, industry studies suggest that more than 60% of the sharing economy models have been launched starting from 2010 [34].

3.2 Technological Drivers

Besides the crucial role of the financial crisis and the consequent increasing attention to less expensive product solutions, the rapid popularity gained by sharing economy has been enormously facilitated and accelerated by the technological developments. In particular, many studies agree on the fact that the developments in information technology, such as Web 2.0, social networking, mobile devices, mobile apps platforms, online payment systems, have significantly contributed to the diffusion of the sharing economy globally in such a short period. Some of these studies suggest that the role of technological development has been even more impactful than that of the financial crisis. For instance, Botsman and Rogers [7] argue that Internet, and in particular the increasing ubiquitous and real-time access to it, has been the key element driving the large-scale diffusion of sharing economy practices. The improvement of broadband Internet connection, the development of apps and the evolution toward the Web 2.0 and more peer-to-peer architectures have given rise to new networks of consumers sharing resources of any kind, and thus becoming prosumers (i.e., producers and consumers at the same time). Nowadays users have the opportunity to interact among each other, exchange information with a multitude of other people globally distributed. This provides the incentive to offer their underutilized and local resources to a global market of peers. A considerable risk in transacting among peers is related to the information asymmetry between seller and buyer about the quality of the product or service to be transacted. Indeed, the absence of professionalism and the lack of reputation at least initially entailed in the sharing economy model could have been a hurdle to the diffusion of this economy model. However, this problem has been eliminated in the context of sharing economy by the adoption of rating and review mechanisms, facilitated once again by the increasing orientation towards the Web 2.0. Indeed, the idea of using rating systems originally adopted by the e-commerce platform eBay in 1995 to rate vendors has been fully adopted by the majority sharing economy platforms in a bi-directional way. That is, both resource providers and resource users can rate and review each other and make this information available to the community accessing the given sharing economy platform. For instance, this occurs in peer-to-peer lodging platforms such as Airbnb and Couchsurfing, where both renters and asset owners can review and rate various aspects of the accommodation, thus mitigating the information asymmetry of future users and providers [12]. Finally, the development of online payment systems has significantly contributed to the rapid growth of the sharing economy. The possibility to pay and access a service through a mobile device using a simple mobile application has been essential to reach a critical mass of platform users. Consider, for instance, car transportation service platforms such as Uber. Consumers needing a ride can simply book it using an app on their smartphone whenever and wherever they are. The easiness, speed and ubiquity of transactions have clearly favored the adoption of sharing economy practices online. Moreover, in the near future peer-to-peer payment methods, e.g., those supported by blockchains, may overcome the traditional systems based on financial intermediaries, and further contribute to the development of the sharing economy.

3.3 Social Drivers

Social changes, supported by technological developments, have also played an important role in the diffusion of the sharing economy. In particular, the desire to be part of a community and the higher attitude to altruism have been identified as important social determinants [7]. The need of connection and belonging to a community has strongly emerged among online users following to the diffusion of social networks. As the term "sharing" suggests, many sharing economy platforms (even those for profit), e.g. CouchSurfing or crowdfunding platforms such as Kickstarter, emphasize these aspects: the transaction is part of a broader process encompassing the creation of relationships, trust, shared experiences, which strengthen the sense of belongingness of users to the given sharing economy community [28]. In this regard, it has been reported that the great majority of people view the sharing economy as a model that helps build stronger communities [28]. Consequently, we have observed the rise of different approaches to consumption, where needs are satisfied not simply by the purchase of a product, but to a greater extent by the search of experiences and situations of consumption that make consumers feel part of a community. These social changes may have been activated also in response to the financial crisis, which has determined a rejection among many people of the traditional (concentrated) capitalistic model and, at the same time, has favored an attempt to head toward more sustainable economic models. These behaviors have been so rooted in the sharing economy that some authors have even prospected the end of employment and the rise of crowd-based capitalism, where micro-entrepreneurship, rather than multinational firms, becomes the core of the economy [35].

3.4 Environmental Drivers

Finally, there are also factors related to the environment that have contributed to the rapid growth of the sharing economy. The financial crisis has also amplified and accelerated the trend of moving toward a more environmentally sustainable society, which has emerged earlier as consequence of the increasing concerns related to the negative effects of an excessively consumerist society. Indeed, the increasing people's awareness of the importance of a sustainable economy is pushing them to pay more attention to the environment through a more rational use of resources by sharing, re-using, recycling products, and finding alternative economic models that can better address the environmental concerns characterizing the current scenario, such as climate change, finite natural resources, pollution. Recent studies suggest that sharing economy may have a positive effect on the environmental issues as it helps rationalize the use of resources and reduces wastes. In this regard, according to PWC (2015), most consumers believe that sharing economy can be useful to reduce wastes. In line with this view, many sharing economy platforms position themselves as "green" and promote the sharing activity as a sustainable solution for reducing wastes (or underutilization) of resources and polluting practices. For instance, a study conducted by Cleantech Group (2014) indicates that home sharing has generated a reduction of energy consumption (63% less in the US and 78% less in Europe) greater than that observed in hotels or similar hospitality providers, as well as a lower waste of water (12% less in the US and 48% less in Europe), and has contributed to a greater consciousness regarding environmental issues. It is noteworthy, however, that the fact that under sharing economy resources are used with higher parsimony does not imply a reduction in the life quality. Rather, it suggests a lifestyle more respectful of the environment and the society as a whole.

4 The Four Business Principles of the Sharing Economy

We discuss four guiding principles of the sharing economy that have determined its disruptive power in the economic scenario [7]. The main idea is to improve the utilization of available resources. Operationally, this occurs through the application of new forms of consumption and behaviors aimed at sharing goods and services to lead to an overall improvement of the society. The first principle consists of reaching the so-called critical mass, i.e., the minimum level of users necessary to start up a given phenomenon, which in this case is the threshold group of initial adopters of sharing economy models [15]. This aspect is crucial and represents the most difficult obstacle to overcome because network-based models such as the sharing economy can be successful and effective only if they are widely known and utilized. Consider, for instance, Airbnb, i.e., the main sharing economy platform for hospitality. Initially, not many people used this service. Travelers need accommodation everywhere, and therefore the higher the geographical area reached by the service, the higher the value that can be generated. To attract a large number of early adopters, i.e., pioneers in the consumption of goods and/or services not yet widespread in the market, firms adopt strategies such as free pricing or discounts when consumers bring new customers to the company. Airbnb, instead, exploited the large diffusion of the advertisement portal Craigslist suggesting the hosts to post their offering not only on Airbnb but also on Craigslist to increase the visibility [8]. This strategy turned out to be successful, as Airbnb has grown incredibly fast in less than ten years reaching a valuation of more than $30 billions. Reaching a critical mass is important as it generates network effects that increase exponentially the number of platform users due imitation and bandwagon effects [15]. The second principle relates to the availability of numerous tangible and intangible resources that a multitude of people worldwide possess, accumulate and do not fully utilize. These underutilized resources are the engine that has allowed many sharing economy platforms to bloom. In the recent past product ownership was much more frequently preferred to product access. That is, consumers were more inclined to buy a multitude of products and then use them only partially or rarely. For instance, products with low frequency of usage, such as home/gardening tools, vacation houses, big consumer electronics, etc., as well as products used for short periods in life and then stored in garages, such as toys for kids, books, clothes, fall into to the category of products with high underutilization, and thus are suitable for sharing economy. Estimates suggest that 80% of products in UK and US are used only once at month. Sharing economy's main goal is to find an opportunity for higher utilization or reuse to these products matching resource owners with new consumers willing to use these products. Platforms such as Depop for used clothes or Splinlister for sports equipments, or even Movieswap for DVDs are example of how a reuse opportunity can be found for many different products. This trend toward higher utilization of resources is not only related to tangible products, but also to intangible assets: for instance, through platforms such as www.Taskrabbit.com individuals can offer their spare time, their knowledge, as well as services under a peer-to-peer logic. The third principle is derived as a consequence of the second principle as it refers to the use of common (shared) resources. That is, the same resource, e.g., cars, bikes, houses, etc., though often owned by a single individual, is used simultaneously or sequentially by many other individuals. Moving from single ownership to economy models where more "commons" become central requires cohesion and trust among users in the community. Lacking these requirements may put in danger the operations of sharing economy models as for instance witnessed by the theft in San Francisco of 4 cars made available by HiGear, the luxury car rental platform, which was forced to stop operations temporarily. Another problem influencing common (shared) products is free-riding, i.e., an opportunistic behavior aimed at exploiting a good produced collectively without contributing effectively to its creation and maintenance. Common (shared) products can be efficiently managed if the governance is well designed. In the Airbnb model, the shared products are owned by single individuals and a third platform (i.e., Airbnb) realizes the matching between supply and demand. Alternatively, if the ownership is distributed among a multitude of individuals, then trust among them becomes extremely relevant and the governance must be designed taking into account this aspect to favor better performance. And here we come to the fourth principle guiding sharing economy, which is trust and mechanisms to support it. Industry reports suggest that 67% of interviewees are reluctant to share, 30% of them because of damages to their own assets/products, 23% of them because they do not trust strangers, and 14% of them because of privacy concerns (Campbell Mithun, 2012). Trust is, indeed, the biggest challenge for sharing economy: if people want to share and cooperate among each other, they have to trust each other. Therefore, mechanisms to nurture trust become necessary. As we have anticipated, rating and review mechanisms have been largely adopted by sharing economy platforms (e.g., Airbnb, Couchsurfing, Blablacar, etc.) to increase trust, favor the creation of reputation, and thus reduce information asymmetry between providers and consumers. In addition, users must adhere to rules for joining these platforms, presenting products, behaving inside the communities (e.g., the case of the crowdfunding platform Kickstarter), which further favor the emergence of trust.

5 The Five Key Sectors of the Sharing Economy

The growth of the sharing economy has been pulled by five main sectors: collaborative finance, peer-to-peer lodging, peer-to-peer transportation services, on demand home services, and on demand professional services. We discuss these sectors, in turn:

– *Collaborative finance.* The main collaborative finance models are crowdfunding and peer-to-peer (or social) lending. Crowdfunding is a form of micro-financing allowing single individuals or organizations to obtain funding from a multitude of small funders in order to develop projects of different nature, such as entrepreneurial, cultural or humanitarian projects. Two forms of crowdfunding currently dominate the scene: reward-based (or product-based) crowdfunding and equity-based crowdfunding (Belleflamme et al., 2014). In the first form, e.g., Kickstarter, entrepreneurs solicit individuals to fund their projects in exchange for rewards commensurate with the level of funding provided. Typical rewards comprise the product that will be commercialized by the entrepreneur if the project is successful. In the second form of crowdfunding, e.g., Crowdfunder, entrepreneurs ask individuals to finance the project in exchange for a share of equity securities

[31, 32]. Peer-to-peer (or social) lending is, instead, a form of lending occurring among private individuals by means of a platform where there is no involvement of traditional financing channels, such as banks [34].

- *Peer-to-peer lodging.* Together with the transportation services, the hospitality is the industry where the sharing economy has been most successful. As a matter of fact, players such as Airbnb and Homeaway already represent 50% of the global market related to vacation accommodation rental [12]. As discussed earlier, these platforms allow individual to monetize on room, apartments or entire houses by renting out them for short or longer periods. Due to the discussed cost advantages, these peer-to-peer lodging platforms become very attractive to a large segment of travelers offering lower prices, large variety (in terms of geographical location heterogeneity, amenities, etc.), as well as a true local experience. There exist also platforms with a more democratic and social footprint, where lodging services do not require payments but simple reciprocal hospitality (this was the case of Couchsurfing, at least initially).
- *Peer-to-peer transportation services.* The rapid urbanization, the changes in the preferences of individuals toward mobility, and the introduction of economic incentives to reduce emissions and traffic congestion are the main factors favoring the exponential growth of the sharing economy models for transportation services. Specifically regarding car transportation services based on sharing economy, we can identify three main models: car sharing, ride sharing and private car hire [12]. Car sharing allows sharing the use of a private car among individuals according their necessities. The ride sharing (e.g., BlaBlaCar) allows sharing the same ride in a car among travelers going to the same destination. Finally, the private car hire (e.g., Uber, Lyft) is a service where individuals offer rides on demand to other individuals using their own car. All these models are naturally supported by platforms matching supply and demand.
- *Home services on demand.* In this sector, there are several platforms matching supply and demand between users having skills and time to carry out a domestic work with users needing to do this type of works, but they do not have time or possess adequate skills. Activities involved in these types of services include furniture assembly, home cleaning, cooking, relocation, gardening, etc. Platforms, such as Taskrabbit, operate exactly this supply-demand matching specifically focusing on manual labor and not particularly qualified jobs.
- *Professional services on demand.* Although under a stringent view professionalism is not a characteristic of the sharing economy, these professional services on demand mediated by third party platforms can also be considered as part of the sharing economy because they are characterized by a multitude of geographically dispersed providers. There are many different types of professional service on demand. For instance, makerspaces put in touch users and providers of services such machining, 3D Printing, and manufacturing at micro-business level. Moreover, Massive Open Online Courses (MOOCs) are courses provided by individuals and organizations with some certified competences and offered to a multitude of users by means of online platforms.

6 Business Models Classification in Sharing Economy

Osterwalder and Pigneur [27] argue that the business model is a conceptual tool explaining how a company creates, appropriates and offers values through the key element of the company. Within the sharing economy environment, we re-interpret the classification provided by Botsman and Rogers [7] as a categorization of business models for the sharing economy. Indeed, although this classification does not entirely describe the business models, the systems identified in it are used to categorize sharing economy platforms in terms of different business approaches. In particular, Botsman and Rogers [7] identify three systems: product service systems, redistribution markets, and collaborative lifestyles.

- *Product Service Systems.* This model is based on the fact that an increasing number of heterogeneous individuals are willing to pay for gaining benefits from the use of the product without acquiring the ownership of it. This model is changing the paradigm of traditional industrial systems based on the logic of selling the product, rather than selling the use of the product. Platforms enabling product service systems allow sharing of a number of products owned by single individuals among a number of different users. Peer-to-peer car transportation services are certainly an example of product service systems.
- *Redistribution Markets.* This model consists of creating a market for redistribution of goods. Re-distributional exchanges can occur for free, in exchange for other goods or upon payment. The benefits generated by this type of model are naturally the reuse and the resale of the existing products. This model has changed the relationship between producers, distributors, and consumers as well as the logic of buying of new products.
- *Collaborative Lifestyles.* According to this model, object of sharing can be time, spaces, knowledge, money contributing to the creation of a collaborative lifestyle from a general viewpoint. Once again trust, social relationships and belongingness to a community are key elements in this peculiar case.

According to this classification, for instance, Beepi, the platform for reselling used cars, falls into the category of redistribution markets, Uber falls into the category of product service systems, whereas Airbnb can be seen as a hybrid between a product service system and a collaborative lifestyle. However, all these business models have a common ground emphasizing resource accessibility over resource ownership. Such business models have been utilized for long time in some business-to-consumers markets (e.g., car rentals) or business-to-business markets (e.g., service outsourcing). However, the key and disruptive innovation represented by sharing economy is that it enables these business models under a peer-to-peer logic. That is, sharing economy platforms facilitate the access of consumers to products, competences, and, more in general, resources owned by other consumers. These platforms can favor the creation of peer-to-peer markets potentially for all products and services exchangeable among consumers. For these reasons, the sharing economy is viewed as a disruptive competitor of traditional businesses providing solutions for rental, mobility, lodging, and similar types of services as it is able to satisfy at least the same needs at much lower prices. Moreover, as widely discussed, sharing economy allows consumers to monetize on their resources and competences, providing them with opportunities of micro-entrepreneurship and reducing the cost of such ownership. PWC [28] defines as sharing economy firms those companies implementing business models based on product accessibility among peers. According to this definition, all well-known sharing economy platforms such as Uber, AirBnB, TaskRabbit, Beepi, etc., have the same business model. Therefore, we can conclude that there may be different categorizations of sharing economy by looking at the perspective of business models.

7 Revenue Streams

In discussing the business models in sharing economy, one crucial aspect is related to how platforms can generate revenue, i.e., revenue streams. In this regard, first of all we can distinguish between for profit and no-profit organizations. Regarding for profit sharing economy platforms, there exist various revenue streams allowing to cover the operational costs and generate profits. In particular, these revenue streams include: transaction fees, subscription fees, advertising, and voluntary donations.

- *Transaction fees.* The percentage on transactions is the most popular revenue stream in sharing economy. It consists of retaining a certain percentage of the price set for the product/service offered through the platform. This percentage varies across different types of platforms and products/services being transacted and shared. For instance, Airbnb retains 3% of the price set by the hosts to manage the operational costs related to the payment 3% and retains an amount between 6% and 12% to the guests to cover the service costs. On the other hand, the transaction fee retained by Uber is about 20% of the price of the ride. Considering that the operational costs of a platform are mainly fixed costs, using a transaction fee revenue stream is very profitable because the revenue increases much more with the volume of transactions than the fixed costs incurred by the platform.
- *Subscription fees.* The subscription to gain access to a service is another very popular source of revenue for sharing economy platforms. Users (possibly from both provider and consumer sides) to have access to the sharing economy platform's service may need to subscribe to the service for certain periods (e.g., yearly or monthly) or based on packages guaranteeing certain utilization levels and/or characteristics. Although this model tends to generate constant revenue levels, it may fail to generate high amounts given that there is low or even no correlation with transaction volumes. This type of revenue stream is quite popular in music and video streaming platforms or in the case of exclusive social networks such as ASmallWorld.
- *Advertising.* Advertisements have been a central element of revenue generation for Internet-based businesses such as websites, portals, social networks, etc. However, it is not particularly adopted by sharing economy firms as its declining effectiveness

does not allow generating high revenues, and more importantly it may reduce users' incentives to use the sharing economy service.
– *Donations.* This type of revenue stream is utilized mostly for collaborative and sharing initiatives not for profit. This allows offering the service for free, and, at the same time, covering the operational costs. This model has been shown to work even in the case a very large mass of users utilize the collaborative service, e.g., the case of the free encyclopedia Wikipedia.

8 Conclusions

In this chapter, we have presented the business perspective of the sharing economy, discussing the major drivers behind the rise and the diffusion of this socio-economic phenomenon, the major business principles and the key sectors where sharing economy is evolving exponentially. We have also analyzed the business models and the revenue streams typically adopted in this context, emphasizing how according to different perspectives there is not univocal classification of business models utilized in the sharing economy. We believe this piece can be considerably useful in the first place to frame the sharing economy and understand its major characteristics and dynamics from a business perspective. It can be definitely considered as a starting point for researchers and practitioners willing to engage with this disruptive phenomenon. To conclude, we provide some suggestions for future research directions. First, we believe that much more research must be done to understand whether and how the growing presence of sharing economy players across several sectors and their changing service characteristics can influence specific traditional businesses' (hotels, taxi companies, etc.) strategies in the short term (e.g., pricing and service innovation strategies), as well as more in the long term (e.g., investment and property development strategies). Second, it is extremely important to understand how the strategic interactions between traditional industry players and sharing economy players affect profitability of both types of players, and consequently shape the market structure. Third, it is undoubtedly worthwhile to unravel the rationales of the above market dynamics by analyzing consumers' choices and behaviors, with specific focus on their preferences and willingness to pay for different services (e.g., preference of Airbnb over hotels or vice versa), their post-experience satisfaction, as well as the main consumer and service characteristics driving such preferences and satisfaction. Finally, to better understand the disruptive impact of the sharing economy, future research should quantify the impact of the growing presence of sharing economy players on the employment level across different industries and on the emergence of micro-entrepreneurial initiatives activated by the rise of the sharing economy.

References

1. Akbar YH, Tracogna A (2018) The sharing economy and the future of the hotel industry: transaction cost theory and platform economics. Int J Hosp Manag 71:91–101
2. Barnes SJ, Mattson J (2015) Understanding current and future issues in collaborative consumption: a four-stage Delphi study. Technol Forecast Soc Chang 104:200–211
3. Belleflamme P, Lambert T, Schwienbacher A (2014) Crowdfunding: tapping the right crowd. J Bus Ventur 29(5):585–609
4. Benjafaar S, Kong G, Li X, Courcoubetis C (2015) Peer-to-peer product sharing: implications for ownership, usage and social welfare in the sharing economy. Working paper
5. Benoit S, Baker TL, Bolton RN, Gruber T, Kandampully J (2017) A triadic framework for collaborative consumption (CC): motives, activities and resources & capabilities of actors. J Bus Res 79:219–227
6. Blal I, Singal M, Templin J (2018) Airbnb's effect on hotel sales growth. Int J Hosp Manag 73:85–92
7. Botsman R, Rogers R (2010) What's mine is yours. The rise of collaborative consumption. Harper-Collins books, New York
8. Brown M (2014) Airbnb: the growth story you didn't know. https://growthhackers.com/growth-studies/airbnb
9. Campbell Mithun (2012) Collaborative consumption: what marketers need to know. http://www.mccannmpls.com
10. Cheng M (2016) Sharing economy: a review and agenda for future research. Int J Hosp Manag 57:60–70
11. Cleantech Group (2014) Environmental impacts of home sharing: phase I report
12. Credit Suisse, The sharing economy. New opportunities, new questions. https://www.credit-suisse.com/media/assets/corporate/docs/news-and-expertise/articles/2016/07/global-investor-2-15-en.pdf
13. Cusumano MA (2015) How traditional firms must compete in the sharing economy. Commun ACM 58:32–34
14. Edelman B, Luca M, Svirsky D (2017) Racial discrimination in the sharing economy: evidence from a field experiment. Am Econ J: Appl Econ 9:1–22
15. Eisenmann TR, Parker GG, Van Alstyne MW (2008) Opening platforms: how, when and why? Harvard Business School Paper
16. Ert E, Fleischer A, Magen N (2016) Trust and reputation in the sharing economy: the role of personal photos in Airbnb. Tour Manag 55:62–73
17. Fang B, Ye Q, Law R (2016) Effect of sharing economy on tourism industry employment. Ann Tour Res 57:264–267
18. Fraiberger SP, Sundararajan A (2017) Peer-to-peer rentals markets in the sharing economy. Working paper
19. Hamari J, Sjoklint M, Ukkonen A (2016) The sharing economy: why people partecipate in collaborative consumption. J Assoc Inf Sci Technol 67:2047–2059
20. Hong S, Lee S (2018) Adaptive governance, status quo bias, and political competition: why the sharing economy is welcome in some cities but not in others. In: Government information quarterly, forthcoming
21. Jiang B, Tian L (2018) Collaborative consumption: strategic and economic implications of product sharing. Manag sci
22. Katz V (2015) Regulating the sharing economy. Berkeley Technol Law J 30:1067–1126
23. Malhotra A, Van Alstyne M (2014) The dark side of the sharing economy… and how to lighten it. Commun ACM 57:24–27
24. Martin CJ (2016) The sharing economy: a pathway to sustainability or a nightmarish form of neoliberal capitalism? Ecol Econ 121:149–159
25. Miller SR (2016) First principles for regulating the sharing economy. Harv J Legis 147:147–202
26. Mohlmann M (2015) Collaborative consumption: determinants of the satisfaction and the likelihood of using a sharing economy option again. J Consum Behav 14:193–207

27. Osterwalder A, Pigneur Y (2010) Business model generation: a handbook for visionaries, game changers, and challengers. John Wiley & Sons Inc
28. PWC, The sharing economy (2015) https://www.pwc.fr/fr/assets/files/pdf/2015/05/pwc_etude_sharing_economy.pdf
29. PWC (2016) Europe's five key sharing economy sectors could deliver €570 billion by 2025. https://press.pwc.com/News-releases/europe-s-five-key-sharing-economy-sectors-could-deliver--570-billion-by-2025/s/45858e92-e1a7-4466-a011-a7f6b9bb488f
30. Quattrone G, Proserpio D, Quercia D, Capra L, Musolesi M (2016) Who benefits from the "sharing" economy of Airbnb. In: Proceedings of the 25th international conference on world wide web. pp 1385–94
31. Roma P, Messeni Petruzzelli A, Perrone G (2017) From the crowd to the market: the role of reward-based crowdfunding performance in attracting professional investors. Res Policy 46:1606–1628
32. Roma P, Gal-Or E, Chen RR (2018) Reward-based crowdfunding campaigns: informational value and access to venture capital. Inf Syst Res 29(3):679–697
33. So KKF, Oh H, Min S (2018) Motivations and constraints of Airbnb consumers: findings from a mixed-methods approach. Tour Manag 67:224–236
34. Stokes K, Clarence E, Anderson L, Rinne A (2014) Making sense of the UK collaborative economy. Nesta. https://media.nesta.org.uk/documents/making_sense_of_the_uk_collaborative_economy_14.pdf
35. Sundararajan A (2016) The sharing economy: the end of employment and the rise of crowd-based capitalism. MIT Press, USA
36. Tian L, Jiang B (2018) Effects of consumer-to-consumer product sharing on distribution channel. Prod Oper Manag 27:350–367
37. Walsh B (2011) Today's smart choice: don't own. share. time. 17 Mar. http://content.time.com/time/specials/packages/article/0,28804,2059521_2059717_2059710,00.html
38. Weber TA (2014) Intermediation in a sharing economy: insurance, moral hazard, and rent extraction. J Manag Inf Syst 31:35–71
39. Zervas G, Proserpio D, Byers JW (2017) The rise of the sharing economy: estimating the impact of Airbnb on the hotel industry. J Mark Res 54:687–705

Distributed Algorithms for Internet-of-Things-Enabled Prosumer Markets: A Control Theoretic Perspective

Syed Eqbal Alam, Robert Shorten, Fabian Wirth and Jia Yuan Yu

Abstract In many sharing economy scenarios, agents both produce as well as consume a resource; we call them *prosumers*. A community of prosumers agrees to sell excess resource to another community in a prosumer market. In this chapter, we propose a control theoretic approach to regulate the number of prosumers in a prosumer community, where each prosumer has a cost function that is coupled through its time-averaged production and consumption of the resource. Furthermore, each prosumer runs its distributed algorithm and takes only binary decisions in a probabilistic way, whether to produce one unit of the resource or not and to consume one unit of the resource or not. In the proposed approach, prosumers do not explicitly exchange information with each other due to privacy reasons, but little exchange of information is required for feedback signals, broadcast by a central agency. In the proposed approach, prosumers achieve the optimal values asymptotically. Furthermore, the proposed approach is suitable to implement in an IoT context with minimal demands on infrastructure. We describe two use cases; community-based car sharing and collaborative energy storage for prosumer markets. We also present simulation results to check the efficacy of the algorithms.

Keywords Distributed optimization · Internet-of-Things (IoT) · Optimal control · Optimal allocation · Prosumers · Sharing economy · Prosumer markets

---

The work is partly supported by Natural Sciences and Engineering Research Council of Canada grant no. RGPIN-2018-05096, Danish ForskEL programme (now EUDP) through the Energy Collective project (grant no. 2016-1-12530), and by Science Foundation Ireland grant no. 16/IA/4610.

S. E. Alam (✉) · J. Y. Yu
Concordia Institute for Information Systems Engineering, Concordia University,
Montreal, QC, Canada
e-mail: syed.eqbal@iiitb.net

R. Shorten
Dyson School of Design Engineering, Imperial College London, London, UK
e-mail: r.shorten@imperial.ac.uk

F. Wirth
Faculty of Computer Science and Mathematics, University of Passau, Passau, Germany

© Springer Nature Switzerland AG 2020
E. Crisostomi et al. (eds.), *Analytics for the Sharing Economy:
Mathematics, Engineering and Business Perspectives*,
https://doi.org/10.1007/978-3-030-35032-1_9

1 Introduction and Setting

Recently, consumers across a range of sectors have started to embrace shared ownership of resources and services with guaranteed access, as opposed to more traditional business models that focus on sole-ownership only. The reasons for this trend are multi-faceted and range from societal issues, such as the need to reduce wastage, and more general environmental concerns [1–4], to pure monetary opportunities arising from increased connectivity (and the ability that this gives to advertise the availability of unused resources and services) [5]. Well-known examples of successful companies building *sharing economy* products include Airbnb (hospitality), Lyft (ride sharing) [6], Bird and Lime (scooter sharing), Mobike (bike sharing) [3], and Google (Google reviews—information sharing).

Roughly speaking, several types of sharing application classes are discerned (as described in [5]).

A. **Opportunistic sharing:** Services based on opportunistic sharing of resources exploit large-scale availability of either unused resources or obsolete business models or both. Examples of products in this area include the parking application JustPark (www.justpark.com) and the peer-to-peer car sharing services Getaround (www.getaround.com). The key enablers for such products are mechanisms for informing agents of available resources, their delivery, and payments.
B. **Federated negotiation and sharing:** Here, groups of agents come together to negotiate better contracts with utilities (electricity, gas, water, health), or to provide mutually beneficial services such as collaborative storage of energy. The key enablers for such products are mechanisms for grouping communities and for enforcing contractual obligations for federations of like-minded consumers.
C. **Bespoke sharing:** In this case, products are designed with the specific objective of being shared, rather than for sole ownership. A basic example of such systems is devices and services that allow sharing of a single electric charge-point by several users. Other examples include time-shared apartments or cars that are owned by several people rather than a single person [5].
D. **Hybrid sharing:** Finally, opportunities also exist for sharing economy to support the regular economy. We readily find examples of such systems in the hospitality industry, where ad-hoc sharing economy infrastructure (spare rooms in local houses) can be used as a buffer to accommodate excess demand in the regular economy (hotels).

The common characteristic in all of the above application classes is the ability for community-wide communication and actuation, both to enable services to be bought and sold, and so that contracts can be enforced.

The development of sharing economy applications [7, 8] is facilitated by Internet-of-Things (IoT). For example, IoT helps to perform secure payments, to track the location and the condition of an object, to list a few. Interested readers can find several IoT-based applications in [9, 10] and the papers cited therein. Therefore, while the value of the sharing economy is not in question [2, 11] and while many of the essential infrastructural elements needed for the deployment of such systems are being developed rapidly, there is an additional requirement for structured platforms to enable distributed *community-wide* buying and distributed *community-wide* selling. Currently, such platforms are at a very early stage of development with significant opportunities for improvement.

Our objective in this chapter is to address this deficit partially and to develop tools to support the design of community-based prosumer markets. We define *prosumers* as agents that both produce and consume a resource [12]. Specifically, we are interested in developing *light* algorithms that can easily be deployed on modest IoT platforms, and that can be used to support distributed community-wide buying and selling of resources. Here, by *light*, we mean algorithms that place low demands on the infrastructure, both in terms of computational power, and actuation and connectivity requirements of individual prosumers. A fundamental requirement is also that such algorithms are scale-free in the sense that they can operate across a range of community sizes; from small communities of a few prosumers to larger communities made up of very many prosumers. These constraints are directly related to the challenges associated with uncertainties that arise in the context of sharing economy problems. Typically, at any time instant, one does not know how many prosumers are participating in the sharing scheme; whether prosumers can or are willing to communicate with each other (perhaps due to privacy considerations), and whether enough computational power is available to the whole network to allocate resources in real-time optimally. A further complication is that we would like any scheme that we develop to be backward compatible with old IoT platforms that support only essential interaction between prosumers and infrastructure. Thus, there is considerable interest in developing *light* algorithms that place only modest demands on infrastructure, yet can be used to implement complex policies in the face of the uncertainties mentioned above.

Given this context, we are particularly interested in situations where communities come together to purchase and sell related commodities simultaneously. Such systems arise, for example, in energy systems where agents (prosumers) both produce and consume energy [12–14].

2 Prosumer Markets and Communities

Prosumers are the agents that both produce and consume resources [12, 15]. We are interested in prosumer markets that facilitate a community of prosumers for distributed production and consumption. Such markets are emerging rapidly in energy sector [16, 17], but also in other areas such as shared mobility [18]. Parag and Sovacool [19] classify prosumer markets according to three network architectures.[1]

---

[1] In the network architectures, *p* represents a prosumer.

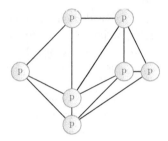

Fig. 1 Prosumer-to-prosumer model. Here *p* represents a prosumer. Figure adapted from [19]

(i) *Prosumer-to-prosumer model:* In this peer-to-peer model, prosumers interact (buy or sell resource) directly with each other as depicted in Fig. 1. This model is widespread. For example, consider the case of car sharing platform Turo. Here, car owners list their cars on Turo sharing platform, and the riders book the cars of their choice through this platform for a certain period with a fee. A similar, peer-to-peer model is proposed in [20] for energy trading in microgrids.

(ii) *Prosumer-to-firm model:* In this model, prosumers interact with a local firm directly. There are two types of prosumer to firm models, *prosumer-to-interconnected-firm* model, and *prosumer-to-isolated-firm* model. In the prosumer-to-interconnected-firm model, prosumers are connected to a local firm, which may be connected to the main firm as presented in Fig. 2a. For example, suppose that prosumers produce energy from renewable sources and are connected to a microgrid. A prosumer satisfies its energy needs from the microgrid and the energy it produces. If the prosumer produces more energy than it needs; then, it can return excess energy to the microgrid, this microgrid may be connected to the main grid. Whereas, in the prosumer-to-isolated-firm model, prosumers are connected to the local firm, which works in isolation as depicted in Fig. 2b. Example of the prosumer-to-isolated-firm model is Island microgrid [21] in which prosumers and microgrid work together to fulfill the energy need of prosumers in the Island.

(iii) *Community-based prosumer model:* In this model, prosumers are located in the same geographic location who have similar resource needs and resource production pattern; more generally, they share common goals and interests. These prosumers are grouped to interact with each other and efficiently manage the resource needs of the community, as depicted in Fig. 3. In this case, communities may also exchange resources with each other. A recent example of community-based trip sharing is found at [22] in which the algorithm clusters commuters in communities to optimize car usage. We clarify that, for simplicity, we consider a single prosumer community that interacts with another community (*external community*) in the rest of the chapter unless otherwise stated.

While work on analytics to help design prosumer markets is still in its infancy, somewhat surprisingly few papers have begun to deal with some of the complex market design issues associated with such systems [13, 23–25]. Roughly speaking,

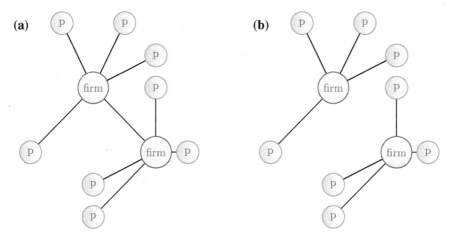

Fig. 2 Prosumer-to-firm models—**a** prosumer-to-interconnected-firm model, **b** prosumer-to-isolated-firm model. Figure adapted from [19]

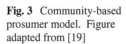

Fig. 3 Community-based prosumer model. Figure adapted from [19]

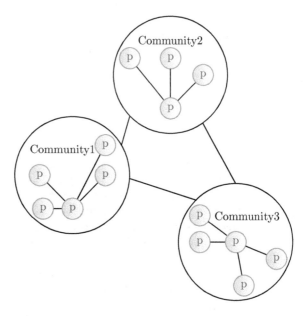

these papers deal with two main issues; (i) the existence of market equilibria, and (ii) methods to allocate resources amongst (competing) prosumers. It is in this latter context that this present work is placed. Generally speaking, resource allocation algorithms for prosumer markets have until now been formulated in an optimization context and can be categorized as is traditionally done for classical optimization models. Namely, resources are either allocated centrally [26] as in the case of Uber, whereby drivers are assigned to the passengers; or in a distributed fashion as in the case of Airbnb, whereby guests and hosts choose each other; or using hybrid of the above two, as in the case of Didi Chuxing [1]. As we have said, typically, these allocation problems are formulated in an optimization context, paying particular attention to the certain constraints that arise in the sharing economy. These include privacy of the individual, fair allocation of resources, the satisfaction of service level agreements, and ever-increasing regulatory constraints (for example, in the case of Airbnb).

Our contribution in this chapter is to address these problems using a different approach. Namely, we shall consider these problems (in a control theoretic context) as regulation problems with optimality constraints. In particular, we are interested in applications where prosumer communities both buy and sell resources from or to one, or more external entities. Thus, we take the view that such communities have a contract to both buy and sell a pre-specified level of a resource at a given time instant, and the objective of the allocation algorithm is to ensure that these levels of demand are met. Given this basic setting, we ask the question as to whether this can be done without any explicit exchange of information between individual prosumers in situations where prosumers take only binary decisions (buy or sell), and whether given these constraints, optimal use of the resource can be realized. We shall see in the next section that it is indeed possible to formulate the problem and develop distributed algorithms that achieve all of these properties, and which can be implemented in an IoT context with minimal demands on infrastructure.

2.1 Prior Work

Recent analytics work on sharing economy has primarily followed two directions; (i) market design and equilibria, and (ii) optimal allocation of resources. Here, we give a very brief picture of some of the most recent work. In [25], Georgiadis et al. propose three types of resource allocation mechanisms; centralized, coalition-based (game-theoretic approach), and peer-to-peer. In [23], Gkatzikis et al. propose a collaborative consumption mechanism to minimize the electricity cost of a community, and in [27], Tushar et al. developed a game-theoretic model for peer-to-peer energy trading. Furthermore, in [13], Moret et al. design a community-based distributed energy collective market model that helps energy prosumers to optimize their energy resources and to achieve social welfare of the community. More recently, Iosifidis and Tassiulas [24], propose optimization techniques for resource exchange and production scheduling for cooperative systems. Courcoubetis and Weber [28], propose mechanisms for the optimal allocation of shared computing resources. In [22], Hasan et al. propose a community-based car sharing model to maximize the trip sharing. To do so, they use mixed-integer programming, graph theory, and clustering techniques. In [20], Zhang et al. propose a game theoretic peer-to-peer energy trading model between local prosumers and distributed energy resources. They show that the model gives rise to an equilibrium between energy production and consumption. In addition to these works, a peer-to-peer equilibrium model for collaborative consumption is proposed by Benjaafar et al. in [29]. Finally, Grijalva et al. [18], discuss architecture for prosumer-based distributed control for the electricity grid. We refer interested readers to a recent review paper by Sousa et al., which covers market design, optimization techniques, and interesting future directions at [30].

2.2 Contribution

We are primarily interested in applications where prosumer communities buy and sell resources via contracts to one or more external entities and must ensure that they meet certain demands in real-time. Thus, the objective of any allocation algorithm is to ensure that these levels of demand are met. Typically, a problem of this nature would be solved in a standard optimization framework. Unfortunately, this approach is not available to us due to the uncertainty that prevails for this system class. For example, the number of prosumers participating at a given time instant may vary, as may the contracted level of prosumption. Also, for reasons of privacy, prosumers may only communicate with each other or with the infrastructure in a limited fashion, making the communication graph unknown a-priori, from the point of view of algorithm development. Thus, our approach is to formulate these allocation problems in a control theoretic setting, where the effect of these uncertainties and other disturbances can be dealt with using feedback. Our principal contribution is, therefore, to develop distributed control algorithms that can asymptotically achieve optimality.

3 Problem Statement

Let us assume that a prosumer market consists of a prosumer community and an *external community*. The prosumer community has $N \in \mathbb{N}$ prosumers producing and consuming a resource, which agrees to sell its excess resource to the external community. We also assume that the prosumer market has a control unit, which measures the aggregate consumption and production of the resource. It communicates with the prosumer community as well as the external community; we call it a *sharing platform*. Additionally, notice that N is not known to the individual prosumers participating in the scheme. For simplicity, we assume that there is a single resource produced and consumed by all prosumers, though our formulation can easily be extended to multiple resources and multiple prosumer communities. In our model, the process of production and consumption takes place at discrete time instants $t_0 < t_1 < t_2 < \cdots$, where $t_0 = 0$. At each time instant, the overall production and consumption of the previous time instant are evaluated and adjusted. Additionally, we assume that communities and external agencies are contracted to, *on aggregate*, consume and produce a certain amount of the resource at time instant t_k, for $k = 0, 1, 2, \ldots$

We assume that each prosumer has limited actuation and at any time instant t_k, for $k = 0, 1, 2, \ldots$, either it consumes one unit of the resource or it does not consume it. Similarly, each prosumer either produces one unit of the resource at a time instant or it does not produce it. Thus, for each prosumer i, we denote by $x_i(k)$ the amount of the resource consumed, and by $y_i(k)$ the amount of the resource produced at time instant t_k.[2]

We also assume that there are constants $C_x \geq 0$ and $C_y \geq 0$, specifying the aggregate consumption and production bounds. Notice that the constants C_x and C_y are known to the sharing platform, but not to individual prosumers in the market. Thus, at each time instant t_k, we require:

$$\sum_{i=1}^{N} x_i(k) = C_x, \quad \text{and,} \tag{1}$$

$$\sum_{i=1}^{N} y_i(k) = C_y. \tag{2}$$

Our primary objective is to ensure that these prosumption bounds are met at each time instant t_k, for $k = 0, 1, 2, \ldots$ However, we are particularly interested in situations where production and consumption are coupled together. For example, in communities that are formed to produce energy, the time-averaged production and consumption of energy might be coupled through battery storage requirements. Furthermore, in the community-based car sharing prosumer market, the average number of delivered and received "cars" might be coupled through the desired value of utilization of cars; and similarly, production of a resource might depend on its consumption. Thus, in these situations, we would like to ensure that consumption and production bounds (1) and (2) are met asymptotically. To formulate this as a long-term requirement, we introduce the time-averaged consumption:

$$\bar{x}_i(k) \triangleq \frac{1}{k+1} \sum_{\ell=0}^{k} x_i(\ell), \quad \text{for } i = 1, \ldots, N, \tag{3}$$

with the time-averaged production $\bar{y}_i(k)$ defined analogously. Now, let $T_i \in \mathbb{R}_+$ be the desired value of utilization of the resource, for $i = 1, \ldots, N$. Thus, depending on the application (refer Sect. 5, cf. (25)), we might require:

$$\lim_{k \to \infty} \bar{x}_i(k) + \bar{y}_i(k) = T_i, \quad \text{for } i = 1, \ldots, N, \tag{4}$$

with some additional constraints on $\bar{x}_i(k)$ and $\bar{y}_i(k)$. In several applications, we might require for $\alpha_i \in [0, 1]$:

---

[2] Depending on the application, the processes of production and consumption may be more appropriately modeled on a continuous time-scale. In this case, we interpret time instants t_k at those times in which the prosumption over the interval $(t_{k-1}, t_k]$ is accounted for.

$$\lim_{k\to\infty} \bar{x}_i(k) = \alpha_i T_i, \quad \text{and,} \quad \lim_{k\to\infty} \bar{y}_i(k) = (1-\alpha_i)T_i, \quad \text{for } i = 1,\ldots,N. \quad (5)$$

Fortunately, in problems that we consider, there is flexibility in satisfying these constraints and it is enough to satisfy that for sufficiently large k, we have:

$$\bar{x}_i(k) + \bar{y}_i(k) \approx T_i, \quad \text{and,} \quad (6)$$

$$\bar{x}_i(k) \approx \alpha_i T_i, \quad \text{and,} \quad \bar{y}_i(k) \approx (1-\alpha_i)T_i, \quad (7)$$

for $i = 1, \ldots, N$. To formulate this mathematically, we associate a cost $g_i : (0, 1]^2 \to \mathbb{R}_+$, $(x_i, y_i) \mapsto g_i(x_i, y_i)$ to the deviation of the actual long-term prosumptions.

Assumption 1 (*Cost function*) The cost function $g_i(\cdot)$ is strictly convex, strictly increasing in each variable, and is continuously differentiable, for all i.

Given this basic setting, we are interested in solving the following optimization problem:

Problem 1 (*Optimization*)

$$\min_{\xi_1,\ldots,\xi_N,\eta_1,\ldots,\eta_N} \sum_{i=1}^{N} g_i(\xi_i, \eta_i), \quad (8)$$

$$\text{subject to} \quad \sum_{i=1}^{N} \xi_i = C_x, \quad (9)$$

$$\sum_{i=1}^{N} \eta_i = C_y, \quad (10)$$

$$\xi_i \geq 0, \quad (11)$$

$$\eta_i \geq 0, \quad \text{for } i = 1, 2, \ldots, N. \quad (12)$$

We assume that for this optimization problem an optimal point $(\xi^*, \eta^*) \in \mathbb{R}_+^{2N}$ exists. The optimal point is unique by Assumption 1 of strict convexity of the cost function $g_i(\cdot)$. As our problem is timed, we aim to design a distributed algorithm determining, for each time instant t_k, values of $x_i(k)$ and $y_i(k)$, such that for the long-term averages, we have:

$$\lim_{k\to\infty} \bar{x}_i(k) = \xi_i^*, \quad \lim_{k\to\infty} \bar{y}_i(k) = \eta_i^*, \quad \text{for } i = 1,\ldots,N. \quad (13)$$

Also, it is desirable that the constraints of the optimization problem are satisfied by $\bar{x}_i(k), \bar{y}_i(k)$, for every k, where $\bar{x}_i(k)$ takes the role of the optimization variable ξ_i and $\bar{y}_i(k)$ that of η_i.

Typically, a problem of this nature can be solved in a standard optimization framework. Unfortunately, this approach is not available to us for many reasons:

(i) The number of prosumers N in the prosumer community, and the constraints C_x and C_y may vary with time, making an offline computation of an optimal solution difficult.
(ii) To preserve privacy, we also assume that prosumers do not necessarily communicate with each other, and they only communicate with the infrastructure in a limited fashion. Thus, even the communication graph is unknown a-priori for this particular problem class. In particular, individual cost function g_i is often private and is not shared by prosumers.
(iii) We are interested in algorithms that self organize and converge to an optimal solution even in the presence of disturbances in the state information.

Our approach, therefore, is to treat the above problem as a feedback (stochastic) control problem, which changes the formulation in a minor way; namely, we allow:

$$\sum_{i=1}^{N} x_i(k) \approx C_x, \text{ and } \sum_{i=1}^{N} y_i(k) \approx C_y, \text{ for all } k. \tag{14}$$

In other words, we allow the instantaneous prosumption to undershoot or overshoot the reference values by a small amount. Then, given this background, and from Assumption 1 for the cost function g_i, for all prosumers i, we shall demonstrate that an elementary feedback control algorithm can be devised to solve an approximate version of Problem 1. As we shall see, this algorithm requires only a few bits of message transfer as intermittent feedback from a control unit (sharing platform) to prosumers in the prosumer market, but no inter-prosumer communication is required.

4 Algorithms for Community-Based Prosumer Markets

The algorithms proposed in this section are motivated by the following elementary argument. We consider only the case in which at time instant t_k, either 0 or 1 unit of the resource is consumed or produced, for $k = 0, 1, 2, \ldots$ For the sake of argument, we only consider the case of pure consumption in this preamble.

Let $z(k)$ denote the number of times an agent (consumer) consumes a unit resource until time instant t_k, where $k = 0, 1, 2, \ldots$ Let $\bar{z}(k) \triangleq \frac{1}{k+1} z(k)$, denote the time-averaged consumption of the resource until time instant t_k. We assume that the consumption at time instant t_k is decided by a stochastic procedure, where the probability of consumption of one unit of the resource is $p(\bar{z}(k))$. Additionally, we assume that this probability conditioned on $\bar{z}(k)$ is independent of the previous history of the process. Then:

$$z(k+1) = z(k) + w(k), \tag{15}$$

where $w(k)$ is a random variable taking the value 0 or 1. Thus, the following holds:

$$\bar{z}(k+1) = \frac{k+1}{k+2}\bar{z}(k) + \frac{1}{k+2}w(k) \tag{16}$$

$$= \bar{z}(k) + \frac{1}{k+2}\big(w(k) - \bar{z}(k)\big). \tag{17}$$

Recall that $p(\bar{z}(k))$ denotes the probability that $w(k) = 1$ at time instant t_k, for $k = 0, 1, 2, \ldots$ Then, we rewrite (17) as:

$$\bar{z}(k+1) = \bar{z}(k) + \frac{1}{k+2}\big(p(\bar{z}(k)) - \bar{z}(k)\big) + \frac{1}{k+2}\big(w(k) - p(\bar{z}(k))\big).$$

Note that systems of this form are discussed extensively in [31]. In particular, the term $\frac{1}{k+2}(w(k) - p(\bar{z}(k)))$ is a martingale difference sequence and is treated as noise. It is shown in [31] that under mild assumptions, $\bar{z}(k)$ converges almost surely. The basic idea of the remainder of this section is to construct stochastic feedback algorithms that mimic this argument. In particular, our basic idea in the sequel is to choose the probability distribution $p(\bar{z}(k))$, so that the stochastic system both solves a regulation problem and also optimization problem of the form of Problem 1, simultaneously.

4.1 Optimality Conditions

In this subsection, we briefly discuss the optimality conditions for Problem 1, using Lagrangian multipliers. These optimality conditions lead to the state dependent probabilities that we alluded to in the preamble.

Let $x = (x_1, x_2, \ldots, x_N)$ and $y = (y_1, y_2, \ldots, y_N)$. Also, let μ^1, μ^2 and $\lambda^1 = (\lambda^1_1, \ldots, \lambda^1_N)$, $\lambda^2 = (\lambda^2_1, \ldots, \lambda^2_N)$ be the Lagrange multipliers corresponding to the equality constraints (9), (10) and the inequality constraints (11), (12), respectively. The Lagrangian of Problem 1 is defined as $\mathcal{L} : \mathbb{R}^{2N} \times \mathbb{R}^2 \times \mathbb{R}^{2N} \to \mathbb{R}$, where

$$\mathcal{L}(x, y, \mu^1, \mu^2, \lambda^1, \lambda^2) = \sum_{i=1}^{N} g_i(x_i, y_i) - \mu^1 \left(\sum_{i=1}^{N} x_i - C_x \right) - \mu^2 \left(\sum_{i=1}^{N} y_i - C_y \right)$$
$$+ \sum_{i=1}^{N} \lambda^1_i x_i + \sum_{i=1}^{N} \lambda^2_i y_i.$$

We assume that the optimal value of Problem 1 is obtained for positive values of consumption and production. We, therefore, let $x_i^*, y_i^* \in (0, 1]$ denote the optimal point of Problem 1. By this assumption, the inequality constraints are not active, and it follows that the corresponding optimal Lagrange multipliers are $\lambda^{*1} = 0 = \lambda^{*2}$.

Additionally, let μ^{*1}, μ^{*2} be the optimal Lagrange multipliers for the equality constraints. The first order optimality condition is that the gradient vanishes, and inspection shows that the gradient condition decouples. Recall that $\nabla_x g_i(.)$ denotes the partial derivative of $g_i(.)$ with respect to x_i and $\nabla_y g_i(.)$ denotes the partial derivative of $g_i(.)$ with respect y_i. Then, we arrive at the following conditions:

$$\nabla_x g_i(x_i^*, y_i^*) = \mu^{*1}, \quad \text{for } i = 1, 2, \ldots, N,$$

and,

$$\nabla_y g_i(x_i^*, y_i^*) = \mu^{*2}, \quad \text{for } i = 1, 2, \ldots, N.$$

In other words, we have:

$$\nabla_x g_i(x_i^*, y_i^*) = \nabla_x g_j(x_j^*, y_j^*), \text{ for } i, j \in \{1, 2, \ldots, N\}. \tag{18}$$

The same consensus condition holds for $\nabla_y g_i(x_i^*, y_i^*)$, $i = 1, \ldots, N$. We find that the optimal values satisfy all the Karush–Kuhn–Tucker (KKT) conditions, which are necessary and sufficient conditions for optimality of differentiable convex functions (Chap. 5.5.3 [32]). Hence, the derivatives of the cost functions of all prosumers with respect to consumption as well as production must reach consensus at the optimal point.

This type of consensus condition has been used in [33, 34] (single resource case) and [35] (multi-resource case) to derive place-dependent probabilities that ensure convergence to the consensus condition and thus, to the optimal point.

4.2 Algorithm for Consumption

Here, we briefly describe the distributed algorithm proposed in [34] for allocating a single resource to consumers. In this model, no production is taking place, hence the corresponding variables are omitted. Suppose that there are N consumers in a community. We assume that consumer i of the community has a cost function $g_i : (0, 1] \to \mathbb{R}_+$, which is strictly convex, strictly increasing in each variable, and continuously differentiable, for $i = 1, 2, \ldots, N$. The random variable $x_i(k) \in \{0, 1\}$ denotes the consumption of the unit resource for consumer i at time instant t_k, for all i and k. As before, let $\bar{x}_i(k)$ be the time-averaged consumption of consumer i until time instant t_k, for all i and k.

The idea is to choose probabilities so as to ensure convergence to the social optimum and to adjust overall consumption by the community to the reference value (capacity) C_x, by applying a *feedback signal* $\Omega(k)$ to the probabilities. At each time instant t_k, the control unit updates $\Omega(k)$ using a gain parameter $\tau > 0$, the past aggregate consumption of the resource, and the capacity C_x as described in (19) and then, broadcasts the new value to all consumers in the community:

$$\Omega(k+1) \triangleq \Omega(k) - \tau\left(\sum_{i=1}^{N} x_i(k) - C_x\right). \tag{19}$$

After receiving this signal, consumer i responds in a probabilistic way. The probability distribution $\sigma_i(\cdot)$ is calculated using the time-averaged consumption $\bar{x}_i(k)$ and the derivative $g'_i(\cdot)$ of the cost function $g_i(\cdot)$, as follows:

$$\sigma_i(\Omega(k), \bar{x}_i(k)) \triangleq \Omega(k) \frac{\bar{x}_i(k)}{g'_i(\bar{x}_i(k))}. \tag{20}$$

Notice that the cost function g_i is chosen as increasing function in each variable so that the probability $\sigma_i(\cdot)$ is in the valid range, for all i. Now, consumer i updates its resource consumption at each time instant t_k, either by consuming one unit of the resource or not consuming it, as follows:

$$x_i(k+1) = \begin{cases} 1 & \text{with probability } \sigma_i(\Omega(k), \bar{x}_i(k)); \\ 0 & \text{with probability } 1 - \sigma_i(\Omega(k), \bar{x}_i(k)). \end{cases}$$

Empirical results show that the time-averaged consumption $\bar{x}_i(k)$ converges to the optimal value x_i^* asymptotically.

Remark 1 (*Integral control action*) Equation (19) defines what is called an integral control action in tracking control. The overall consumption by the community should "track" the available capacity, i.e., approximate it. In other words, the objective of the integrator is to ensure that the tracking error $e(k) = \sum_{i=1}^{N} x_i(k) - C_x \approx 0$ asymptotically.

Remark 2 (*Consensus of derivatives*) The policy for $\sigma_i(\Omega(k), \bar{x}_i(k))$ in (20) is to ensure that asymptotically, $g'_i(\bar{x}_i(k)) = g'_j(\bar{x}_j(k))$, for all consumers i and j. As is discussed in [34], the convergence of the above algorithm and strict convexity of all cost functions (and of course assuming the existence of a feasible solution in the constraint set) is enough to imply that Problem 1 is solved asymptotically for the consumption case.

4.3 Algorithm for Coupled Prosumption

The case of *coupled prosumption* of a single resource is characterized by the constraint (4), that is, consumption and production are coupled through the desired value of utilization of the resource. The algorithm follows the case of exclusive consumption closely taking into account the cost function as discussed in the problem formulation (Sect. 3). Before proceeding, note that the following discussion extends to an arbitrary number of resources, but is presented for a single resource, both to aid exposition and to be consistent with the application class that is the principal consideration in this chapter. Interested readers can look at [36] for preliminary results on multi-resource allocation. With this background in mind, following the discussion for consumption, let us assume that in a prosumer market there is a community of N prosumers. The prosumer community sells the excess resource to an external community. Furthermore, let τ_x, τ_y denote the gain parameters for consumption and production, $\Omega_x(k)$, $\Omega_y(k)$ denote the *feedback signals* for both processes, and C_x, C_y represent the respective contract (capacity) constraints. We assume the existence of a centralized control unit (sharing platform) in the prosumer market that can measure the aggregate response of the prosumer community and broadcast a feedback signal in the prosumer market at each time instant t_k, for both consumption and production type. Specifically, the sharing platform updates the feedback signal $\Omega_x(k)$, as follows:

$$\Omega_x(k+1) \triangleq \Omega_x(k) - \tau_x\left(\sum_{i=1}^{N} x_i(k) - C_x\right), \quad \text{for all } k, \qquad (21)$$

with $\Omega_y(k)$ updated analogously. After receiving a feedback signal prosumer i's algorithm responds in a probabilistic manner. The probability that prosumer i responds to the feedback signal is given by:

$$\sigma_{i,x}(k) \triangleq \Omega_x(k) \frac{\nabla_x g_i(\overline{x}_i(k), \overline{y}_i(k))}{\overline{x}_i(k)}, \quad \text{for all } i \text{ and } k. \qquad (22)$$

Here, $\sigma_{i,x}(k)$ denotes the probability of prosumer i, responding to a demand for consumption of the resource, at time instant t_k; similarly, $\sigma_{i,y}(k)$ denotes the probability of prosumer i, responding for production of the resource at time instant t_k, for all k. Similar to the consumption case, the cost function g_i is chosen as increasing in each variable, to keep the probabilities $\sigma_{i,x}(k)$ and $\sigma_{i,y}(k)$ in the valid range. However, the definition of $\sigma_{i,x}(k)$ and $\sigma_{i,y}(k)$ is slightly different from the consumption case, described previously. Now, prosumer i updates its consumption and production of the resource at each time instant in the following ways:

$$x_i(k+1) = \begin{cases} 1 & \text{with probability } \sigma_{i,x}(k); \\ 0 & \text{with probability } 1 - \sigma_{i,x}(k), \quad \text{and,} \end{cases}$$

$$y_i(k+1) = \begin{cases} 1 & \text{with probability } \sigma_{i,y}(k); \\ 0 & \text{with probability } 1 - \sigma_{i,y}(k), \end{cases}$$

for all i and k. Empirically, we observe that the time-averaged consumption $\overline{x}_i(k)$ converges to the optimal consumption value x_i^*, and similarly, the time-averaged production $\overline{y}_i(k)$ converges to the optimal production value y_i^* asymptotically, for all prosumers in the community. We describe the algorithm of the sharing platform in Algorithm 1 and the algorithm of prosumers of the community in Algorithm 2. We make the following remarks.

Remark 3 (*Communication overhead and privacy*) There is no explicit communication between prosumers. Thus, the algorithm is low cost in terms of communication and is private.

Remark 4 (*Probability bounds*) The gain parameters τ_x, τ_y are small constants chosen to ensure $\sigma_{i,x}(k)$ and $\sigma_{i,y}(k)$ are the probabilities; namely, are in $[0, 1]$, for all i and k.

Remark 5 (*Consensus of partial derivatives*) The well-posedness of our algorithm follows from the assumption of strict convexity of $g_i(\cdot)$, and that the constraint sets are closed and bounded; namely, that there exists unique optimal solutions to Problem 1. To show that the long-term average values converge to the optimal values, we use the consensus of (partial) derivatives of the cost functions of prosumers as described previously.[3] That is:

$$\lim_{k\to\infty} \nabla_x g_i(\overline{x}_i(k), \overline{y}_i(k)) = \lim_{k\to\infty} \nabla_x g_j(\overline{x}_j(k), \overline{y}_j(k)),$$

and similarly,

$$\lim_{k\to\infty} \nabla_y g_i(\overline{x}_i(k), \overline{y}_i(k)) = \lim_{k\to\infty} \nabla_y g_j(\overline{x}_j(k), \overline{y}_j(k)),$$

for all $i, j \in \{1, 2, \ldots, N\}$.

Algorithm 1: Algorithm of control unit (sharing platform).

1 Input: $C_x, C_y, \tau_x, \tau_y, x_i(k), y_i(k)$, for $k = 0, 1, 2, \ldots$ and $i = 1, 2, \ldots, N$.
2 Output: $\Omega_x(k+1), \Omega_y(k+1)$, for $k = 0, 1, 2, \ldots$
3 Initialization: $\Omega_x(0) \leftarrow 0.06$ and $\Omega_y(0) \leftarrow 0.06$,
4 **foreach** $k = 0, 1, 2, \ldots$ **do**
5   calculate $\Omega_x(k+1)$ and $\Omega_y(k+1)$ as follows and broadcast them in the prosumer community;

$$\Omega_x(k+1) \leftarrow \Omega_x(k) - \tau_x\left(\sum_{i=1}^{N} x_i(k) - C_x\right), \text{ and,}$$

$$\Omega_y(k+1) \leftarrow \Omega_y(k) - \tau_y\left(\sum_{i=1}^{N} y_i(k) - C_y\right).$$

6 **end**

---

[3] We initialize $\Omega_x(0)$ and $\Omega_y(0)$ with positive real numbers.

Algorithm 2: Algorithm of prosumer i.

1 Input: $\Omega_x(k)$, $\Omega_y(k)$, for $k = 0, 1, 2, \ldots$
2 Output: $\bar{x}_i(k+1)$, $\bar{y}_i(k+1)$, for $k = 0, 1, 2, \ldots$
3 Initialization: $x_i(0), y_i(0) \leftarrow 1$ and $\bar{x}_i(0) \leftarrow x_i(0)$ and $\bar{y}_i(0) \leftarrow y_i(0)$.
4 **while** *prosumer i is active at $k = 0, 1, 2, \ldots$* **do**
5

$$\sigma_{i,x}(k) \leftarrow \Omega_x(k) \frac{\nabla_x g_i\left(\bar{x}_i(k), \bar{y}_i(k)\right)}{\bar{x}_i(k)};$$

$$\sigma_{i,y}(k) \leftarrow \Omega_y(k) \frac{\nabla_y g_i\left(\bar{x}_i(k), \bar{y}_i(k)\right)}{\bar{y}_i(k)};$$

calculate outcome of the random variables;

$$x_i(k+1) \leftarrow \begin{cases} 1 & \text{w. p. } \sigma_{i,x}(k) \\ 0 & \text{w. p. } 1 - \sigma_{i,x}(k); \end{cases}$$

$$y_i(k+1) \leftarrow \begin{cases} 1 & \text{w. p. } \sigma_{i,y}(k) \\ 0 & \text{w. p. } 1 - \sigma_{i,y}(k); \end{cases}$$

update $\bar{x}_i(k+1)$ and $\bar{y}_i(k+1)$ as follows;

$$\bar{x}_i(k+1) \leftarrow \frac{k+1}{k+2}\bar{x}_i(k) + \frac{1}{k+2}x_i(k+1);$$

$$\bar{y}_i(k+1) \leftarrow \frac{k+1}{k+2}\bar{y}_i(k) + \frac{1}{k+2}y_i(k+1);$$

6 **end**

5 Use Cases

We now describe two use cases for community-based prosumer market. The first is a transportation example and concerns a car sharing prosumer market. The second example is from the energy sector and considers a prosumer market, where produced and consumed energy is coupled through storage constraints.

5.1 Community-Based Car Sharing

In this use case, we consider a community of N households (prosumers) with several cars, not all of which are required each day. We assume that cars are pooled and shared amongst community members to multiplex and monetize the excess capacity, but that the average aggregate daily community demand for cars is known. We let T_i denote the average number of cars desired to be used by each household i, for all i. Suppose that C_x cars are required within the community each day, and an excess of

C_y cars are made available to an external community each day. For simplicity, we assume that C_x and C_y are fixed. Notice that a household is a prosumer in the sense that it requires cars for its transportation needs and supplies excess car-days to an external community. For each prosumer i, let $x_i(k) \in \{0, 1\}$ denote that prosumer i requires a car at day k or not, and let $y_i(k) \in \{0, 1\}$ denote that prosumer i supplies a car to an external community at day k or not. Thus, we assume that aggregated over the entire community, the demand for *shared cars* on a given day k is:

$$\sum_{i=1}^{N} x_i(k) = C_x, \tag{23}$$

leaving the *excess* capacity so that C_y cars can be supplied to an external community:

$$\sum_{i=1}^{N} y_i(k) = C_y. \tag{24}$$

Thus, $\sum_{\ell=0}^{k} x_i(\ell)$ is the number of days a car was required by prosumer i over k days, and $\sum_{\ell=0}^{k} y_i(\ell)$ is the number of days that the same prosumer made a car available to an external community. Thus, $\sum_{\ell=0}^{k}(x_i(\ell) + y_i(\ell))$ is the total number of days that a car was used as a result of prosumer i. Over some days, prosumer i will require that the time-averaged of this number be equal to the desired value of utilization of cars, T_i. For example, if a prosumer has two cars, thus, over seven days period (a week) it has 14 car-days. Now, suppose that the prosumer only needs access to 10 car-days over a week. Then, this prosumer might choose $\sum_{\ell=0}^{6} x_i(\ell)$ and $\sum_{\ell=0}^{6} y_i(\ell)$, such that $\sum_{\ell=0}^{6}(x_i(\ell) + y_i(\ell)) = 12$. In this case, the prosumer sells access to two excess car-days over the week (two car-days rather than the maximum of four to provide some margin in case that a car is required for personal use more than expected). Let $\bar{x}_i(k)$ denote the average number of days prosumer i requires a car over k days, and let $\bar{y}_i(k)$ denote the average number of days prosumer i makes a car available to an external community over k days, for all i and k. Thus, over a long period, this constraint can be scaled by the number of days to yield:

$$\bar{x}_i(k) + \bar{y}_i(k) \approx T_i, \tag{25}$$

with the cost of not achieving this goal captured by a penalty function $g_i(x_i + y_i - T_i)$. For sufficiently large k, we might also require that $\bar{x}_i(k) \approx \alpha_i T_i$ and $\bar{y}_i(k) \approx (1 - \alpha_i)T_i$ with $\alpha_i \in [0, 1]$, for all i. This latter constraint can be formulated in terms of a cost via a penalty function. For example, in residential areas in Ireland; two car households are common, meaning that households can in principle both consume and produce cars simultaneously, hence act as prosumers. However, this may not always be possible, and prosumers may be required to make alternative arrangements or pay penalties, should they not be able to meet contractual demands. To formulate the cost function, we associate costs $h_i : (0, 1] \to \mathbb{R}_+$, $x_i \mapsto h_i(x_i)$ to the deviation from $\alpha_i T_i$ and $l_i : (0, 1] \to \mathbb{R}_+$, $y_i \mapsto l_i(y_i)$ to the deviation from $(1 - \alpha_i)T_i$, for all i. Then, the aim of the sharing platform is to minimize:

$$\sum_{i=1}^{N} \left(g_i(x_i + y_i - T_i) + h_i(x_i - \alpha_i T_i) + l_i(y_i - (1-\alpha_i)T_i) \right), \tag{26}$$

subject to the additional constraints listed in Problem 1.

5.2 Collaborative Energy Storage

In this use case, we again assume that there are N households in a community participating in a prosumer market, with each household connected to the grid, and also has installed solar panels. Every household has batteries to store the energy either from the solar panel or the grid. The households act as prosumers in the sense that they can consume stored energy as well as sell excess energy for monetary benefits. Let $x_i(k) \in \{0, 1\}$ denote that household i consumes stored energy at day k or not, and let $y_i(k) \in \{0, 1\}$ denote that household i sells stored energy to an external community at day k or not, for all i. Let $\bar{x}_i(k)$ denote the time-averaged amount of stored energy consumed by household i over k days, and let $\bar{y}_i(k)$ denote the time-averaged amount of stored energy sold by household i to an external community over k days, for all i. Now, we assume that $\sum_{i=1}^{N} x_i(k) = C_x$, be the aggregated consumption of stored energy over the entire community on a given day k; whereas, the excess energy C_y is sold to an external community, with $\sum_{i=1}^{N} y_i(k) = C_y$. Furthermore, we assume that each household may have a constraint on the amount of energy stored in order to realize the above strategy. As in the previous use case, this *soft* constraint is captured as follows. Let T_i be the desired amount of energy stored by household i. Then, on average, over a sufficiently large given period k, household i may expect to store the following desired amount of energy temporarily:

$$\bar{x}_i(k) + \bar{y}_i(k) \approx T_i, \tag{27}$$

with the deviation from this goal captured by a penalty function $g_i(x_i + y_i - T_i)$. We might also require that roughly speaking, a certain amount of storage is reserved for consumption and a certain amount to sell (production). Hence, again, the objective of the sharing platform is to minimize:

$$\sum_{i=1}^{N} \left(g_i(x_i + y_i - T_i) + h_i(x_i - \alpha_i T_i) + l_i(y_i - (1-\alpha_i)T_i) \right),$$

subject to the constraints listed in Problem 1. Notice that the definition of $h_i(\cdot)$ and $l_i(\cdot)$ is same as the previous use case.

6 Simulation Results

In this section, we present simulation results for 99 prosumers participating in a prosumer market. We assume that these $N = 99$ in the prosumer market are grouped into two prosumer communities. Prosumers 1 to 50 are grouped in Community 1, and Prosumers 51 to 99 are grouped in Community 2. Additionally, each community has a cost function type, and prosumers of a particular community use the cost function type of that community, but with randomized parameter values. Recall that prosumers can produce the resource as well as consume it. Let time instants $t_0 < t_1 < t_2 < \cdots$, represent days, and let the capacity constraints be $C_x = 90$ and $C_y = 80$. Furthermore, let the cost factors $a_i \in \{1, 2, \ldots, 10\}$ and $b_i \in \{1, 2, \ldots, 15\}$ be drawn from uniformly distributed random variables. Additionally, let $\delta = 11.75$ and $\gamma = 11.65$. Notice that the values of δ and γ are chosen in such a way that the cost function $g_i(\cdot)$ is increasing in each variable. Recall that this is done to keep the probabilities $\sigma_{i,x}$ and $\sigma_{i,y}$ in $[0, 1]$. The cost functions are presented as follows:

$$g_i(\overline{x}_i(k), \overline{y}_i(k)) = \begin{cases} \delta(\overline{x}_i(k) + \overline{y}_i(k)) + \frac{1}{4}a_i(\overline{x}_i(k) + \overline{y}_i(k) - T_1)^2 + \\ \frac{1}{2}(\overline{x}_i(k) - \frac{1}{2}T_1)^2 + \frac{1}{2}(\overline{y}_i(k) - \frac{1}{2}T_1)^2 \quad \text{Community 1,} \\ \gamma(\overline{x}_i(k) + \overline{y}_i(k)) + \frac{1}{8}a_i(\overline{x}_i(k) + \overline{y}_i(k) - T_2)^2 + \\ \frac{5}{4}b_i(\overline{x}_i(k) + \overline{y}_i(k) - T_2)^4 + \frac{1}{2}(\overline{x}_i(k) - \frac{1}{2}T_2)^2 + \\ \frac{1}{2}(\overline{y}_i(k) - \frac{1}{2}T_2)^2 \quad \text{Community 2.} \end{cases}$$

Now, we present the simulation results and show the convergence of the long-term average of prosumption by each prosumer in Fig. 4. Figure 4a shows the time-averaged consumption and Fig. 4b shows the time-averaged production over 200 days. In the context of car sharing, these are the average of cars used by a prosumer and the average of cars shared with another person by the same prosumer over 200 days, respectively. Figure 5 illustrates the average prosumption on the 200th day by every

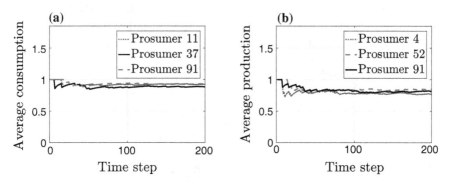

Fig. 4 a Evolution of time-averaged consumption of the resource, and b evolution of time-averaged production of the resource

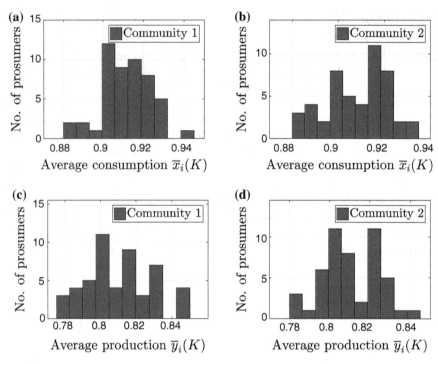

Fig. 5 At time index $K = 200$—**a** average consumption $\bar{x}_i(K)$ by prosumers of Community 1, **b** average consumption $\bar{x}_i(K)$ by prosumers of Community 2, **c** average production $\bar{y}_i(K)$ by prosumers of Community 1, and **d** average production $\bar{y}_i(K)$ by prosumers of Community 2

prosumer of a particular community. Furthermore, the absolute difference between the desired value of utilization T_i of cars and the actual utilization $\bar{x}_i(k) + \bar{y}_i(k)$ by prosumer i for a certain period is shown in Fig. 6. Figure 6a illustrates the evolution of the absolute difference between T_i and $\bar{x}_i(k) + \bar{y}_i(k)$ for individual prosumers. Here, we observe that gradually the difference comes closer to zero. Additionally, in Fig. 6b, c, we observe that the absolute difference between the quantities is close to zero for most of the prosumers.

Now, we analyze the derivatives of the cost functions and see, whether they gather close to each other to make consensus over time or not. Recall that the derivative of the cost function with respect to consumption is $\nabla_x g_i(.)$ and with respect to production is $\nabla_y g_i(.)$, that are shown in Fig. 7 for a single simulation. We plot the shaded errorbars as depicted in Fig. 7a, b. It is observed that in both the cases the derivatives gather close to each other over time. Therefore, we say that the derivatives make consensus asymptotically in their respective prosumer communities, which is a necessary and sufficient condition for optimality as described in Sect. 4.1. Notice that because both the derivatives (with respect to consumption and production) are same, therefore, we illustrate here just one of them. We clarify here that because of the chosen initial values, the probability $\sigma_{i,x}(k)$ may overshoot at the start of the algorithm; to keep it

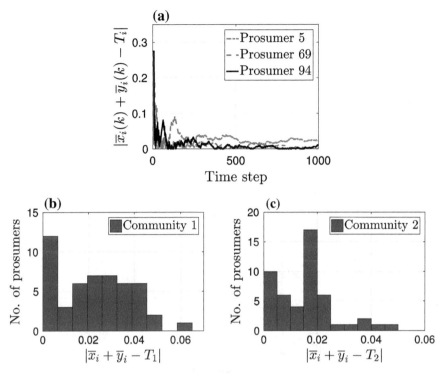

Fig. 6 a Evolution of absolute difference between desired value of utilization T_i and actual utilization of the resource $\bar{x}_i(k) + \bar{y}_i(k)$ of individual prosumers, **b** absolute difference between desired value of utilization and actual utilization $|\bar{x}_i(K) + \bar{y}_i(K) - T_1|$ of prosumers of Community 1, here $T_1 = 1.74$, and **c** absolute difference between desired value of utilization and actual utilization $|\bar{x}_i(K) + \bar{y}_i(K) - T_2|$ of prosumers of Community 2, here $T_2 = 1.725$ and time index $K = 200$

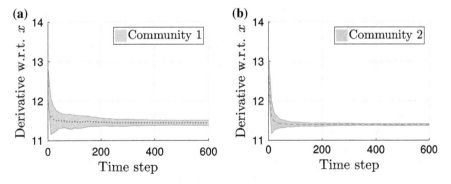

Fig. 7 a Evolution of derivatives of $g_i(.)$ w. r. t. x for prosumers of Community 1, and **b** evolution of derivatives of $g_i(.)$ w. r. t. x for prosumers of Community 2

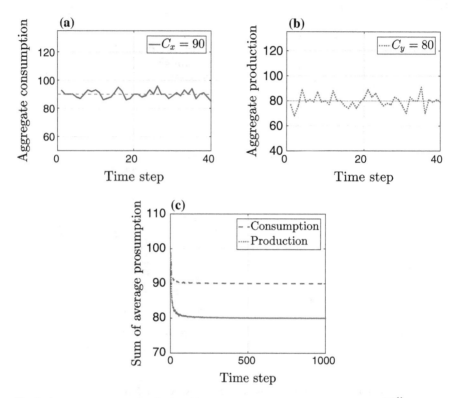

Fig. 8 Aggregate prosumption for last 40 time instants—**a** aggregate consumption $\sum_{i=1}^{N} x_i(k)$, **b** aggregate production $\sum_{i=1}^{N} y_i(k)$, and **c** evolution of sum of time-averaged prosumption

in the valid range, we use $\min\left\{1, \Omega_x(k) \frac{\nabla_x g_i(\bar{x}_i(k), \bar{y}_i(k))}{\bar{x}_i(k)}\right\}$. Similar step is used to keep $\sigma_{i,y}(k)$ in the valid probability range.

Now, we analyze the aggregate consumption $\sum_{i=1}^{N} x_i(k)$ and production $\sum_{i=1}^{N} y_i(k)$ by the prosumer communities. The aggregate consumption $\sum_{i=1}^{N} x_i(k)$ is presented in Fig. 8a for last 40 time instants (days), and similarly, the aggregate production by the communities $\sum_{i=1}^{N} y_i(k)$ is shown in Fig. 8b. Notice that the aggregate prosumption is close to the respective capacity constraints C_x and C_y; overshoots and undershoots are due to the assumption of soft constraints, as described previously. Additionally, Fig. 8c shows the time-averaged consumption $\sum_{i=1}^{N} \bar{x}_i(k)$ by all prosumers in the prosumer market until 1000 time instants, and similarly, the time-averaged production $\sum_{i=1}^{N} \bar{y}_i(k)$ by all prosumers in the prosumer market for the same period, these averages are approximately equal to the respective capacities, satisfying the capacity constraints of Problem 1. In addition to the above results, we observe in Fig. 9a, b that most of the time the aggregate consumption $\sum_{i=1}^{N} x_i(k)$ and the aggregate production $\sum_{i=1}^{N} y_i(k)$ are close to their respective capacities C_x

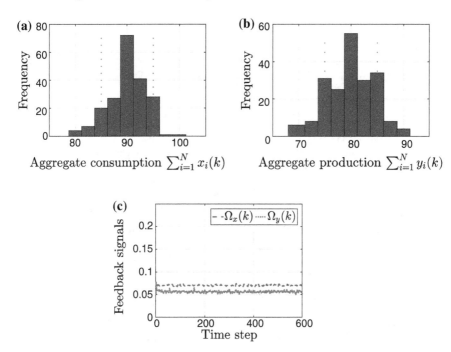

Fig. 9 Frequency of prosumption—**a** frequency of aggregate consumption, **b** frequency of aggregate production for last 200 time instants, and **c** evolution of feedback signals $\Omega_x(k)$ and $\Omega_y(k)$

and C_y. Furthermore, the convergence of feedback signals Ω_x and Ω_y are shown in Fig. 9c.

7 Conclusion and Future Directions

We have proposed distributed control algorithms to solve regulation problems with optimality constraints for community-based prosumer market. Our algorithms are based on ideas from stochastic approximation but formulated in a control theoretic setting. The algorithm reaches optimality asymptotically, while simultaneously regulating instantaneous contract constraints. To do so, the algorithm does not require communication between prosumers, but little communication with the sharing platform. Additionally, the algorithm is light and is suitable to implement in an Internet-of-Things (IoT) context with minimal demands on infrastructure. Two applications are described and simulation results presented to demonstrate the efficacy of the algorithms. Future work will explore the theoretical aspects of the algorithm (convergence properties), new applications and use cases, and the development of policies to reach more complicated equilibria.

References

1. Narasimhan C, Papatla P, Jiang B, Kopalle PK, Messinger PR, Moorthy S, Proserpio D, Subramanian U, Wu C, Zhu T (2018) Sharing economy: review of current research and future directions. Cust Needs Solut 5(1):93–106
2. Hamari J, Sjoklint M, Ukkonen A (2016) The sharing economy: why people participate in collaborative consumption. J Assoc Inf Sci Technol 67(9):2047–2059
3. Lan J, Ma Y, Zhu D, Mangalagiu D, Thornton TF (2017) Enabling value co-creation in the sharing economy: the case of mobike. Sustainability 9(9):1–20
4. Chen TD, Kockelman KM (2016) Carsharing's life-cycle impacts on energy use and greenhouse gas emissions. Transp Res Part D: Transp Environ 47:276–284
5. Crisostomi E, Shorten R, Studli S, Wirth F (2017) Electric and plug-in hybrid vehicle networks: optimization and control. CRC Press (Taylor and Francis Group), Boca Raton
6. Fraiberger S, Sundararajan A (2015) Peer-to-peer rental markets in the sharing economy. SSRN Electron J
7. Huckle S, Bhattacharya R, White M, Beloff N (2016) Internet of things, blockchain and shared economy applications. Procedia Comput Sci EUSPN 98:461–466
8. Kortuem G, Bourgeois J (2016) The Internet of things for the open sharing economy. In: Proceedings of the ACM international joint conference on pervasive and ubiquitous computing, pp 666–669
9. Atzori L, Iera A, Morabito G (2010) The Internet of things: a survey. Comput Netw 54(15):2787–2805
10. Al-Fuqaha A, Guizani M, Mohammadi M, Aledhari M, Ayyash M (2015) Internet of things: a survey on enabling technologies, protocols, and applications. IEEE Commun Surv Tutor 17(4):2347–2376
11. Goudin P (2016) The cost of non-Europe in the sharing economy: economic, social and legal challenges and opportunities. Technical Report, European Parliamentary Research Service
12. Ritzer G, Jurgenson N (2010) Production, consumption, prosumption: the nature of capitalism in the age of the digital 'prosumer'. J Consum Cult 10(1):13–36
13. Moret F, Pinson P (2019) Energy collectives: a community and fairness based approach to future electricity markets. IEEE Trans Power Syst 34(5):3994–4004
14. Agnew S, Dargusch P (2015) Effect of residential solar and storage on centralized electricity supply systems. Nat Clim Chang 5(315)
15. Patel S, Rajagopal R (2017) The value of distributed energy resources for heterogeneous residential consumers. arXiv:1709.08140 [cs.SY]
16. Inderberg TJ, Tews K, Turner B (2018) Is there a prosumer pathway? Exploring household solar energy development in Germany, Norway, and the United Kingdom. Energy Res Soc Sci 42:258–269
17. Schill W, Zerrahn A, Kunz F (2017) Prosumage of solar electricity: pros, cons, and the system perspective, vol 1637. DIW, Berlin
18. Grijalva S, Costley M, Ainsworth N (2011) Prosumer-based control architecture for the future electricity grid. In: IEEE international conference on control applications, pp 43–48
19. Parag Y, Sovacool BK (2016) Electricity market design for the prosumer era. Nat Energy 1
20. Zhang C, Wu J, Zhou Y, Cheng M, Long C (2018) Peer-to-peer energy trading in a microgrid. Appl Energy 220:1–12
21. de Souza Ribeiro LA, Saavedra OR, de Lima SL, de Matos JG (2011) Isolated micro-grids with renewable hybrid generation: the case of Lencois Island. IEEE Trans Sustain Energy 2(1):1–11
22. Hasan MH, Hentenryck PV, Budak C, Chen J, Chaudhry C (2018) Community-based trip sharing for urban commuting. In: AAAI conference on artificial intelligence, pp 6589–6597
23. Gkatzikis L, Iosifidis G, Koutsopoulos I, Tassiulas L (2014) Collaborative placement and sharing of storage resources in the smart grid. In: International conference on smart grid communications, pp 103–108
24. Iosifidis G, Tassiulas L (2017) Dynamic policies for cooperative networked systems. In: Workshop on the economics of networks, systems and computation, Ser. NetEcon, pp 1–6

25. Georgiadis L, Iosifidis G, Tassiulas L (2017) On the efficiency of sharing economy networks. arXiv:1703.09669 [cs.GT]
26. Einav L, Farronato C, Levin J (2016) Peer-to-peer markets. Annu Rev Econ 8:615–635
27. Tushar W, Yuen C, Mohsenian-Rad H, Saha T, Poor HV, Wood KL (2018) Transforming energy networks via peer-to-peer energy trading: the potential of game-theoretic approaches. IEEE Signal Process Mag 35(4):90–111
28. Courcoubetis C, Weber R (2012) Economic issues in shared infrastructures. IEEE/ACM Trans Netw 20(2):594–608
29. Benjaafar S, Kong G, Li X, Courcoubetis C (2018) Peer-to-peer product sharing: implications for ownership, usage, and social welfare in the sharing economy. Manag Sci
30. Sousa T, Soares T, Pinson P, Moret F, Baroche T, Sorin E (2018) Peer-to-peer and community-based markets: a comprehensive review. arXiv:1810.09859 [cs.CY]
31. Borkar VS (2008) Stochastic approximation. Cambridge University Press, Cambridge
32. Boyd S, Vandenberghe L (2004) Convex optimization. Cambridge University Press, New York
33. Wirth F, Stuedli S, Yu JY, Corless M, Shorten R (2019) Nonhomogeneous place-dependent Markov chains, unsynchronised AIMD, and network utility maximization. J ACM 66(4)
34. Griggs WM, Yu JY, Wirth FR, Hausler F, Shorten R (2016) On the design of campus parking systems with QoS guarantees. IEEE Trans Intell Transp Syst 17(5):1428–1437
35. Alam SE, Shorten R, Wirth F, Yu JY (2018) Communication-efficient distributed multi-resource allocation. In: IEEE international smart cities conference, pp 1–8
36. Alam SE, Shorten R, Wirth F, Yu JY (2018) On the control of agents coupled through shared resources. arXiv:1803.10386 [cs.SY]

Negotiation Approaches for Sharing Systems

From Pool-Based to Peer-to-peer

Pierre Pinson, Fabio Moret, Thomas Baroche
and Athanasios Papakonstantinou

Abstract Sharing systems require some form of organization and negotiation for optimal access to the resource and price discovery. Various forms of organizations are introduced and discussed here, from pool-based to peer-to-peer, also considering a community-based approach. These various organizations are motivated, the related negotiation mechanisms described and their properties discussed. The example of electric energy sharing is used as a basis for discussion and illustration.

Keywords Negotiation mechanisms · Distibuted optimization · Electric energy

1 Introduction

Sharing systems rely on the idea that users collaborate for the use of a finite resource or a set of resources, possibly under a given social contract. This is in line with some of the modern views of economics that aim to reconcile resource allocation with environmental and human considerations e.g. [9]. As for any collaboration to be successful, the rules of the game need to be clear upfront, being in terms of access to the resource, negotiation processes, settlement if payment for usage of the resources, and penalization in case agreements are not respected. Taking the example of a bike-sharing system, access to the resource may be at centralized storage points like for the Velib' in Paris (France), or at any location where a bike has been left as for the many

---

P. Pinson (✉) · F. Moret
Department of Electrical Engineering, Technical University of Denmark, Lyngby, Denmark
e-mail: ppin@elektro.dtu.dk
e-mail: fmoret@dtu.dk

T. Baroche
Department of Mechatronics, Ecole Normales Supérieure Rennes, Rennes, France
e-mail: thomas.baroche@ens-rennes.fr

A. Papakonstantinou
Department of Management Engineering, Technical University of Denmark, Lyngby, Denmark
e-mail: athpapa@dtu.dk

bike-sharing systems in Beijing (China). Negotiation for those bikes may simply take the form of access on a first-arrived first-served basis, as no early booking is possible. Settlement uses a flat price per minute, while finally penalization consists in a fee for not returning the bike or freeing it on time. In this Chapter, emphasis is placed on negotiation approaches for sharing systems, conditional to some organizational models.

The way negotiation is to take place fundamentally relates to some of the characteristics of the resource to be shared. If having to take turns for accessing the resource (as for bike sharing), one ends up with using concepts of queuing theory for such system design. When this time dimension of the access to the resource can be overlooked, the design simplifies to considering a general resource allocation problem with distributed agents, which is the case we will further develop in the following. The nice features of recent views on sharing systems include the fact that these may accommodate and implement an agreed-upon social contract, allow users to express preferences, while being solved in a fully decentralized manner even though aiming to reach a social global optimum.

To make this Chapter easier to follow, we base its thread on the practical case of electric energy sharing. In the view of the scale taken by sharing systems over the past years, the idea that such sharing system can be applied to electric energy has gained incredible momentum over the last few years, see e.g. [5, 7]. As some have argued, electric energy systems comprise one of the long-lasting and most successful sharing system, since consisting of a resource most often available through a network, for which all producers and consumers somehow have to collaborate for optimal usage of that resource. It is also a highly constrained sharing system owing to the need for constant balance between production and consumption, as well as other physical constraints stemming from electric power flows on networks.

In that context, even though not using time indices in the following in order to lighten the notations, we restrict ourselves to problems where a set of agents (being suppliers and consumers of that resource) have to negotiate and eventually settle at time t on usage of the resource at time $t + k$. For simplicity, inter-temporal constraints are disregarded. These may readily be accounted for at the expense of modeling and computational complexity. In the following, our aim is to introduce various models for sharing by focus on the negotiation process, also discussing settlement, going from high-level concepts like organization of and interaction among agents to the mathematical description of the mathematical problems at hand. Fundamentally, all those negotiation approaches rely on the formulation of an optimization problem which aims at maximizing the utility of all agents involved. These may eventually be solved in centralized or distributed manner, depending upon the way the agents are organized. Since aiming to concentrate on the key features of organizational aspects and negotiation approaches, the more technical challenges related to network constraints, power balance and other operational aspects are overlooked here.

Alternative approaches to sharing and their motivations are first covered in Sect. 2 using electric energy an illustrative example. Opportunity is taken to introduce the setup for a running example that is to be revisited throughout the Chapter. Subsequently, we go through a set of typical organizations to discuss setup and properties as well as providing extensive mathematical formulations. This will go from the pool-based approach in Sect. 3, through the peer-to-peer approach in Sect. 4, and finally to community-based approach in Sect. 5. The Chapter ends in Sect. 6 with a set of general insights and perspectives for further developments in that field.

2 Organizational Basis for Negotiation in Sharing Systems

Sharing can be based on various levels of decentralization. The basic intuition behind representative models for sharing systems organization and negotiation approaches is discussed and motivated here.

2.1 Sharing System Organizations and Their Motivation

Knowing that a resource is limited, that various agents have different possibilities to supply and consume (also with different appraisal of the resource value), that some other individual and commonly agreed preferences may step in, what are relevant, efficient and fair ways to optimally share that resource? This is a fundamental question at the heart of the design and operation of sharing systems. For example, the problem of how to share electric energy has been around for a long time, see e.g. a recent review in [8]. Sharing electric energy in an open market environment is agreed upon, but how such a market, negotiation mechanisms and settlement should look like is still an open design question.

Today, pools form the basis for how electricity is exchanged at the wholesale level, with retailers acting as interface between this wholesale level and actual consumers (i.e., actual users of the electric energy resource). Negotiation is based on quantity-price products in their simplest form, with a number of more advanced products aiming to reflect flexibility or constraints of the agents. Nevertheless, market access is typically restricted to power companies, utilities and industrial sized producers or consumers. Such a practice is challenged by recent advancements such as the deployment of distributed energy resources at a larger scale, advances in ICT (Information and Communication Technologies) and new business models inspired by sharing economy and collaborative commons.

In this context, a view at bike sharing systems allows envisaging how current electricity markets could adapt towards a more consumer-centric design, hence rethinking completely the existing centralized and hierarchical organization models. While the pool offers a convenient starting point, it is a reference centralized approach for sharing systems, with a market operator centralizing all communication and resource allocation tasks. Various levels of decentralization may then yield peer-to-peer approaches and community based approaches to sharing systems. These three alternative approaches, to be extensively introduced and analyzed in the remainder of this Chapter, are illustrated in Fig. 1. An overview of the motivations for such

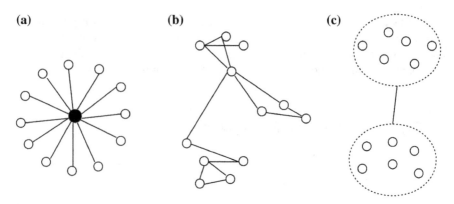

Fig. 1 Set of alternative organizations for sharing systems as a basis for negotiation, resource allocation and settlement, from pool based (**a**) to peer-to-peer (**b**) and community-based (**c**)

alternative organizations and related business models was recently covered in [7] for the case of electric energy. Typically here, the motivations may not be related to some form of optimality in economic terms only, but certain preferences, environmental considerations and other social components.

In practice, the peer-to-peer approach allows each and every agent to directly negotiate with each other through bilateral contracts in the absence of third-party supervisor. There may be several variations of this model with consumers being linked to producers through a series of multiple bilateral contracts [11], coordination of fleets of electric vehicles [1] or even interconnection with the distribution grid and the wholesale markets [6]. On the other hand, the community-based approach is more structured and may be seen as a flexible way to be somewhere between pool-based and peer-to-peer approaches. Agents there gather in communities, based on a given social contract which may involve localization, family of friendship-based ties, economic considerations e.g. shared investment, etc. And, instead of having a for-profit third party acting as an interface between that community and the outside world, this role is given to a not-for-profit node which we refer to as "community manager". This node is there also to coordinate interaction among agents within the community and implementation of the social contract that binds them together.

2.2 An Example Setup to Illustrate Further Developments

To make the understanding of concepts and workings of negotiation approaches for sharing systems easier, let us introduce here an example setup which will be re-visited regularly throughout this Chapter. It is for the case of sharing electric energy and is inspired by the 2 village setup of [11]. 12 agents, seen as part of 2 communities when relevant (which could be seen as two villages, as in the original example of [11]), are to engage in sharing electric energy. It is assumed that all network and operational

Table 1 Characteristics of the agents for our electric energy sharing example problem. These agents have various types of assets and cost functions, while belonging to two communities (the first 6 with the first community and the following 6 with the second one)

Agent	P/C	Type	$[P_i^{min}, P_i^{max}]$ [kWh]	a_i (c€/kWh)	b_i (c€)
1 (Comm. 1)	P	Wind	[30, 30]	0	0
2 (Comm. 1)	C	Household	[−20, 0]	0.4	25
3 (Comm. 1)	P	Solar	[20, 20]	0	0
4 (Comm. 1)	C	Household	[−30, 0]	0.5	20
5 (Comm. 1)	C	Industrial	[−120, −6]	0.3	40
6 (Comm. 1)	P	Wind	[50, 50]	0	0
7 (Comm. 2)	C	Household	[−15, 0]	0.2	28
8 (Comm. 2)	C	Household	[−25, 0]	0.3	22
9 (Comm. 2)	P	Fossil	[0, 40]	0.7	15
10 (Comm. 2)	P	Fossil	[20, 90]	0.6	10
11 (Comm. 2)	C	Industrial	[−120, −10]	0.4	55
12 (Comm. 2)	P	Fossil	[15, 105]	0.3	6

aspects are such that this can be done freely without considering constraints for the underlying infrastructure. The list and characteristics of agents are collated in Table 1.

Agents are either producers (P) or consumers (C) for simplicity, while gathering a set of wind and solar producers, fossil fuel generators, as well as household and industrial consumers. They are all assumed to have quadratic cost functions, and hence linear marginal cost functions f_i, over a feasible range $[P_i^{min}, P_i^{max}]$, i.e.,

$$f_i(p) = a_i p + b_i, \quad p \in [P_i^{min}, P_i^{max}], \quad i = 1, \ldots, 12 \qquad (1)$$

Additional information about the setup and the agents characteristics, e.g. their preferences, will be given when necessary when revisiting this example in the remainder of this Chapter.

3 Pool-Based Approach to Sharing Systems

A traditional approach to sharing, and often seen as a benchmark, consists in having a centralized view of the problem. In practice, this translates into having a pool where all agents meet, being both suppliers and consumers, for an optimal allocation of the resource. In the following, its organization and the role of the agents are described, followed by its mathematical formulation and a discussion of its properties and limitations.

3.1 Organization and Agents

Let us consider a set Ω of agents, including a set Ω_s of agents that are willing to supply the resource and Ω_c the set of agents willing to consume, $\Omega = \Omega_s \cup \Omega_c$. This can generalize to the case where all agents are both suppliers and consumers, though not simultaneously, as for the example of electricity prosumers.[1] Those are willing to sell the production surplus from their solar panels (or other generation asset) at certain times, while needing to consume from others at other times since their solar panels do not produce enough for self-consumption. While we still separate suppliers and consumers in the description of the pool-based approach here, all will be seen as prosumers when focusing on communities and peer-to-peer frameworks. This actually relates to one of the practical reasons for shifting from a centralized pool approach to the decentralized one discussed subsequently. Intuitively, a pool best fits a practical setup with a few agents with well defined roles, like having a limited number of hydrothermal power plants supplying electricity to consumers through a limited number of retailers. Though, as the roles of agents become more generic (e.g., all supply and consume) while their number increases, many see decentralized approaches as more suitable. Some of their reasons may be ideological ("local production—local consumption", grass root movements, etc.), or related to system resilience, etc.

Agents are considered as rational in the sense that they have well-defined supply and consumption marginal cost functions, denoted as f^s and f^c respectively, and that they aim to optimize based on those functions [3]. Such marginal cost functions relate quantity to their willingness to be paid or to pay for an additional unit quantity of the resource, respectively. For simplicity, we assume that these functions are convex and $\mathcal{C}_1$, meaning that they are differentiable and with first-order derivatives being continuous. The purpose of this assumption is to have simpler optimization problems to be solved in the following. It may be relaxed in practice, but possibly at the expense of computational complexity while affecting some of the properties of their solution. The agents additionally are non-strategic, in the sense that their decisions are not forward looking [10].

In a pool-based approach, at a given time t and projecting themselves at a later time $t + k$, all agents declare their cost functions to the centralized market mechanism. The clearing of this market has the purpose of finding the optimal allocation for all agents, i.e., how much suppliers will supply and how much consumers will consume, in an overall optimal manner. It also serves as a basis to define a price to be considered for eventual settlement, that is, the price that suppliers will receive and the consumers will pay.

---

[1] "Prosumer" is the generic term for those agents that can both supply and consume. The term "prosumer" literally comes from merging the terms "producer" and "consumer".

3.2 Mathematical Formulation

Eventually, the optimal resource allocation under a pool-based approach translates to solving an optimization problem where the aim is to maximize social welfare, defined as the difference between consumer and supplier costs. This is done under a balance constraint, since supply has to be equal to demand. In practice, writing it as a minimization problem instead of a maximization one, this writes

$$\min_{\Gamma} \quad \sum_{i \in \Omega_s} f^s(p_i^s) - \sum_{j \in \Omega_c} f^c(p_j^c) \tag{2a}$$

$$\text{s.t.} \quad \sum_{i \in \Omega_s} p_i^s - \sum_{j \in \Omega_c} p_j^c = 0 \qquad [\lambda] \tag{2b}$$

where $\Gamma = \{p^s, p^c\}$ is the set of decision variables describing the resource allocation for suppliers ($p^s = [p_1^s, p_2^s, \ldots]$) and consumers ($p^c = [p_1^c, p_2^c, \ldots]$). The balance between supply and demand is imposed by the equality constraint (2b). Its associated dual variable λ is to be seen as the price for the resource at equilibrium.

In practice, these problems are fairly straightforward to solve. For instance, in the case where the cost functions simplify to a single quantity-price offer (or several discretized one), the optimization problem (2) is a Linear Program (LP). Similarly, if costs functions are linear, it becomes a Quadratic Program (QP). In both cases, these may readily be solved by a wealth of available solvers today, using one's favourite programming language.

3.3 Properties and Limitations

Considering first its organization, the centralized approach of the pool makes that each and every agent does not directly interact with the others, which is then supposed to anonymize their information and actions. However, the central operator that clears the market collects all information from all agents, that is, their marginal costs functions. Many of those who support the development of sharing systems, possibly to be seen as collaborative commons, prefer to have a more decentralized view of the resource allocation (and possibly pricing) problem. This may additionally allow to increase resilience, improve privacy, etc.

The price λ is an outcome of the market clearing and is to be seen as the price that supports the resource allocation. This is to be understood in the sense that all suppliers will be happy to supply at that price, since it is greater or equal to their marginal cost function value for that allocated supply quantity. Indeed, if supplier i is to supply p_i^s, then $\lambda \geq f_i^s(p_i^s)$. Inversely, consumers will be happy to consume at that price, since it is lower or equal to their marginal cost function value for their allocated consumption quantity. For consumer j who is to consume p_j^c, then $\lambda \leq f_i^c(p_i^c)$. This nice property supports the choice of uniform pricing in such a pool approach: this single price λ is

Table 2 Resource allocation as an outcome of the pool-based approach

Agent	1	2	3	4	5	6
p_i [kWh]	30	−10.29	20	0	−63.72	50
Revenue/Payment [€]	6.27	−2.15	4.18	0	−13.31	10.44
Agent	7	8	9	10	11	12
p_i [kWh]	−15	−3.72	8.41	20	−85.29	49.61
Revenue/Payment [€]	−3.13	−0.78	1.76	4.18	−17.81	10.36

used for all consumers and suppliers. An alternative (and less-employed) approach is to consider a pay-as bid approach, for which the price to pay (or to be paid) is equal to the cost function value for the quantity allocated. While the simplicity and desired properties of uniform pricing are appealing, they also prevent to accommodate some of the aspects of negotiation in sharing systems e.g. user preferences.

3.4 Example Application Results

The example setup of Sect. 2.2 is revisited here to see what would happen if all agents were to share electric energy and negotiate within a pool-based approach. All agents readily share their cost functions and constraints (in terms of minimum and maximum resource quantity) with the central agent which is the pool operator. Based on that information, problem 2 is solved to obtain the resource allocation and corresponding price for exchange. Under uniform pricing, it is used for all payment and revenue calculations. The resource allocation outcome, as well as revenue and payment, are collated in Table 2. Those are based on a single price of 20.88 c€/kWh.

What happens there is that resource allocation follows a simple merit-order principle, meaning that those who were ready to pay the most are supplied first, and those ready to supply for the lowest cost sell their resource first.

4 Peer-to-Peer Approaches

At the opposite of the pool-based approach introduced in the above lies the peer-to-peer (some times abbreviated P2P) one. After describing its organization and the role of the agents, we provide a mathematical formulation of the negotiation problem. Subsequently, some of the interesting properties of the P2P approach are underlined, e.g., the possibility to readily generalize on the products to be exchange and to account for product differentiation.

4.1 Organization and Agents

Within a peer-to-peer organization, negotiation and settlement are based on bilateral contracts for a direct exchange of a resource between two agents. In principle, bilateral contracts may exist between any two agents, on supply and demand sides. As illustrated later in Sect. 5, community-based exchanges may also be seen as a specific instance of sets of multiple bilateral contracts. This means that such a peer-to-peer setup may be the most general and flexible framework to operate a range of organization structures for sharing systems. In parallel, it does not assume nor require specific communication structures, in the sense that in principle all may have the possibility to exchange and negotiate with all others. This comes in contrast with the pool-based approach, where all communicate with a single central node, and with the community-based approach that will be discussed in the following.

Focusing on the agents, only one type is considered here, who may supply or consume the resource. For the electric energy example, these are the prosumers that we mentioned previously. All agents have cost functions that allow to rationally optimize their supply or procurement. Cost functions can be as general as self-scheduling objective in the case agents have multiple assets. Agents can negotiate with all and select their negotiation partners. The only requirement to form a bilateral contract is that the match must be mutually accepted. At the extreme, agents can trade with every participant but themselves, which constitutes a complete peer-to-peer market. This freedom component comprises one of the main advantages of this organization, besides the fact it is fully decentralized. In fact, once partnerships are formed, agents can still choose to favor or incentivize a partnership over another. This ability to order and prioritize negotiation partners yields a preference mechanism which opens even broader possibilities. For example with electric energy, it makes it possible to have consumers expressing their preferences to buy energy produced in their neighborhood, from a friend or family member, or from solar production assets preferred over nuclear ones.

A peer-to peer approach assumes the use of a fully distributed negotiation mechanism with a communication structure matching the partnerships formed by participants. In this way no information on cost functions needs to be disclosed, not even between partners. To obtain valid bilateral contracts each of them must agree both on quantity and prices, e.g. exchange 1kWh of electric energy at a price of 0.1€. As mathematically formulated below, reciprocity in traded quantities is enforced by a complicating constraint, the dual variable of which yields the price for the trade. Besides the choice of negotiation partners, quantity proposals is the only information that is to be exchanged between agents to drive the negotiation. In practice, this relies on an iterative process where each agent optimizes its own cost function to update quantity values, then communicated to others.

4.2 Mathematical Formulation

The peer-to-peer negotiation mechanism is firstly formulated as an optimization problem. This choice is motivated by driving the negotiation mechanism towards overall social welfare maximization, as for the pool-based approach. The problem is then reformulated to yield a decentralized and iterative negotiation approach.

Let us consider a peer-to-peer setup with a set Ω of agents. These are again seen as rational [3] and non-strategic, i.e. do not anticipate actions and reactions of other agents. Each agent i has a cost function f_i and a set ω_i of negotiation partners. For each trade p_{ij} (resource quantity, positive when selling and negative when buying) with a partner $j \in \omega_i$, agent i can also apply a set of preferences internalized through g_i, a function of vector $P_i = (p_{ij})_{j \in \omega_i}$ the whole set of trades. The overall procurement for all agents, for a single time step, is optimized by total cost minimization as

$$\min_{\boldsymbol{P},(p_i)_{i \in \Omega}} \quad \sum_{i \in \Omega} f_i(p_i) + g_i(P_i) \tag{3a}$$

$$\text{s.t.} \quad \boldsymbol{P} = -\boldsymbol{P}^\mathsf{T}[\Lambda] \tag{3b}$$

$$p_i = \sum_{j \in \omega_i} p_{ij}[\mu_i] \quad i \in \Omega \tag{3c}$$

$$p_i \in \mathscr{P}_i \quad i \in \Omega \tag{3d}$$

where the matrix of all possible power trades $\boldsymbol{P} = (P_i)^\mathsf{T}_{i \in \Omega}$ describes trade power proposals (p_{ij}). Scalar p_i is the power set-point of agent i and $\cdot^\mathsf{T}$ denotes the matrix transpose operator. Overall procurement cost minimization is to be seen as a way to maximize social welfare for all agents.

For each agent we consider technical constraints (3d). Balance equation (3c) ensures that one cannot sell or buy more than physically produced or consumed. Dual variable μ_i is interpreted as the price perceived by agent i for the resource. Reciprocity constraint (3b) ensures a balance for each trade as well as a resulting overall equilibrium between supply and demand. In other words, it enforces that selling and buying quantities are equal for each trade. Matrix dual variable $\Lambda = (\lambda_{ij})_{i,j}$, represents the energy price of every trade. Hence, agent i trades quantity p_{ij} at price λ_{ij} with agent j, while agent j trades quantity p_{ji} at price λ_{ji} with agent i. The objective of the negotiation mechanism is to reach consensus for both variables, i.e. to have $p_{ij} = -p_{ji}$ and $\lambda_{ij} = \lambda_{ji}$.

It is important to note that, even though problem (3) is formulated as centralized optimization problem, it is to be eventually distributed among agents. However, the decomposition of (3) is challenging. In consequence, we reformulate it as an equivalent consensus problem by replacing constraint (3b) with

$$\left(\boldsymbol{C} - \boldsymbol{C}^\mathsf{T}\right)/2 = \boldsymbol{P} \tag{4}$$

where matrix $\boldsymbol{C} = (c_{ij})_{i,j}$ is a decoupling slack variable. Problem (3) can easily be decomposed and solved by distributed optimization approaches like the Alternating

Direction Method of Multipliers (ADMM) [2] when relaxing the complicating constraint (3b) into (4). This is done by first writing the augmented Lagrangian of the new formulation as

$$L_\rho(\boldsymbol{P}, \boldsymbol{C}, (p_i, \mu_i)_{i \in \Omega}, \boldsymbol{\Lambda}) = \sum_{i \in \Omega} L_{i,\rho}(P_i, \boldsymbol{C}, p_i, \mu_i, \Lambda_i) \quad (5)$$

with

$$L_{i,\rho}(P_i, \boldsymbol{C}, p_i, \mu_i, \Lambda_i) = f_i(p_i) + g_i(P_i) + \mu_i \left(\sum_{j \in \omega_i} p_{ij} - p_i \right)$$
$$+ \sum_{j \in \omega_i} \lambda_{ij} \left(\frac{c_{ij} - c_{ji}}{2} - p_{ij} \right) + \rho \left(\frac{c_{ij} - c_{ji}}{2} - p_{ij} \right)^2 \quad (6)$$

where $\rho > 0$ is the regularization parameter to guarantee smoothness of the objective function even if f or g are not, and vector $\Lambda_i = (\lambda_{ij})_{j \in \omega_i}$ lists trade prices of agent i. After simplifications, as proposed in [2], the iterative algorithm to solve (3) is based on two update steps to go from iteration k to iteration $k+1$, which are

$$\{P_i^{k+1}, p_i^{k+1}, \mu_i^{k+1}\} = \arg\min_{P_i, p_i, \mu_i} L_{i,\rho}(P_i, \boldsymbol{P}^k, p_i, \mu_i, \Lambda_i^k) \quad i \in \Omega \quad (7a)$$

$$\boldsymbol{\Lambda}^{k+1} = \boldsymbol{\Lambda}^k - \rho \left(\boldsymbol{P}^{k+1} + \boldsymbol{P}^{\mathsf{T},k+1} \right)/2 \quad (7b)$$

Note that the 2-step updates in (7) are strictly separable among agents. It means that, at step (7a) agents optimize their own costs and preferences independently of each other, and only by means of previous step information on trade proposals of negotiation partners and prices. Then, agents communicate counter proposals to their negotiation partners and update prices (7b) to reach trade reciprocity in terms of resource quantity. The convexity of all functions (both f and g) guarantees that the iterative algorithm reaches the global optimum.

4.3 Expression of Preferences

One of the advantages of the peer-to-peer approach is that each agent can favour negotiation partners following given preferences. As proposed in [11], each agent i can apply a specific linear coefficient g_{ij} for each individual trade made with a partner j, which leads to preference cost functions expressed as

$$g_i(P_i) = \sum_{j \in \omega_i} g_{ij} |p_{ij}|, \quad i \in \Omega \quad (8)$$

where $|\cdot|$ denotes the absolute value operator so that $g_{ij}|p_{ij}|$ is always a positive cost even if buying energy, i.e. when p_{ij} is negative.

The agent can then sort partnerships, or better say, affect trades, based on a set $\mathcal{L}_i$ of preference criteria. The importance of each criterion l is expressed by its associated weight g_i^l, yielding

$$g_{ij} = \sum_{l \in \mathcal{L}_i} g_i^l \gamma_{ij}^l, \quad i \in \Omega, \; j \in \omega_i \tag{9}$$

where γ_{ij}^l is a positive score evaluating the characteristics of partner j with respect to criterion l of agent i. Thinking of the electric energy application, preference criteria could be distance or CO_2 emissions, and possibly the agent's reputation based on online votes or self-determined reputation. Eventually, since these yield a preference component to the original cost functions of the agents, it has the effect of internalizing those preferences. When a criterion-related cost value is positive, the resource provided is perceived as more expensive, and if negative as cheaper.

The main drawback of this formulation of preferences, atomized per agent and per partnership, is that it does not naturally permit for a group of agents to reach common goals. The first way to get around this calls for an additional agent (or node) coordinating groups of agents with common goals, as will be detailed in Sect. 5 with the concept of community manager. Another approach would be to rely on regulatory measures to affect agents' cost functions. For example to reduce carbon emissions the regulator may choose to set a carbon tax, which will then enter the cost function as for the preferences. Similarly, to limit the risk of congestion on power networks, a system operator could impose a network fee, e.g. as a function of the distance between agents, for the use of the infrastructure.

4.4 Implication for Prices

While by construction, the pool-based approach yields a single price for all trades based on a single balance constraint, a peer-to-peer setup fundamentally allows for a different price for each and every trade. By writing Karush–Kuhn–Tucker conditions, one obtains that the energy price μ_i perceived by agent i links (i) the generation or consumption price of the resource, and (ii) prices λ_{ij} of the negotiated trades. In consequence, prices are such that

$$\mu_i = \begin{cases} \frac{\partial f_i}{\partial p_i}(p_i) \\ \lambda_{ij} - \text{sign}(p_{ij}) g_{ij} \; j \in \omega_i \end{cases}, \quad i \in \Omega \tag{10}$$

where $\text{sign}(\cdot)$ returns the sign of the given argument. Consequently, when no preference is expressed, so when all g_{ij} equal zero, perceived prices μ_i are uniform among agents and equal to the energy price of the pool market, denoted λ in Sect. 3.

However, in presence of preferences, the price λ_{ij} negotiated between two agents i and j is impacted by the internalized cost of preferences g_{ij}. For example with electric energy, if a regulator imposes a carbon emission tax $g_i^{CO_2}$, consumer i will have an economical interest to trade with its greenest partners, so with the lowest scores $\gamma_{ij}^{CO_2}$. In fact, in addition to the cost paid to its provider j at price λ_{ij}, consumer i will have to pay the carbon tax at price $g_i^{CO_2}\gamma_{ij}^{CO_2}$ to the regulator.

As mentioned previously, the P2P market allows to apply several individual preferences opening new business opportunities as they can be used as commission fees and taxes. This feature is fundamental when looking at sharing economies aiming at changing natural behaviors of market participants. Finally, the true advantage of the P2P model is that it englobes any bilateral and multilateral market structures.

4.5 Example Application Results

In the peer-to-peer case, further the basic setup information given in Sect. 2.2, preferences are to be defined. These are gathered in Table 3 below. In that example, the rationale is that households prefer to be supplied by renewables (over fossil fuel power plants), industrial consumers are indifferent, while consumers do not trade with other consumers (same for producers).

Based on those preferences, and going through the peer-to-peer negotiation process described above, the eventual resource allocation for all agents is gathered in Table 4. There, summing over each and every row gives the overall resource allocation per agent, which may be readily compared to the pool-based case. One may then verify that the resource allocation is different overall if compared to the pool-based case, even if not the case for each and every agent. For instance, wind power producers still sell their whole production. Similarly, the trade prices are collated in Table 5, where prices can vary on a trade to trade basis, hence affecting the overall revenue and payments of the agents.

Table 3 Preferences for trades among agents. Increasing values at row i and column j indicates lower preference for a trade between agent i and j

Agent	1	2	3	4	5	6	7	8	9	10	11	12
1	0	1	100	1	2	100	1	1	100	100	2	100
2	1	0	1	100	100	1	100	100	3	3	100	3
3	100	1	0	1	2	100	1	1	100	100	2	100
4	1	100	1	0	100	1	100	100	3	3	100	3
5	2	100	2	100	0	2	100	100	2	2	100	2
6	100	1	100	1	2	0	1	1	100	100	2	100
7	1	100	1	100	100	1	0	100	3	3	100	3
8	1	100	1	100	100	1	100	0	3	3	100	3
9	100	3	100	3	2	100	3	3	0	100	2	100
10	100	3	100	3	2	100	3	3	100	0	2	100
11	2	100	2	100	100	2	100	100	2	2	0	2
12	100	3	100	3	2	100	3	3	100	100	2	0

Table 4 Resource allocation on an agent-to-agent basis (in kWh). Overall resource allocation is obtained by summing over each row

Agent	1	2	3	4	5	6	7	8	9	10	11	12
1	0	3.37	0	0	9.02	0	4.79	1.44	0	0	11.38	0
2	−3.37	0	−2.68	0	0	−4.57	0	0	0	0	0	0
3	0	2.68	0	0	5.83	0	3.34	1.26	0	0	6.90	0
4	0	0	0	0	0	0	0	0	0	0	0	0
5	−9.02	0	−5.83	0	0	−13.98	0	0	−2.55	−9.43	0	−16.68
6	0	4.57	0	0	13.98	0	6.88	1.46	0	0	23.11	0
7	−4.79	0	−3.34	0	0	−6.88	0	0	0	0	0	0
8	−1.44	0	−1.26	0	0	−1.46	0	0	0	0	0	0
9	0	0	0	0	2.55	0	0	0	0	0	2.82	0
10	0	0	0	0	9.43	0	0	0	0	0	10.57	0
11	−11.38	0	−6.90	0	0	−23.11	0	0	−2.82	−10.57	0	−25.83
12	0	0	0	0	16.68	0	0	0	0	0	25.83	0

Table 5 Trade prices on an agent-to-agent basis (in c€/kWh)

Agent	1	2	3	4	5	6	7	8	9	10	11	12
1	0	19.75	0	0	20.75	0	19.75	19.75	0	0	20.75	0
2	19.75	0	19.75	0	0	19.75	0	0	0	0	0	0
3	0	19.75	0	0	20.75	0	19.75	19.75	0	0	20.75	0
4	0	0	0	0	0	0	0	0	0	0	0	0
5	20.75	0	20.75	0	0	20.75	0	0	20.75	20.75	0	20.75
6	0	19.75	0	0	20.75	0	19.75	19.75	0	0	20.75	0
7	19.75	0	19.75	0	0	19.75	0	0	0	0	0	0
8	19.75	0	19.75	0	0	19.75	0	0	0	0	0	0
9	0	0	0	0	20.75	0	0	0	0	0	20.75	0
10	0	0	0	0	20.75	0	0	0	0	0	20.75	0
11	20.75	0	20.75	0	0	20.75	0	0	20.75	20.75	0	20.75
12	0	0	0	0	20.75	0	0	0	0	0	20.75	0

In the given example it can be observed that households are able to neglect completely fossil fuel power plants as renewable sources produce enough for all households. However, industrial consumption cannot only be satisfied by renewables, hence they also trade with fossil fuel power plants. In consequence renewable producers face two different trading prices, the lower with households and the higher with industrial consumers, but with the same perceived price as depicted in Sect. 4.4. For example, wind producer 1 trade with household 2 at 19.75 c€/kWh, leading to a perceived price of $19.75 - 1 = 18.75$ c€/kWh, and with industrial consumer 5 at 20.75 c€/kWh, leading to a perceived price of $20.75 - 2 = 18.75$ c€/kWh. Finally, please remember that in absence of preferences the P2P would lead to the same results as the pool-based case but with set-points split in bilateral trades of the same sum.

5 Community-Based Approaches

After looking at two opposite views of negotiation for sharing systems, in the form of pool-based (centralized) and peer-to-peer (fully decentralized) approaches, we concentrate here on a more structured version of the peer-to-peer approach, to be seen as an in-between alternative with interesting prospects and properties. Since using electric energy as an example application throughout our developments, let us introduce here the terms energy communities or energy collectives to refer to a group of agents (all prosumers), who gather to share their excess or lack of energy under a given social contract. Straightforward applications of this structure include situations where by nature it exists a point of common coupling, e.g., behind-the-meter energy management of a building with multiple apartments, energy markets among buses under the same feeder at distribution level, etc. However, being a pure market construct, there is no need to have a physical point of common coupling to create an energy community. Hence agents part of an energy collective do not need to be geographically located in the same area. In the following we first describe organizational aspects, before to give a mathematical formulation and an illustrative example.

5.1 Organization and Agent Description

Two types of agents are to be considered, the community members and a community manager. On the one hand, the community members are prosumers who own and operate a set of assets for production and consumption. On the other hand we refer to as community manager a non-profit third-party, who coordinates consensus among community members while accounting for social contracts, previously agreed among members, in the negotiation mechanism. Social contracts represent the core of a community-based approach. In fact, thinking of what makes a community, it is the fact that its members share one or more common goals or interests. These goals are then translated into a social contract, i.e., an agreement among individuals to secure welfare or to regulate the relations among community members. Further details on how social contracts are defined and handled will be given in the following when getting into mathematical formulations.

A community-based setup can be conceived as a bridge between pool-based and peer-to-peer approaches. As in pool-based setup, agents negotiate with a central node and the resulting price is unique. However, the negotiation relies on distributed (and iterative) mechanisms. This way, no information on cost or utility curves of the agents' assets needs to be disclosed, not even to the community manager. That latter agent coordinates consensus among members and makes sure that the community preferences (common since given by a social contract) are respected by the negotiation outcome. Therefore, each community member communicates directly with the community manager and not to other agents. It is important to note that we refer to communication networks and not physical networks, since communication among two parties is the only requirement for purely economic trades to happen. As described in Sect. 4, peer-to-peer approaches are generally conceived as markets where each agent can directly trade with other peers without a third-party. If one considers the community manager as an agent, whose objectives are the social contracts of the community members, then a community-based setup is nothing but a special configuration of a peer-to-peer approach.

5.2 Mathematical Formulation

The community-based negotiation for a sharing system is first formulated as an optimization problem. Then we show how to reformulate it using distributed optimization, to make it a distributed and iterative negotiation mechanism. ADMM [2] is used, as for the peer-to-peer case in the above.

Let us consider a community with a set Ω_{co} of members, which is a subset of Ω of Sects. 3 and 4. Each member i has a cost function f_i, which could be extended to a self scheduling cost function if several assets are owned and operated. As for the other cases, agents are supposed rational and non-strategic. The community procurement, for a single time step, is optimized by total cost minimization as the following exchange problem

$$\min_{\Gamma} \sum_{i \in \Omega_{co}} f_i(p_i) + g(q_{imp}, q_{exp}) \tag{11a}$$

$$\text{s.t.} \sum_{i \in \Omega_{co}} p_i = q_{imp} - q_{exp} \quad [\lambda] \tag{11b}$$

$$p_i \in \mathscr{P}_i \qquad\qquad i \in \Omega_{co} \tag{11c}$$

$$q_{imp}, q_{exp} \geq 0 \tag{11d}$$

where $\Gamma = \{p, q_{imp}, q_{exp}\}$ is the set of decision variables describing respectively the resource allocation vector of agents ($p = [p_1, \ldots, p_i]$) and the resource sold to or bought from outside the community (q_{imp} and q_{exp}). For each community member we consider technical constraints (11c) while the overall community balance is expressed by (11b), with its associated dual variable λ representing the price for the trades within the community. It does not have to be equal to the price for the resource outside of the community.

Problem (11) can be readily decomposed and solved by ADMM if relaxing the complicating constraint (11b). This can be done by writing the augmented Lagrangian as

$$L_\rho(p, q_{imp}, q_{exp}) = \sum_{i \in \Omega_{co}} f_i(p_i) + g(q_{imp}, q_{exp}) \\ + \lambda \left(\sum_{i \in \Omega} p_i - q_{imp} + q_{exp} \right) + \frac{\rho}{2} \left\| \sum_{i \in \Omega} p_i - q_{imp} + q_{exp} \right\|_2^2 \tag{12}$$

where $\rho > 0$ is the regularization parameter to guarantee smoothness of the objective function even if f or g are not. $\|\cdot\|_2^2$ denotes the squared Euclidean norm operator. The iterative algorithm to solves (11) follows the common ADMM steps, i.e.,

$$p_i^{k+1} = \arg\min_{p_i} L_\rho(p_i, q_{\text{imp}}^k, q_{\text{exp}}^k, \lambda^k) \qquad i \in \Omega_{\text{co}} \qquad (13a)$$

$$\{q_{\text{imp}}^{k+1}, q_{\text{exp}}^{k+1}\} = \arg\min_{q_{\text{imp}}, q_{\text{exp}}} L_\rho(p^{k+1}, q_{\text{imp}}, q_{\text{exp}}, \lambda^k) \qquad (13b)$$

$$\lambda^{k+1} = \lambda^k + \rho \left(\sum_{i \in \Omega_{\text{co}}} p_i^{k+1} - q_{\text{imp}}^{k+1} + q_{\text{exp}}^{k+1} \right) \qquad (13c)$$

The steps in (13) fit the communication architecture of the community. In step (13a) each community member optimizes its procurement only by means of information on the price and the imported and exported resource (for the community as a whole) at the previous iteration. Then the new resource allocation of each member p_i^{k+1} is communicated to the community manager that calculates the new imported and exported resource quantity (13b) and updates the price (13c). The convexity of all functions (both f and g) guarantees that the iterative algorithm reaches the optimal market equilibrium, as discussed in [2].

5.3 Social Contracts

One of the advantages of community-based market is that the community manager is able to coordinate the negotiation process while ensuring that the previously agreed social contracts are respected. This is done by properly defining the function g in (11). In other words, since the community manager represents all the collective members, its objective function needs to be commonly decided among members. The most straightforward example is the market-driven case. In this situation the community members do not have any preference besides optimizing costs. Therefore the objective function of the community manager is simply $g(q_{\text{imp}}, q_{\text{exp}}) = \lambda_{\text{imp}} q_{\text{imp}} - \lambda_{\text{exp}} q_{\text{exp}}$, with λ_{imp} and λ_{exp} respectively the price of buying and selling energy outside the collective. In case of more advanced social contracts, it may be useful to split the power dispatch of each agent p_i into the energy traded within the community, the one contributing to the community import and export. One can therefore extend (11) by introducing a free variable q_i to account for the energy traded (positive if bought) within the community and two positive variables representing agent i share of import (α_i) and export (β_i) to outside the community. The optimization problem then becomes

$$\min_\Gamma \sum_{i \in \Omega_{\text{co}}} f_i(p_i) + g(q_{\text{imp}}, q_{\text{exp}}) \qquad (14a)$$

$$\text{s.t.} \quad p_i + q_i + \alpha_i - \beta_i = 0 \qquad i \in \Omega_{\text{co}} \qquad (14b)$$

$$\sum_{i \in \Omega_{\text{co}}} q_i = 0 \qquad [\lambda] \qquad (14c)$$

$$\sum_{i \in \Omega_{co}} \alpha_i = q_{\text{imp}} \quad [\lambda_{\text{imp}}] \tag{14d}$$

$$\sum_{i \in \Omega_{co}} \beta_i = q_{\text{exp}} \quad [\lambda_{\text{exp}}] \tag{14e}$$

$$p_i \in \mathscr{P}_i \qquad\qquad i \in \Omega_{co} \tag{14f}$$

$$\alpha_i, \beta_i \geq 0 \qquad\qquad i \in \Omega_{co} \tag{14g}$$

This extended formulation implies three complicating constraints, (14c), (14d) and (14e), however the ADMM algorithm as in (13) can still be applied if relaxing all three constraints in the augmented Lagrangian through their dual variables λ, λ_{imp} and λ_{exp}.

Social contracts can be used not only for common objectives, but also in order to enforce fairness strategies within the community. In other words, the collective members may perceive as not fair that the whole community acts as net importer only because of some individuals. Hence, a penalization for the maximum importer can be added to the objective function of the community manager to reduce this phenomenon. This takes the form of an extra cost γ_{imp} times the maximum among the imported energy $\|\alpha\|_\infty$, expressed as the infinity norm of the vector of energy imported by each member. Different types of social contracts can be designed and implemented in such a negotiation framework [4], but the key point remains that they are relatively easily included in the negotiation process among community members. This feature becomes fundamental when looking at sharing systems relying on communities, as may be the case for electric energy for instance, with shared production (e.g., PV) and storage assets.

5.4 Example Application Results

In the community-based case, we then consider this fact that the overall population is split into two communities, which could be villages or communities based on the fact they agreed on a given social contract. Let us say that, in principle those communities aim for autonomy in terms of electric energy sharing, meaning that they would like to produce and consume within their community as much as possible. The two communities can exchange electricity if necessary trade with each other through a bilateral contract. Two types of social contracts are considered to go towards community autonomy: having an extra cost per kWh (say, a penalty) on all who induce electricity import within their community, or alternatively imposing this extra cost on the maximum importer only. In this last case, one expects a ripple effect with all agents in a community trying not to be this max importer who has to pay more than others for electricity. In all cases the penalty is defined as 2 c€/kWh. The resource allocation results, as well as corresponding electricity prices, are gathered in Tables 6 and 7, for the penalization of all imports and max importer only, respectively.

First of all, the resource allocation results and prices are different from the pool-based and peer-to-peer cases. Those are also influenced by the choice of a social

Table 6 Resource allocation of the community-based approach with penalty on importation

Agent	1	2	3	4	5	6	7	8	9	10	11	12
p_i [kWh]	30	−16.16	20	−2.93	−71.54	50	−15	0	10.77	20.89	−81.16	55.12
q_i [kWh]	−30	16.16	−20	12.30	71.54	−50	15	0	−10.77	−20.89	71.79	−55.12
α_i [kWh]	0	0	0	0	0	0	0	0	0	0	9.37	0
β_i [kWh]	0	0	0	9.37	0	0	0	0	0	0	0	0

(a) Agent level

Community	q_{imp} [kWh]	q_{exp} [kWh]	λ [c€/kWh]	λ_{imp} [c€/kWh]	λ_{exp} [c€/kWh]
1	0	9.37	18.54	22.54	18.54
2	9.37	0	22.54	22.54	18.54

(b) Community level

Table 7 Resource allocation of the community-based approach with max importer penalization

Agent	1	2	3	4	5	6	7	8	9	10	11	12
p_i [kWh]	30	−10.83	20	0	−64.44	50	−15	−3.33	8.57	20	−84.99	50.01
q_i [kWh]	−30	10.83	4.74	0	64.44	−50	10.88	−0.80	−12.70	−24.12	80.87	−54.13
α_i [kWh]	0	0	0	0	0	0	4.12	4.12	4.12	4.12	4.12	4.12
β_i [kWh]	0	0	24.74	0	0	0	0	0	0	0	0	0

(a) Agent level

Community	q_{imp} [kWh]	q_{exp} [kWh]	λ [c€/kWh]	λ_{imp} [c€/kWh]	λ_{exp} [c€/kWh]
1	0	24.74	20.67	20.67	20.67
2	24.74	0	21.00	20.67	20.67

(b) Community level

contract. Here the overall impact on exchange between communities is such that it is lowest when penalizing all imports, though it should not be seen as something to be expected generally. On an agent-per-agent basis, the resource allocation and related costs and revenues are also affected, in direct relation with their flexibility in terms of range of kWh, but also based on their cost functions.

6 Conclusions and Perspectives

Alternative approaches to organization and negotiation for sharing systems were presented, though relying on a fairly specific and simplified setup, also illustrated for the case of electric energy sharing. Even though this type of resource allocation problem is representative for a wide range of sharing systems, additional aspects ought to be considered.

First of all a number of assumptions made here ought to be relaxed if aiming to consider sharing systems more generally with their characteristics and constraints.

One mainly think of inter-temporal constraints, being in terms of usage of the resource (as electric energy consumers, in view of their inter-temporal flexibility) but also in terms of access to the resource, e.g., queues in bike-sharing systems. This then requires recasting the negotiation approaches and individual optimization problems in a control or dynamic programming framework. Similarly, a number of sharing systems have inherent uncertainty in demand and availability of the resource, hence requiring to rethink negotiation in a framework of optimization under uncertainty e.g. stochastic programming and chance-constrained optimization. More fundamentally, the organization and negotiations approaches rely on synchronicity of negotiation on well defined products (quantity-price) for a given time period. In general cases, it might be desirable that negotiation may be asynchronous or that product types and characteristics are flexible. Finally, the fact that access and use of the resource may also require the use of an underlying infrastructure (like the power network for electric energy) was set aside here. Accounting for such infrastructure and its constraints, pricing for its usage, as well as design of incentives for fair access and optimal usage, are all current research problems considered for a wealth of applications.

Thinking of mechanism design and the agents involved, gaming and strategic aspects, coalition formation, as well as fairness-driven mechanism design, are all timely problems to make sure we design and operate sharing system in a socially acceptable and sustainable manner.

Acknowledgements The authors are partly supported by the Danish Innovation Fund and the ForskEL programme through the projects '5s'—Future Electricity Markets (12-132636/DSF), CITIES (DSF-1305-00027B) and The Energy Collective (grant no. 2016-1-12530). In addition, the authors thank Etienne Sorin, Lucien Bobo, Tiago Sousa, Roman Le Goff Latimier, Hamid Ben Ahmed and many others for inspiring discussions and input on sharing electric energy.

References

1. Alvaro-Hermana R, Fraile-Ardanuy J, Zufiria PJ, Knapen L, Janssens D (2016) Peer to peer energy trading with electric vehicles. IEEE Intell Transp Syst Mag 8(3):33–44. https://doi.org/10.1109/MITS.2016.2573178
2. Boyd S, Parikh N, Chu E, Peleato B, Eckstein J (2011) Distributed optimization and statistical learning via the alternating direction method of multipliers. Found Trends Mach Learn 3(1):1–122. https://doi.org/10.1561/2200000016
3. Day RH (1971) Rational choice and economic behavior. Theory Decis 1(3):229–251. https://doi.org/10.1007/BF00139569
4. Moret F, Pinson P (2018) Energy collectives: a community and fairness based approach to future electricity markets. IEEE Trans Power Syst 99:1–1. https://doi.org/10.1109/TPWRS.2018.2808961
5. Morstyn T, Farrell N, Darby SJ, McCulloch MD (2018) Using peer-to-peer energy-trading platforms to incentivize prosumers to form federated power plants. Nat Energy 3(2):94–101. https://doi.org/10.1038/s41560-017-0075-y
6. Morstyn T, Teytelboym A, McCulloch MD (2018) Bilateral contract networks for peer-to-peer energy trading. IEEE Trans Smart Grid 99:1–1. https://doi.org/10.1109/TSG.2017.2786668
7. Parag Y, Sovacool BK (2016) Electricity market design for the prosumer era. Nat Energy 1:16,032. https://doi.org/10.1038/nenergy.2016.32

8. Peter C (2017) Electricity market design. Oxf Rev Econ Policy 33(4):589–612
9. Raworth K (2017) Doughnut economics - seven ways to think like a 21st-century economist. Cornerstone, San Diego
10. Reuben E, Suetens S (2012) Revisiting strategic versus non-strategic cooperation. Exp Econ 15(1):24–43. https://doi.org/10.1007/s10683-011-9286-4
11. Sorin E, Bobo LA, Pinson P (2017) Consensus-based approach to peer-to-peer electricity markets with product differentiation. Under review

Behaviour Change for the Sharing Economy

Léa Deleris and Pól Mac Aonghusa

Abstract Our understanding of human behavior and how to effect changes in behavior have evolved to the point where behavior change techniques have been applied successfully in areas as diverse as public health, advertising and retail. Despite its successes, behavior change science remains largely unknown among the engineering and technology community. In this chapter we provide a brief introduction to a vast field of knowledge. We describe practical behavior change tools and techniques that can be used to design behavioral interventions as elements of complex projects such as are typical of the sharing economy. Structured design of behavioral change programs is discussed and key concepts discussed followed by an overview of important techniques such as the COM-B Model, the Behavior Change Wheel. The Taxonomy of Behavior Change Techniques we discuss sits in an exciting area of active research using ontology techniques to organize knowledge. This approach lends itself naturally to application of technological tools such as artificial intelligence. Behavior change is a rapidly evolving science with sufficient rigorous and structured underpinnings to support more holistic design of complex projects where human behaviors are critical to sustainability and long-term success. We invite the reader to see this chapter as a practical introduction and to use the accompanying references to explore the literature around this fascinating subject.

Keywords Behavior change · Interventions · Behavior change wheel · COM-B model · Taxonomy

---

L. Deleris
BNP Paribas, Paris, France
e-mail: lea.deleris@bnpparibas.com

P. Mac Aonghusa (✉)
IBM Research Ireland, Dublin, Ireland
e-mail: aonghusa@ie.ibm.com

© Springer Nature Switzerland AG 2020
E. Crisostomi et al. (eds.), *Analytics for the Sharing Economy: Mathematics, Engineering and Business Perspectives*,
https://doi.org/10.1007/978-3-030-35032-1_11

1 An Introduction to Behaviour Change

What do you think of when you hear the term "behaviour change"? The term seems to crop up everywhere—from images of mind-control in science fiction to manipulation of election results through "false news". Practical reality, as is often the case, is alas less sensational.

The science of behaviour change is a vast field with established roots in areas from changing health outcomes such as smoking and obesity management to nudging retail consumers to buy more. In this introductory presentation, we provide a single "slice" through the behaviour change literature with the objective of giving the reader an overview of practical tools and techniques that the authors have found useful in practice. We will necessarily skip over details and we invite the reader who wishes to dig deeper into this fascinating field to explore the bibliography for this chapter.

Our focus is on *Human Behaviours* and how they can be changed—so a natural question is "what do we mean by a behaviour" and especially behaviours we can reasonably hope to change? We choose the following formal definition to suit our purpose ..."anything a person does in response to internal or external events. Actions may be overt (motor or verbal) and directly measurable or, covert (activities not viewable but involving voluntary muscles) and indirectly measurable; behaviours are physical events that occur in the body and are controlled by the brain" [4]. The key idea is that a behaviour implies a voluntary action in response to a stimulus. This is important because *voluntary actions* are realistically the only things where we can reasonably expect to change and a *response to a stimulus* is something we can use to measure change before, and after, the stimulus is applied.

Our definition helps to clarify what is in scope—and perhaps more importantly what is not! For example, *attitudes* are not behaviours in our definition. An attitude is a way of *thinking* about someone or something, for example, your attitude towards sharing your personal resources with people from a neighbourhood with a high crime rate may be positive if asked as part of a survey. A behaviour is how you *act*, for example, refusing to share your car with someone from a high-crime neighbourhood is a negative behaviour. Polling professionals are very used to the phenomenon where subjects indicate a particular attitude publicly—such as voting for a political party—but may behave differently privately. This is why we should always read the footnotes below survey results to check if they are surveying attitudes or behaviours—and be careful that we target *behaviours* we wish to change! Research shows that changing attitudes alone has limited effect on behaviours and understanding attitudes alone does not help reliably predict behaviours, [1]. In fact, research shows that it is often the case that the reverse is true—behaviours we experience appear to influence our attitudes, [17].

The term *Sharing Economy* has been in popular use sufficiently long to have become an umbrella term with multiple meanings, [8]. Measuring the effectiveness of behaviour change interventions requires us to be specific about *what* behaviour is to be changed. For the purposes of this chapter, we take the sharing economy to mean "the set of resource circulation systems which enable consumers to both obtain and

Behaviour Change for the Sharing Economy

provide, temporarily or permanently, valuable resources or services through direct interaction with other consumers or through a mediator", [21].[1] The characteristic identifying behaviours of participants in this definition of a sharing economy we wish to reinforce through behaviour change are; dynamic switching of individual roles between producer and consumer, and, active collaboration among participants to organise the exchange of resources.

Once we have identified a behaviour that we wish to change a key question we often ask is, "*what* behaviour change intervention works, *when*, *how* and for *whom*"? Reasoning formally, let $\mathcal{B}$ be a measurable event denoting the occurrence of behavioural event we wish to encourage in a sample cohort $\mathcal{P}$ in the general population. Given a set of available behaviour change interventions $\mathcal{I} = \{i_1, i_2, \ldots, i_n\}$ we could apply to $\mathcal{P}$, we wish to identify behaviour change interventions solving

$$\sup_{i \in \mathcal{I}} \frac{\mathbf{Pr}(\mathcal{B} \mid I = i, \mathcal{P})}{\mathbf{Pr}(\mathcal{B} \mid \mathcal{P})} \qquad (1)$$

where I is a random variable over $\mathcal{I}$ denoting choice of behaviour change intervention. Expression (1) is appropriate when we wish to increase prevalence of $\mathcal{B}$, when considering a negative behaviour we wish to discourage, we change (1) by substituting inf for sup.

Attempting to solve (1) in practice is an empirical rather than mathematical exercise. It requires us to think carefully about context and to answer each of the questions …what, when, how and who …when designing a behaviour change program. A large part of the content of this chapter is dedicated to developing the vocabulary and structured tools to model context around (1).

1.1 Behaviour Change and the Sharing Economy

Why do people participate in the sharing economy? After all, the classic model of individualistic consumption is familiar and seems to work on average in Western society. For discussion purposes we introduce a hypothetical use case that we develop as we proceed.

> **CarFree**
> Our setting is a city of population 50,000 located in the remote Mid-west of the United States. High daily pollution levels from car exhausts around key locations such as schools and the central business district have long been a source of dissatisfaction among citizens. A recent study of pollution levels indicated that schools were particularly badly effected by exhaust emissions

---

[1] The authors use the term the *collaborative consumption* instead of Sharing Economy.

> when parents dropped off and collected children. The city recently underwent an internet upgrade and boasts one of the highest participation rates in social media in the State, across all demographics of the population. The mayor is facing re-election next year and has announced a car sharing program—called CarFree—intended to reduce car usage in the city area. Citizens who sign up for CarFree agree to pool their personal car in return for incentives such as tax breaks and free use of public transport. Cars enrolled in CarFree will also be allowed to enter key areas of the City without paying a congestion toll, which cars not enrolled in CarFree will have to pay. The vision is that CarFree will encourage people to reduce personal car usage in order to maximise benefits from CarFree participation. The mayor has asked you to explore behaviour change interventions that can support the CarFree vision.

In reality, behaviours are influenced by other behaviours, and individuals are influenced by other individuals, [16]. Individual knowledge and values change, in short there is no single demographic and no universal behaviour change technique. We do not try to provide exhaustive suggestions to the CarFree use case therefore. Instead we encourage readers to explore and develop their own approaches based on their own experiences and research to gain familiarity with the tools presented in this chapter.

2 Structuring a Behaviour Change Project

2.1 Expectation Management

The most important thing we can do to achieve sustainable behaviour change is to plan. In our experience, projects that fail to plan and prepare in a structured way are unlikely to meet stakeholder expectations—achieving, at best, partial and temporary results. A structured approach will help manage stakeholder expectations and avoid some common pitfalls in undertaking a behaviour change program.

It is not simple There is a belief that changing behaviour is a matter of getting the message right. In fact there is ample evidence that seeing tangible benefits and the associated feelings, rather than analysing information and thinking rationally, are much more powerful motivators of change, [9, 22].

People are neither fully rational nor fully irrational People act with reference to the *situation they are in* rather than the *way they are*, [9]. People do things that their political, economic, social and technological environment allows them to do. For example, if charging points are not readily available for electric cars people concerned about driving range are less likely to switch to electric vehicles.

Prediction is difficult Unintended consequences from interventions are common. The true sustainability of car sharing schemes is complex to assess, [13], while some users reduced their car usage, for others access to vehicles caused carbon footprints. It is very difficult to say with high confidence how people will behave in a given circumstance, and even more difficult to predict behaviour over long time frames, [12].

We suggest keeping these pitfalls in mind when setting stakeholder expectations at the initial stages of the project.

> **CarFree: Setting Expectations**
> You are preparing for an initial review with the mayor to discuss CarFree. On consulting the behaviour change literature you learn that, at population level, participation in collaborative or shared consumption appears to be motivated by factors such as environmental sustainability, enjoyment of the activity and expected economic gains, [21]. At individual level, behaviours are more complex—for example—in [21] environmental sustainability as a motivating factor was not observed to be directly associated with participation—*unless* it was also associated with positive attitudes to shared consumption.
>
> This suggests that when designing behaviour change interventions around CarFree that promoting a positive image of shared consumption is important. What other factors did you uncover in your research? Would children be more influenced by enjoyment or financial gain? What about 25–45 year old professionals? Would a sense of legacy play a role in influencing the over-65 population with families?

2.2 The Intervention Design Process

The Intervention Design Process (IPD), [16], summarised in Fig. 1 is a structured approach to the design and planning of a behaviour change intervention strategy.

The first step in the IDP is to identify the target behaviour *precisely*, meaning who needs to do what, when, where, and how. For instance, in the CarFree case, rather than say the target behaviour is not just about sharing of cars in the general population—a more specific target is to say sharing of 80% of all carbon fuel cars of residents living near key areas such as schools during peak hours.

Based on a clear description of the target behaviour, the first step for disciplined behaviour change analysis is to perform a behavioural diagnosis, using for instance the COM-B model in Fig. 1. A good behavioural diagnosis is more likely to lead to effective interventions, the second step of the IDP. Consequently, we recommend you consider the full range of options open to you and to target many levels simultaneously. Finally the third, and equally challenging, step of a rigorous behaviour change analysis is to evaluate interventions. An important aspect of the evaluation is

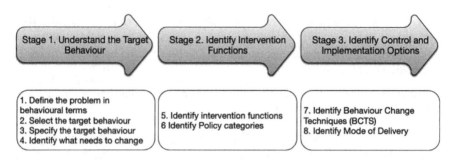

Fig. 1 The intervention design process

the definition of how the intervention is actually undertaken which is referred to as *Mode of delivery*.

> **CarFree: Unintended Consequences**
>
> It is also important to recognise that behaviours are part of a system of other behaviours within and between people and may not be considered independently. Building on the example of CarFree, we must also pay attention to unintended consequences from transfer towards other modes of transport. What if carbon fuel vehicles not included in CarFree have to pay a congestion toll, while more expensive electric vehicles do not?
>
> Might that introduce an unintended socio-economic bias in favour of those who could afford a second electric car for short drives around town—also resulting in more total vehicles on the road? How would you control against this kind of unintended consequence?

The Centre for Behaviour Change at University College London has produced a practical guide to applying the IDP that is available online as a PDF document, [16].

2.3 Behaviour Change Intervention

A sizeable portion of the effort in a behaviour change program is spent in the design and evaluation of various Behaviour Change Intervention (BCI) scenarios to assess what works, when, how and for whom. A BCI scenario is defined as a sequence or development of events consisting of an *intervention*, its target behaviours, and factors that influence the *outcome* of the BCI in relation to the *target behaviour* (Fig. 2). Specifying a BCI in this way helps to complete the first 5 steps in the IDP (Fig. 1).

Behaviour Change for the Sharing Economy

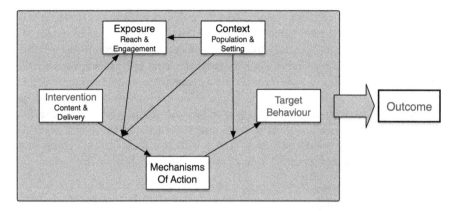

Fig. 2 Upper level entities in the BCI scenarios and their causal connections

Outcome and target behaviours in Fig. 2 relate to the type(s) of behaviour that the BCI seeks to change (e.g., participation in car sharing) together with a collection of attributes (e.g., duration, frequency or incidence). Together these specify types of outcome measure (e.g., Self-reporting of participation in the CarFree scheme for 6 months supported by an average of at least 2 CarFree bookings per week evidenced by receipts in the online booking system and measured at the final follow up point). For an example of this in action in the case of smoking cessation see [23].

Intervention in Fig. 2 is defined as a set of types of policies, activities, services or products that are intended to result in a specified outcome in relation to the target behaviour. The intervention is specified in terms of summary descriptors (e.g., "brief opportunistic advice on CarFree from local newspapers") together with detailed descriptions of "content" such as the techniques used (e.g., financial support, verbal persuasion about benefits etc.), and "delivery" (e.g., monthly single session, verbal, face-to-face, during a routine consultation, by a city CarFree officer). The term intervention is also used to refer to any comparator in a BCI evaluation (e.g., intensive, usual level).

Context in Fig. 2 is defined as any factor (consisting of characteristics of the population and setting) not directly connected with the intervention that may influence the effect of the intervention.

The entities in the BCI scenario interact in specific ways, as showed by the arrows in Fig. 2. The content and delivery of an intervention influences the target behaviour through one or more mechanisms of action. The context moderates the influence of (1) the intervention on the mechanism of action and (2) the mechanism of action on the behaviour. Exposure moderates the influence of the intervention on the mechanism of action and is itself influenced by the intervention and context.

To illustrate with a medical example if a doctor prescribes nicotine replacement therapy (intervention) to smokers interested in stopping (population), as part of a routine consultation in a doctor's surgery in the UK (context), and 60% of smokers obtain the medication and start the treatment, and 50% take it as prescribed (exposure), this may reduce cigarette cravings (mechanism of action) and so lead to at least 6 months of abstinence (outcome behaviour) in 15% (outcome value) of cases [14].

If one were to conduct a study to assess the effect of GPs prescribing nicotine replacement therapy, this scenario would be compared with a BCI scenario such as GP advice without the offer of a prescription. The comparison would have a number of features relating to study design (for example a Randomised Control Trial), sample recruitment and selection, sample size, baseline and outcome measures etc. The comparison of outcomes between the two scenarios would constitute the "effect" of the prescription intervention relative to advice without a prescription, expressed in terms of an odds ratio or risk ratio with a corresponding confidence interval.

> **CarFree: Interventions**
>
> How does the medical example above translate into the formal model in Eq. 1?
>
> You want to design a behaviour change study to assess the effect of advice on participation in the CarFree scheme. The available modes of delivery are social media, local newspapers and posters displayed on public transport.
>
> Use the material of this section to formalise the interventions in the proposed study. Pay attention to how you would define a measurable outcome and how you would compare the relative effectiveness of each mode of delivery. How would you select a sample population for each model of delivery for valid comparison later? What specific techniques would you use to gather data for each mode?

3 Tools of Behaviour Change

3.1 COM-B Model

The COM-B model serves as a framework to analyze behaviour in context and identify targets for change [14]. The COM-B models seeks to articulate the interactions between capability, opportunity and motivation on one side and the targeted or observed behaviour as represented on Fig. 3. The model can be used for instance as a starting point for intervention design, asking questions to focus groups leveraging the categories outlined in Table 1 and helping frame what the target behaviour is.

In the case of the CarFree example, the target behaviour is to increase the use of CarFree programs among city dwellers. Here motivation can come in many forms, from cost and time savings to seeking social interactions or acting with social responsibility to wanting to try something new.

Capabilities to support the behaviour and motivation would be the physical ability to ride in any kind of car. Some people for instance people that use a wheelchair or people that need to carry a large volume of equipment to work may not be able to participate. Others may not be able to accept the fact that different people may be

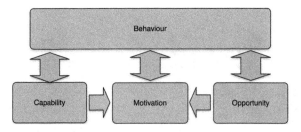

Fig. 3 The COM-B model

Table 1 Elements of the COM-B model with descriptions

Component	Definition	Example
Physical capability	Physical skill to enact the behaviour	Having the skill to take a blood sample
Psychological capability	The capacity to engage in the necessary though processes to enact the behaviour (comprehension and reasoning)	Awareness of the impact of CO_2 on the environment
Reflective motivation	Reflective processes, involving evaluations and plans	Deciding to buy a fridge based on its energy consumption profile
Automatic motivation	Automatic processes, involving emotions and impulses	Deciding to buy a car because you liked the advertising campaign
Physical opportunity	Opportunity afforded by the environment	Being able to go for a run because you own running shoes
Social opportunity	Opportunity afforded by the cultural context and associated social norms	Being able to smoke in the house of someone who smokes but not in a professional meeting

driving their cars and the possibility of damage to their car each time they use the service. Finally, only people living in neighbourhoods where the service is available can participate.

The opportunity aspect also includes the ease with which one can register for the service and then request rides. If it is supported by a mobile app, then non technologically savvy populations may not have the opportunity to engage with the program.

3.2 Behaviour Change Wheel

The Behaviour Change Wheel (BCW), [14], illustrated in Fig. 4, is a comprehensive framework designed for defining the intervention strategy and the behaviour change techniques comprising the designed intervention. The BCW is a visual tool based on the synthesis of 19 frameworks to classify interventions (health, environment,

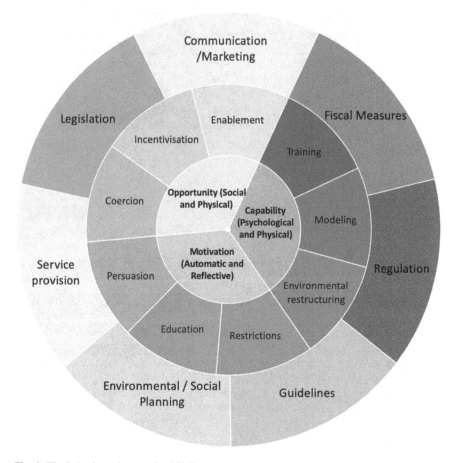

Fig. 4 The behaviour change wheel [16]

culture change and social marketing) organised into 3 rings. The inner ring represents intervention functions addressing the purposes the interventions serve, i.e., what aspects of the COM-B model do they enable or disable. The middle ring outlines categories of actions which influence different aspects of the COM-B model. Those categories of actions are defined below:

- Education: Increasing knowledge and Understanding
- Restrictions: Using rules to reduce the opportunity to engage in the target behaviour
- Environmental Restructuring: Changing the physical or social context
- Modelling: Providing an example for people to aspire to or imitate
- Enablement: Increasing means/reducing barriers to increase capability (beyond education and training) or opportunity (beyond environmental restructuring)
- Training: Imparting skills
- Coercion: Creating an expectation of punishment or cost

- Incentivization: Creating an expectation of reward
- Persuasion: Using communication to induce positive or negative feelings or stimulate action

Finally, the outer ring aligns with policy categories, i.e., actions taken at the system level whether community, county or state:

- Environment and Social Planning: Designing and/or controlling the physical or social environment
- Guidelines: Creating documents that recommend or mandate practice. This include all changes to service provisions
- Fiscal Measures: Using the tax system to reduce or increase the financial cost.
- Regulation: Establishing rules or principles of behaviour or practice
- Service Provision: delivering a service
- Legislation: Making or changing laws
- Communication/Marketing: Using print, electronic, telephonic or broadcast media.

The wheel is designed to help behavioural scientists build a broad strategy for behaviour change, going from the source of behaviour to the intervention function to policy level considerations. While access to policy levers may not be relevant in some sharing economy use cases, it is essential to keep them in mind as they are powerful elements of a behaviour change strategy.

Returning to the CarFree example, then relevant intervention functions (center ring) are education (e.g., informing people of the existence of the program) along with training (e.g., explaining how to register and use the car sharing program) and persuasion (e.g., outlining the benefits for the individuals using the program). Furthermore, the program also relies on incentivisation through the tax rebates and free public transport usage. Finally, at policy level, the Financial measures, Communication and Marketing and Environment and Social Planning represent the natural levers to influence positively the adoption of the CarFree program.

3.3 Taxonomy of Behaviour Change Techniques

Traditionally, reports of evaluations of behaviour change interventions barely described the actual intervention, making it very difficult to identify the most effective methods from past experiences. This was increasingly recognised in the late nineties and early twenty-first century, where behaviour change methods gained increasing popularity. A need thus emerged to provide a standardised method of classifying intervention content. Michie and colleagues have leveraged the combined expertise of 400 experts from 12 countries to develop the behaviour Change Technique (BCT) taxonomy [15] which includes 93 items that allow the active ingredients of interventions to be systematically described, reviewed, and replicated. The 93 BCTs[2] are

---

[2]http://www.ucl.ac.uk/health-psychology/bcttaxonomy.

organised around 16 groupings and addressing the COM-B elements of capability, opportunity, and/or motivation.

A key task in behavioural science can be seen as understanding the extent to which BCTs contribute to the effectiveness of interventions of which they form a part. To date the BCTs have been largely applied in the area of health behaviours and examples of such endeavours in the domain of smoking cessation are presented in [2, 19] looking respectively at the case of pregnant women and water-pipe smokers.

Going further, some researchers have long advocated the use of analytical theories, and specifically Control Theory, as conceptual framework to understand the mechanisms behind human behaviour changes [3]. Recently, one specific project proposed to use Control Theory to understand which combinations of BCTs can be effective for a particular problem and why such a combination was effective. More specifically, they have found that the BCTs that were congruent with Control Theory, i.e., that had a conceptual link with Control Theory concepts (e.g., self-regulation), were more effective to reduce alcohol consumption [7].

4 Examples of Domains with Active Behaviour Change Research

In this section, we discuss briefly three domains that have been and often still are the focus of active behaviour change research. Our intention is to guide readers the reader towards settings that can be potentially translated into a specific sharing economy challenges. The first domain is recycling which seeks to influence a change of behaviour requiring a personal effort for the benefit of society. The second domain is smoking, the quintessential problem of encouraging people to abstain from a personally harmful yet enjoyable behaviour. Other similar domains are alcohol and drug cessation. Finally, the third domain, patients adherence, is relevant to contexts where the motivation of the human is obvious but practical aspects prevent him or her to actually follow through with the intended behaviour.

4.1 Recycling

One common behaviour that has progressed steadily over the past half century is our effort as consumer to sort our waste in order to recycle materials. Progression from almost no recycling to an average of 35% in the US[3] and 39% in Europe[4] in recent years. Recycling is also a good example in the sense that behaviour change

---

[3]https://www.epa.gov/sites/production/files/2018-07/documents/2015_smm_msw_factsheet_07242018_fnl_508_002.pdf.

[4]https://www.eea.europa.eu/about-us/competitions/waste-smart-competition/recycling-rates-in-europe/image_view_fullscreen.

intervention campaign to encourage recycling are typically led locally, which implies a disparity of approaches and some variability in the outcome. Specifically, San Francisco boasts about 77% of recycling and composting of its waste while New York City is at about 15%.[5] Participation rate in recycling is affected by a multitude of aspects, in particular whether the collection of recyclable is so called single-stream-recycling (all recyclables in one recycling bin). Recently a study using techniques such as the Behaviour Change Wheel [6] described in this chapter has explored in-depth recycling behaviour. User groups studies led to the realization that people motivation was affected by the physical opportunity to recycle (i.e. availability and proximity of resources for recycling) and their psychological capability to recycle in the present context (i.e. knowing which items can be recycled and where). Such findings were then used to determine which intervention functions were the best aligned with the change of behaviour needed was rather to focus on making it possible (bins and training/support in determining into what goes where) than to convince people that recycling was positive.

4.2 Smoking

Smoking cessation is an area of health behaviour change that has received a significant amount of interest over the past 50 years. Tobacco smoking contributes to millions of premature deaths each year worldwide and is highly addictive, with more than 95% of unaided attempts at cessation failing to last 6 months [10]. Success in quitting is increased by behavioural support and a range of pharmacotherapies [5]. However, a majority of the smokers in the world live in countries in which the average household income is less than 200 per week and in which smoking-cessation medications are much more expensive than smoking. Such context challenges researchers further in including the cost element in the search for effective interventions [24].

4.3 Patients

Simple, lightweight interventions can sometimes have a significant effect on patient's behaviour. In the case of medication adherence in stroke survivors, providing patients with two discussion sessions on medication-taking routine and addressing mistaken beliefs about the treatment had significant effect on their pill taking behaviour, increasing from 87% in the control group to 97% in the intervention group [18]. In fact a recent systematic review on medication adherence has highlighted that the effective intervention were associated with two main BCTs specifically, increasing knowledge and self-efficacy, i.e., an individual's belief in his or her innate ability to achieve goals [11]. Another study looked at the challenge of patients movements

---

[5]https://www.scientificamerican.com/article/has-recycling-lived-up-to-its-promises/.

while doing MRI examinations. Movement causes motion artefacts on the images that may reduce the diagnostic accuracy of a scan. In this experiment, the intervention group was presented with an information DVD about the procedure while the control group simply received the default leaflet. The results showed that 35 participants in the intervention group out of 41 completed the scan and had no motions against 23 out of 42 in the control group [20].

5 Conclusion

The intersection of the sharing economy and behaviour change is a developing area with exciting opportunities to apply knowledge gained from areas such as health behaviour to encourage participation in the shared economy. Indeed without considering human behaviours we suggest that wide-spread adoption of shared economy models is unlikely to reach, and sustain, critical mass in the general population.

The field of behaviour change is vast and growing in application and importance as a tool to support societal programs. Our introduction is necessarily brief and intended to orient the reader for further exploration. We have outlined some of the key considerations, approaches and tools used in constructing behaviour change programs. We invite the reader to jump in and explore this fascinating area.

References

1. Ajzen I, Fishbein M (2005) The influence on attitudes on behavior. The handbook of attitudes, pp 173–221
2. Campbell KA, Fergie L, Coleman-Haynes T, Cooper S, Lorencatto F, Ussher M, Dyas J, Coleman T (2018) Improving behavioral support for smoking cessation in pregnancy: what are the barriers to stopping and which behavior change techniques can influence these? Application of theoretical domains framework. Int J Environ Res Public Health
3. Carver CS, Scheier MF (1982) Control theory: a useful conceptual framework for personality-social, clinical, and health psychology. Psychol Bull
4. Davis R, Campbell R, Hildon Z, Hobbs L, Michie S (2015) Theories of behaviour and behaviour change across the social and behavioural sciences: a scoping review. Health Psychol Rev 9(3):323–344
5. Fiore M, Jaen C, Baker T, Bailey W, Benowitz NL, Curry S, Dorfman S (2008) Treating tobacco use and dependence: 2008 update. Clin Pract Guidel. arXiv:gr-qc/9809069v1
6. Gainforth HL, Sheals K, Atkins L, Jackson R, Michie S (2016) Developing interventions to change recycling behaviors: a case study of applying behavioral science. Appl Environ Educ Commun 15(4):325–339
7. Garnett CV, Crane D, Brown J, Kaner EFS, Beyer FR, Muirhead CR, Hickman M, Beard E, Redmore J, de Vocht F, Michie S (2018) Behavior change techniques used in digital behavior change interventions to reduce excessive alcohol consumption: a meta-regression. Ann Behav Med
8. Hamari J, Sjöklint M, Ukkonen A (2015) The sharing economy: why people participate in collaborative consumption. J Assoc Inf Sci Technol 67(ID 2271971):2047–2059
9. Heath C, Heath D (2010) Switch: how to change things when change is hard. Self-improvement

10. Hughes JR, Keely J, Naud S (2004) Shape of the relapse curve and long-term abstinence among untreated smokers
11. Kahwati L, Viswanathan M, Golin CE, Kane H, Lewis M, Jacobs S (2016) Identifying configurations of behavior change techniques in effective medication adherence interventions: a qualitative comparative analysis. Syst Rev
12. Kelly MP, Barker M (2016) Why is changing health-related behaviour so difficult? Public Health
13. Martin EW, Shaheen SA (2011) Greenhouse gas emission impacts of carsharing in North America. IEEE Trans Intell Transp Syst. arXiv:0803.1716
14. Michie S, van Stralen MM, West R (2011) The behaviour change wheel: a new method for characterising and designing behaviour change interventions. Implement Sci
15. Michie S, Richardson M, Johnston M, Abraham C, Francis J, Hardeman W, Eccles MP, Cane J, Wood CE (2013) The behavior change technique taxonomy (v1) of 93 hierarchically clustered techniques: building an international consensus for the reporting of behavior change interventions. Ann Behav Med. arXiv:1011.1669v3
16. Michie S, Atkins L, West R (2014) The behaviour change wheel. A guide to designing interventions, 1st edn. Silverback Publishing, London, p 329
17. Myers DG (2009) Social psychology, 10th edn. arXiv:1011.1669v3
18. O'Carroll RE, Chambers JA, Dennis M, Sudlow C, Johnston M (2014) Improving medication adherence in stroke survivors: mediators and moderators of treatment effects. Health Psychol: Off J Div Health Psychol Am Psychol Assoc
19. O'Neill N, Dogar O, Jawad M, Kellar I, Kanaan M, Siddiqi K (2018) Which behavior change techniques may help waterpipe smokers to quit? An expert consensus using a modified Delphi technique. Nicotine Tob Res
20. Powell R, Ahmad M, Gilbert FJ, Brian D, Johnston M (2015) Improving magnetic resonance imaging (MRI) examinations: development and evaluation of an intervention to reduce movement in scanners and facilitate scan completion. Br J Health Psychol
21. Taeihagh A (2017) Crowdsourcing, sharing economies and development. J Dev Soc 33(2):191–222. arXiv:1707.06603
22. Webb A (2010) Making the emotional case for change: an interview with chip heath. McKinsey Q. arXiv:1011.1669v3. ISSN 00475394 (Interview)
23. West R, Hajek P, Stead L, Stapleton J (2005) Outcome criteria in smoking cessation trials: proposal for a common standard
24. West R, Zatonski W, Cedzynska M, Lewandowska D, Pazik J, Aveyard P, Stapleton J (2011) Placebo-controlled trial of cytisine for smoking cessation. N Engl J Med

Platforms and New Use Cases

Emanuele Crisostomi, Bissan Ghaddar, Florian Häusler, Joe Naoum-Sawaya, Giovanni Russo and Robert Shorten

1 Introduction

Use-cases drive excitement and interest justifies allocation of resources, and resources drive innovation and research, which in turn drives even more exciting use-cases. This has certainly been true in the sharing economy. Even though the sharing economy has only recently been the subject of interest in the academic community, already many exciting and very disruptive use cases have emerged that are changing the nature of products and services in several domains. In this part of the book we give a snapshot of a very limited number of these, to illustrate just

E. Crisostomi
University of Pisa, Pisa, Italy
e-mail: emanuele.crisostomi@unipi.it

B. Ghaddar · J. Naoum-Sawaya
Ivery Business School, Canada
e-mail: bghaddar@ivey.ca
e-mail: jnaoum-sawaya@ivey.ca

F. Häusler
Moovel, Germany
e-mail: florian.haeusler@gmail.com

G. Russo
Department of Information & Electrical Engineering and Applied Mathematics,
University of Salerno, Fisciano, Salerno, Italy
e-mail: giovarusso@unisa.it

R. Shorten
Dyson School of Design Engineering, Imperial College London, London, UK
e-mail: r.shorten@imperial.ac.uk

some of the work and ideas that are emerging in areas such as energy, transport and advanced manufacturing. Roughly speaking, sharing systems come in all *sorts and sizes*. For example, in opportunistic sharing systems, services are based on opportunistic sharing of resources to exploit large scale availability of underutilized resources, and/or outdated business models. Examples of successful sharing economy ventures include Airbnb[1], Uber[2], Parkatmyhouse[3], and Vrumi[4]. Other very important realizations exist beyond these mainstream sharing economy examples. As discussed in Chap. 12, sharing data is one such important realization of the sharing economy. The new business dynamics of data sharing and open source concepts present several challenges to industries such as transportation which have traditionally safeguarded data secrets. At the heart of the sharing economy is also the role of technology that enables data availability and data parsing. Connected vehicles for instance is a technology platform that permits ubiquitous data sharing and is currently being monetized by several services. As such, Chap. 13 outlines the concept of a parsing engine that makes sense of relevant information and makes it possible for a connected vehicle to become a *cognitive body* that can interact with a driver in a dynamic and continuously changing environment. Chapter 14 presents mobility-on-demand which is another realization of the sharing economy that not only has reshaped in recent years how goods and passengers are transported, but also how cities and communities are designed and served. Optimal resource allocation is a major challenge for many sharing systems. Chapter 15 outlines the efforts of Citi Bike to develop and implement a data driven system to aggregate and optimize the rebalancing efforts of New York City Bikeshare to reduce customer dissatisfaction and enable seamless system operation. Chapter 16 discusses another realization of the sharing economy which is peer-to-peer energy trading that has also emerged with the increasing availability of energy from renewable sources and the development of new technologies for energy consumption management and storage. While transportation and energy have been leading innovations in the sharing economy, Chap. 17 considers the healthcare field which has remained cautious about adopting such new business models and it is not until recently that healthcare firms started experimenting with business models for healthcare cost and medical assets sharing among others. Finally as the sharing economy is transitioning the manufacturing industry from physical assets that are manufactured and sold to assets that are manufactured and consumed as services, Industry 4.0 has emerged as the fourth industrial revolution and Chaps. 18 and 19 discuss its role in enabling the sharing economy.

---

[1] www.airbnb.com.
[2] www.uber.com.
[3] www.justpark.com.
[4] www.vrumi.com.

Sharing Data in Automotive Applications

Joachim Taiber

Abstract As we are approaching a new mobility age where zero-emission and highly automated vehicles become the foundation of a mobility as a service ecosystem, sharing data in real time via V2X to optimize safety, security and energy efficiency as well as operational service efficiency becomes essential. Giving the history of proprietary system development in the automotive industry and the vehicle-centric business model, sharing in particular safety and security relevant data across the complete smart mobility ecosystem poses a big hurdle and requires significant standardization efforts between the different stakeholders involved. Low latency, high bandwidth and blockchain/DLT-based standardized data protocols are critical factors to be considered in the problem resolution. It is also important that testbed infrastructure is shared among the ecosystem participants both in the development stage as well as in the certification stage, test scenarios are standardized and test data is shared.

Keywords Data sharing · Blockchain/DLT · V2X · Standardization · Mobility as a service ecosystem

1 Introduction

The automotive industry comes from a tradition where product specifications and its data are typically safeguarded secrets as the competition in the fight for market share between Automotive manufacturers is fierce. On the other hand, the automotive industry also successfully developed standards which are shared among all competitors and which require the sharing and exchange of data, for example standards around vehicle safety.

The latest McKinsey Race 2050 report [1] outlines how automotive industry players need to seek new forms of cooperation and alliances for example in domains such

J. Taiber (✉)
International Transportation Innovation Center, Greenville, SC, USA
e-mail: Joachim.Taiber@itic-sc.com

Fig. 1 1920 Anderson

as autonomous and connected vehicles, infrastructure development for connected and electrified vehicles, battery cell production, supply chain topics, and data-enabled mobility service platforms.

If we reflect how automotive manufacturers originally built cars (see Fig. 1, we will find that parts were manufactured with special tools that were typically custom made so that parts could not even be shared among vehicles of the same manufacturer. The standardization of product component design and tool design was essential to achieve parts sharing across the vehicles of the same make. The next step was the standardization of manufacturing processes which allowed the mass production of vehicles which also allows the comparison of performance metrics among car plants that produce the same vehicle—based on shared data and based on shared production systems.

When it comes to the operation of the vehicle, the sharing of data can become essential in particular both from a safety as well as from a security perspective. But it can also be relevant with respect to energy efficiency as well as the time efficiency. As vehicles become connected and automated, we do have the opportunity to communicate and share driving trajectories as well as observed behavior of objects participating in traffic activities such as pedestrians or bicyclists. Sharing the operational behavior of moving objects within a well-defined framework of driving scenarios lead to new forms of traffic optimization (see Fig. 2). The higher the level of driving automation, the more predictable the operational behavior becomes which results in better optimization results.

One of the most significant use cases of blockchain/distributed ledger technology in the automotive world could be the real time data sharing of vehicle behavior within heterogeneous vehicle networks. In this chapter we will explain in more detail what is necessary to implement such a system and how it could revolutionize the way traffic is currently organized and how it could become an important component in mobility-as-a-service platforms.

Fig. 2 Autonomous Shuttle "Olli" in Seaside Community, Florida

2 Sharing Data Within an Organization and Sharing Data with Your R&D Partners

As previously indicated, many car companies come from a culture that data needs to be protected—even within a company which then leads to data silos. This is of course true for data such as styling data, product development knowledge, information about suppliers and their pricing or market research data about customers and competitors. Critical data is linked to people and through fluctuation of company experts to competitors as well as through the use of contracted experts, knowledge is often difficult to be kept exclusive and it comes with a price to acquire, maintain and grow.

The open source movement had a significant impact on the way R&D knowledge is shared and additional knowledge is being added through a developer ecosystem. From a functionality point of view, software applications based on open source operating systems can be competitive to software applications based on proprietary operating systems. The ecosystem of application developers benefits from a robust kernel system which attracts the most competent experts to provide source code. At the same time, the cost of development can be reduced and the source code can be accessed by anybody.

If we consider the current situation in the development of automated driving, some companies such as Waymo do not share the source code of the algorithms used in the controls systems that are being provided to Original Equipment Manufacturer (OEM) R&D partners whereas Baidu has an open source Autonomous Vehicle (AV) development platform called Apollo [2] which can be accessed by the whole developer ecosystem. Waymo does not share the data that is accumulated by driving its pilot vehicle fleet whereas Baidu decided to share data [3]. But we do need to understand the underlying business model. Waymo started early to invest in automated driving with its own development resources and wants to lead the market of highly automated mobility services itself where the profit is being generated on a per mile revenue basis through a fleet operator. The OEM basically has the role of a supplier which provides the vehicles for the mobility service but the integration of the automated driving system (HW, SW) is controlled by Waymo which carefully selects OEM partnerships from a system integration perspective. Furthermore Waymo wants to get access to as much OEM vehicle data as possible to optimize the fleet operation it is responsible for. Baidu has a very different approach as it came much later to the game. The Apollo program is focused on generating a large developer base working on different vehicle types and it supports different OEM's via an ecosystem where knowledge needs to be built up quickly and where knowledge can actually be shared among developers working on competing products and services. Waymo on the other hand wants to monopolize its knowledge on driving automation and operating automated vehicle fleets and NOT share it.

From an OEM perspective a key problem is how to create a critical mass of in-house developers that can work on a proprietary AV system that is superior to other alternatives on the market including the open source options. Engineering groups of large AV system developers or AV divisions of large OEMs or Tier 1 suppliers can easily reach a workforce of 1,000 people or more but a start-up which provides an elegant open source kernel system which attracts hundreds of highly capable software developers that use it and share the source code contributions could be competitive and pose a threat to those proprietary investments. Nevertheless open source initiatives typically require organizational forms of consortiums which need to be supported by large companies and players in the ecosystem. This indicates the power of data sharing.

Many universities demonstrated that automated vehicles can be built based on open source solutions with commercial of-the-shelf components, see Fig. 3.

However, the real challenge is to manufacture highly automated vehicles in large volumes and at repeatable quality levels. This requires substantial knowledge both in high volume hardware manufacturing as well as in high quality software manufacturing. Many Automotive OEMs do have the hardware manufacturing competence but not necessarily the software manufacturing competence.

The real challenge for the OEM is on one hand to develop a highly automated product at cost that are acceptable for the user, which could also be a fleet operator, and on the other hand to ensure that the product and service quality meets customer expectations at all times. If a mobility service operator such as Uber provides robo-taxis to its customer, it is ultimately liable for the quality of the service delivery and it needs to select the OEM based on its capability to delivery a product that enables Uber to meet customer expectations. As pointed out before, it is essential how the

Fig. 3 Automated Nissan Leaf—Clemson University International Center of Automotive Research at the enclosed testbed of the International Transportation Innovation Center

knowledge to build and operate a highly automated vehicle and mobility service is accumulated to determine the cost per mile for the end user. The knowledge can be built up by the mobility service operator on an exclusive base such as in the Waymo example where Waymo brings in the knowledge about the sensors, the operating system, the map data as well as the fleet operation and where the Automotive OEM is basically a system supplier with limited access to the value chain and the inherent profit pool which depends on the economy of scale of the service. The strategic decision of Waymo is not to build a car but to do everything else to provide a highly automated mobility service which includes the process of controlling the integration of the vehicle as a system into the mobility service platform. Another option for the mobility service operator is to work with OEMs that utilize a shared open source platform where system functionalities are standardized in the ecosystem and where the quality of hardware and software integration is the key role of the OEM and not the mobility service operator. This model is followed by Baidu as a mobility ecosystem provider which drives the cost down for the mobility service operator if a critical mass of OEM adoption can be reached to support the ecosystem in a plug and play fashion. In this case the large mobility service operators such for example Didi in China can focus on the development of the fleet management system and do not need to deal with the vehicle system integration on the device level itself. Thus the OEM integration becomes "commoditized" and the cost of market entry for OEMs into the operation of mobility services as a supplier goes down. Many Automotive OEMs are currently struggling with the strategic decision in which role they want to enter into the mobility service market, as a supplier to a mobility service operator on a proprietary or standardized open source platform (or both) or as a provider of mobility services to end customers.

From a consumer perspective, the balance between functionality, cost but also quality is key. Economy of scale is essential on the cost side. The larger the AV fleet of the mobility service operator, the better the development cost of sensors and software can be distributed. The more customers the mobility service has, the better the development cost for the fleet management software can be distributed. It is to be expected that the market forces play in favor of large mobility service operators that make prudent strategic choices on OEM partnerships that can deliver the quality that is expected by the customer.

3 Sharing Data in Mobility-as-a-Service Ecosystems

A mobility-as-a-service ecosystem has a very different business philosophy compared to the traditional business dynamics of the automotive world.

The current business model of the automotive industry is based on private car ownership and the principle to sell as many new cars as possible with a profit, and quite often a substantial part of the profits are generated via vehicle financing instruments such as loans or lease deals. The majority of vehicles sold to private customers are practically underutilized and are often driven only with single occupancy. As the market is saturated, the profit margins on the product get thinner and in the extreme case new vehicles are sold at cost or even with a loss and money is only earned with the maintenance service. The new car sales business is cyclical in nature and car replacement cycles of privately used vehicles are long.

The cost of owning a car is significant, in particular if the owner lives in an urban environment with limited parking space. As a consequence urban citizens often use public transportation and/or on demand shared mobility services and also bicycles. As so-called last mile option shared scooters became more recently popular. Thus the replacement of private cars requires multi-modal transportation systems (Fig. 4) which are often organized in public private partnership constellations.

The RethinkX Sector Disruption Report [4] predicts that by 2030 95% of US passenger miles will be provided by a Transportation as a Service model where mobility is provided by fleets of electric automated vehicles and where cost savings for consumers compared to owning a private vehicle is assumed to be the primary driver of adoption.

What is described in [5] are three different scenarios towards the future of mobility in 2030 considering five key evaluation criteria of mobility systems which are availability, affordability, efficiency, convenience and sustainability. By implementing connectivity, autonomy, sharing and electrification correctly the model of seamless mobility could be realized which has the potential to allow for 30% more traffic and a 10% travel time reduction. With seamless mobility private cars and robotaxis could

Fig. 4 Multi-modal transportation concept in New Urbanism Communities—Seaside, Florida

provide for 30% of passenger miles in 2030 compared to 40% provided by private cars today.

Mobility service providers such as Uber or Lyft provide an attractive alternative of owning a car in particular in dense urban environments but it competes with public transportation and it adds to traffic congestion (see Fig. 5).

What Uber and Lyft do well is to match customers with mobility demands, with drivers that supply the transportation tool. The Schaller Report [6] makes an interesting statement that Transportation Network Companies doubled in size since 2012 and provide a major source of urban transportation to surpass local bus ridership by end of 2018. A mobility-as-a-service (MaaS) ecosystem takes all transportation options into consideration and connects them in an intelligent way to meet the transportation demand at all times in the most efficient way. Such an ecosystem works much better if data is being shared between the different mobility service providers. Blockchain/Distributed Ledge Technology (DLT) is an interesting technology to be used in context of matching buyers and sellers of mobility services. In [7] the different options are being described how Blockchain/DLT models can be utilized for mobility-as-a-service platforms. What needs to be pointed out that latency of data transactions and scalability of the platform are key for successful deployment of MaaS platforms and here the practical experience with Blockchain/DLT models is limited. The interaction between public transportation agencies and private mobility service operators as well as the interaction among private mobility service operators in terms of data sharing could be organized principally via scalable Blockchain/DLT platforms. Conceptually this has been demonstrated by multiple start-ups but it requires adoption

Fig. 5 Dualism of public and private transportation on shared roads in city of Zurich, Switzerland

from large mobility service operators to support Blockchain/DLT protocols which in return requires standardization via organizations such as the Society of Automotive Engineers or the Institute of Electrical and Electronics Engineers.

Mobility-as-a-service ecosystems can be developed both around people transportation as well as around goods transportation. A key problem in goods transportation is often last mile delivery as well as the optimal allocation of transportation resources. Due to the competition between package delivery services such as Fedex or Amazon which optimize their operations individually, it is questionable whether delivery vehicles are used in an optimal way and whether they could be organized in a mobility-as-a-service ecosystem to use less energy and transport the goods more efficiently in terms of time or cost.

In general there is a significant change to be observed in shopping behavior towards e-commerce. As described in [8] online sales in the US averaged more than 15% year-over-year since 2010, Amazon being the primary driver of this growth. It has been shown in the example of Seattle that geocoding of private truck load/unload bays and loading docks needs to be complemented by public truck load/unload spaces which can be shared via the city curb data layer. In [9] it projected that 80% of parcel deliveries will ultimately be done by autonomous vehicles (with locker capabilities). In [10] it is described how curbside zones can be used in a flexible manner and how the usage can be priced. It is also outlined how users of the street and vehicles generate data and how this data can be shared via data-driven policies.

An interesting aspect in context of the utilization of blockchain/DLT for mobility services is the tokenization of the transportation economy. What this basically means is to eliminate traditional car ownership and to replace it with a usage right for a variety of mobility services which can be linked to a set of constraints and which has a specific monetary value. Reference [11] describes how a token-based mobility ecosystem can be organized which ultimately results in a solution where different mobility services can be utilized under a common approach of user interaction and payment considering transportation route optimization in a multi-modal mobility system from point to point.

4 Sharing Data to Validate and Certify Highly Automated Mobility Services

The use of highly automated mobility services will change the dependency of the driver on paying attention to the road and it also provides the option for the future to live without the need of owning a car. A key problem for a vehicle driver is to anticipate the intent of other vehicles, bicyclists or pedestrians correctly even under difficult environmental circumstances. In order to understand human driver behavior patterns, intensive data analysis is needed and the collected data can be shared among the automotive safety systems development community. On the other hand, a huge variety of driving scenarios needs to be validated for highly automated vehicles to understand potential reaction patterns of the engaging actors—whether machine-based or human. This can be done both in a virtual as well as in a physical environment as well as in mixed reality configurations as shown in Figs. 6, 7 and 8.

The current approach of many OEMs as well as start-up companies is to develop their own test scenarios for automated driving and to NOT share test data as they are in a competitive environment. Larger OEMs and Tier 1 suppliers test their highly

Fig. 6 SUNTRAX—dedicated automated vehicle physical proving ground in Florida

Fig. 7 Virtual smart city development in China including transportation layout

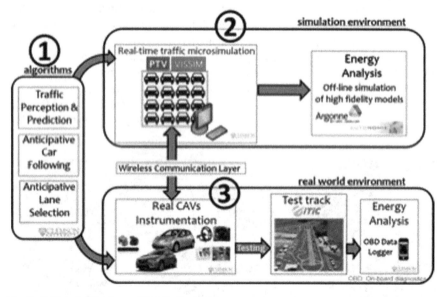

Fig. 8 Vehicle-In-The-Loop (VIL) Testbed—Clemson University and International Transportation Innovation Center (ITIC)

automated driving systems typically in their own testbeds, however the underlying driving scenarios being used may differ. Regulators are very much interested in standardized system behavior in standardized driving scenarios which needs to be demonstrated in certified testbeds designed as shared facilities and in shared test data to be used to validate safety and security relevant aspects.

Universities and research institutions play an important role to demonstrate the power of new algorithmic methods or sensor technologies in testbeds and their data is typically shared in the research community.

From an Automotive OEM perspective, safety and security regulations lead to higher product and service cost (similar to emission regulations). If those regulations need to be met by all OEMs there is first no competitive advantage linked to it unless benchmark tests organized by independent test organizations allow a quality rating (as for example applied in crash related tests). High quality ratings can be utilized by the OEMs from a sales and marketing perspective which may lead to higher sales of products or services. Communicating unbiased benchmark data to consumers is an effective instrument to channel market demand towards products that achieve the best results but at the same time compliance to standards guarantee that only products and services are approved to be offered on the market that fulfill a minimum set of safety and security requirements.

The next step in safety evolution is to enhance the "vision" of highly automated vehicles by sharing sensor data in swarm networks [12] which basically requires a V2X approach where data is being communicated via standardized communication interfaces. As latency needs to be very low and bandwidth very high, suitable network technologies such as DSRC or 5G need to be utilized. Blockchain/DLT technologies can help to share the data among different OEMs and to avoid data manipulation.

5 Sharing Data to Optimize Mobility Related Infrastructure Usage

Road capacity is usually limited—in particular at peak hours. Thus sharing data regarding travel routes among vehicles can help reduce traffic congestion. Another limitation can be the capacity of electric vehicle charging infrastructure which would require that travel routes as well as the information about charging needs of an electrified fleet have to be shared with the infrastructure operators.

In particular in the context of highly automated driving, it is important to share data about the condition and the capability of the mobility-related infrastructure. Figure 9 shows the different mobility infrastructure configurations in a closed testbed which can be used to study the impact of data sharing. It is very likely in the future that roads and even road lanes need to be classified according to their capability to support highly automated driving (e.g. high resolution map information, supporting data, access to high performance edge and cloud computing). The different data sharing V2X configurations need to be tested in cyberphysical testbeds for optimal

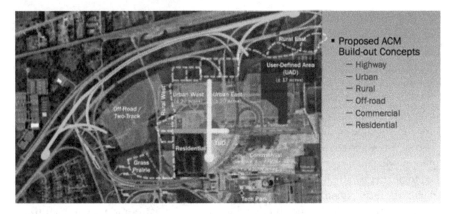

Fig. 9 Infrastructure configurations of American Center of Mobility (ACM)

performance under different environmental conditions as well as different traffic load configurations. What is also highly relevant to understand is the influence of non-automated vehicles on the operation of automated vehicles. Non-automated vehicles are typically connected and can receive information from automated vehicles. But as the percentage of automated vehicles will increase over time, the reaction patterns in specific driving scenarios may change.

In the future road infrastructure layouts can be flexible and sharing data between the different mobility service providers can help utilize the infrastructure in a more efficient way.

6 Societal Impact of Data Sharing—Example Mobility

Data sharing can increase road safety as well as can prevent or fight cybersecurity attacks and road rage or other forms of physical vehicle related attacks. It can improve the traffic flow and reduce travel times and it can also reduce the use of energy. It can furthermore help to protect pedestrians as well as bicycle drivers and contribute to more livable cities and communities.

Both users as well as operators of mobility services need to be aware of the fact that data sharing has a major impact on safety, security, efficiency and sustainability on transportation. Lack of data sharing can lead to more traffic congestion, more use of energy, avoidable accidents and more crime. On the other hand full data transparency can lead to a lack of privacy as well as exposure of critical operational knowledge to competitors. Regulators play an important role to establish the rules of a fair data-driven mobility economy which on one hand protects the users and general citizens and on the other hand stimulates economic growth and prosperity.

7 Technical and Non Technical Hurdles to Overcome with Data Sharing

From a technical perspective we need to improve the ways data can be transmitted in real time between vehicles of different makes in particular in vehicles swarms. Such data sharing might become mandatory when technically feasible at reasonable cost to improve the safety and security level of highly automated driving. The chosen technology to distribute safety and security relevant data in vehicle swarms needs to be scalable to very large numbers of nodes as well as low latency.

We need to consider the legal aspect of data sharing which starts with the question who actually owns the data in a vehicle. If the vehicle is privately owned, you need to deal with the privacy rights of the vehicle owner. If the vehicle is owned by a commercial fleet operator, there could be rules between the operator and the city regarding the data that is being shared to reach a win-win situation.

The German Association of the Automotive Industry developed a data sharing concept for vehicle generated data [13] that can be utilized to create a data market for third parties. The core idea is to share data via neutral servers in the cloud which support a standardized data communication interface with the OEM vehicle data cloud. The vehicle data itself is being transferred from he vehicle to the cloud under full control of the OEM. This approach eliminates security risks however, existing V2V interfaces as well as access to repair and maintenance data via the OBDII port will still be supported by the OEM.

8 Conclusions

Data sharing is a difficult topic for OEMs. On the one end sharing data can lead to higher level of product liability exposure as the shared data can be validated by third party analysis. On the other hand data sharing can contribute to a substantial more efficient operation of mobility-as-a-service platforms which could be organized as a public private partnership effort. The regulators in close collaboration with standards development organizations as well as insurance companies need to define clear guidelines which data needs to be shared on a mandatory base—and still addressing privacy concerns. From an R&D perspective independent test and certification companies need to ensure that sufficient data is being shared on a voluntary collaborative basis by OEMs, mobility service operators and infrastructure developers/operators to validate and certify mobility services in mobility-as-a-service platforms. This can be encouraged by a consortium approach where stakeholders can communicate more freely in a protected space and with the focus on developing solutions for identified problems.

References

1. McKinsey center for future mobility – Race 2050 – A vision for the European automotive industry (2019)
2. http://apollo.auto/
3. http://apolloscape.auto/
4. Rethinking transportation 2020–2030, A Rethink X sector disruption report, James Arbib and Tony Seba (2017)
5. McKinsey center for future mobility – An integrated perspective on the future of mobility, part 3: setting the direction towards seamless mobility (2019)
6. The new automobility: lyft, uber and the future of American cities, Schaller consulting (2018)
7. Blockchain and beyond encoding 21st century transport. Corporate partnership board report, ITF/OECD (2018)
8. The final 50 feet urban goods delivery system, Research scan and data collection project. Seattle Department of Transportation (2018)
9. Parcel delivery – The future of last mile, McKinsey, travel, transport and logistics (2016)
10. Blueprint for autonomous urbanism, module 1. NACTO, Fall (2017)
11. iomob – Internet of mobility – www.iomob.net, Whitepaper (2018)
12. The blockchain: a new framework for robotic swarm systems, MIT Media Lab (2017)
13. Access to vehicle and vehicle generated data –"NEVADA share and secure concept", VDA (2017)

On Parsing Shared Information: An Application from the Connected Car Domain

Rodrigo Ordóñez-Hurtado, Giovanni Russo, Sam Sinnott and Robert Shorten

Abstract Vehicles are becoming connected entities. As a result, a possible scenario is that such entities might be literally bombarded with information from a multitude of devices. In this context, a key challenging requirement for connected vehicles is that they will need to become *cognitive* bodies, able to *parse* information and use only the pieces of information that are relevant to the driver in the context of a given journey. Motivated by this scenario, we propose in this chapter a *parsing engine*, a collaborative service for connected vehicles. The service predicts the likely route/destination of the driver and checks if any obstruction occurs along the likely route. If this happens, recommendations are provided. The recommendations have the goal of regulating the flow of vehicles along the obstructed route (so as not to exceed a given maximum capacity) and to balance the re-routed flow along a set of alternative routes (so that there are no alternative routes that are overloaded). We validate the service via a mixture of numerical and Hardware-in-the-Loop simulations.

---

[1] See www.iota.org for an example of a data market place that monetizes shared data.

R. Ordóñez-Hurtado (✉)
IBM Research Ireland, Dublin, Ireland
e-mail: rodrigo.ordonez.hurtado@ibm.com

G. Russo
Department of Information & Electrical Engineering and Applied Mathematics,
University of Salerno, Fisciano, Salerno, Italy
e-mail: giovarusso@unisa.it

S. Sinnott
University College Dublin, Dublin, Ireland
e-mail: sam.sinnott@ucd.ie

R. Shorten
Dyson School of Design Engineering, Imperial College London, London, UK
e-mail: r.shorten@imperial.ac.uk

© Springer Nature Switzerland AG 2020
E. Crisostomi et al. (eds.), *Analytics for the Sharing Economy:
Mathematics, Engineering and Business Perspectives*,
https://doi.org/10.1007/978-3-030-35032-1_13

1 Introduction

A common use-case to illustrate the commercial importance of the sharing economy concerns the making available and use of shared data.[1] One of the most prominent applications of the use of shared data arises in the connected car space. Already many companies have emerged that attempt to monetize such data; Waze[2] is a very good example of one such company, and there are already many others. As vehicles become more and more connected, applications based on sharing of data are likely to become ubiquitous [7, 10]. It is in this context that the main idea of this present chapter (built upon the preliminary results obtained in [8]) is inspired. Specifically, in order to enable certain services, connected vehicles are likely to be literally *bombarded* in the near future with information from a multitude of devices. In this scenario, it is then of utmost importance that vehicles are able to filter or *parse* this information, so that only the relevant information to the journey is passed to the driver. Motivated by this observation, we propose in this chapter a *parsing engine* for connected vehicles and illustrate how it can be used in the context of congestion management. Specifically, the main contributions of this chapter can be summarized as follows:

- We first introduce a set of algorithms, which are able to predict the most likely route/destination of the driver and, based on this, check if any obstruction is occurring along the route;
- We build upon these algorithms and, in the case where an obstruction is detected, an alternative route is recommended to the driver. In turn, the recommended route is computed via a distributed algorithm, having the goal of regulating the flow of vehicles along the obstructed route (so as not to exceed a given maximum capacity) and to balance the re-routed flow along a set of alternative routes (so that there are no alternative routes that are overloaded);
- Finally, in addition to the use of comprehensive conventional numerical and SUMO-based simulations (similar in nature to those presented in [8]) to test and validate our service, we now make use of the Hardware-in-the-Loop (HIL) platform originally presented in [3]. In particular, in order to test our service, we develop new capabilities of the HIL, so that it can now interface with external IoT objects and Social Media (Twitter®). The use of the HIL platform allows us to embed a few real vehicles into scenarios created using the microscopic traffic simulation package SUMO [5]. As a result, the real car (onto which our algorithm is deployed) is able to interact with simulated vehicles and, in this way, the large scale effects of our service can be effectively validated, as well as driver acceptability. Consequently, during the tests, the driver of the real vehicle experiences how it would actually *feel* to be part of a collaborative service.

The rest of the chapter is organized as follows. In Sect. 2, we introduce the reference scenario for the proposed service. Section 3 details the service presented in this chapter, while the experimental validation is illustrated in Sect. 4. Final remarks are given in Sect. 5.

---

[2]www.waze.com.

2 The Reference Scenario

Consider the scenario where a set of $V > 1$ connected vehicles is traveling within a given geographic area. Each of the vehicles is currently traveling along some route to reach its destination.

At some point, one road becomes partially obstructed (or, simply, obstructed in what follows). In the context of this chapter, an obstruction is an event that affects (i.e. decreases) the capacity of a given road. Obstructions might be due to unexpected events, such as accidents, or to scheduled events, such as road maintenance. The obstruction (or event in what follows) impacts the vehicles whose route would pass through the obstructed road. As a result, the driver of those vehicles would experience a disruption to his/her trip.

The overall goal of this chapter is to address the ubiquitous scenario discussed above. The service we propose to handle this scenario consists of two main macro components: an in-car system and a remote service. Essentially, the in-car system for the ith vehicle can sense environmental information (i.e. the presence of obstructions) and is able to predict the most likely destination and route followed by the driver. Based on this, the in-car system parses the environmental information to check whether an obstruction will impact the trip for the driver. If this is the case, then a notification is sent to a remote service. The remote service is able to monitor the capacity of the road links within the geographic area of interest and, based on this, is able to provide alternative routes to the vehicles that are going to experience the obstruction. In particular, the recommendations are computed via a feedback loop mechanism, which: (i) provides to all vehicles an opportunity to access their preferred route; (ii) balances the vehicles that need to be re-routed across adjacent routes; and (iii) regulates the flow of vehicles along an obstructed road so that the reduced capacity of that road is not exceeded.

Finally, we remark that we consider two classes of road obstructions. Namely:

- *Irregular obstructions*, which typically do not occur at the same location/time and thus can not be predicted. Examples of these are lane closures due to road works or accidents.
- *Regular obstructions*, which are periodic/systematic and thus can be predicted, and resources can be regularly adjusted in advance for the duration of the obstruction. Examples of these are the traffic buildups at opening/closing times of a school or an event at an entertainment venue.

2.1 Notation

We now introduce the notation that will be used in the rest of the chapter. In the context of this chapter, the road network onto which the connected vehicles travel is characterized by a weighted directed graph, $\mathcal{G}_T$, and the associated adjacency (or connectivity) matrix, G_T [2]. The nodes of this graph are specific road locations such as intersections and waypoints. The links of the graph physically represent roads connecting the nodes. A link from node l to node m of $\mathcal{G}_T$ is denoted by (l, m). Also, we denote by $\mathcal{N}^{in}_{(l,m)}$ the set of edges of $\mathcal{G}_T$ which point to edge (l, m) (i.e. pointing to node l) and by $\mathcal{N}^{out}_{(l,m)}$ the set of edges starting from edge (l, m) (i.e. departing from node m). The occupancy limit associated with the link (l, m) of $\mathcal{G}_T$ is denoted by $c_{(l,m)}$ and it represents the maximum capacity for that link, i.e. the maximum number of vehicles that can use that link simultaneously.

Each vehicle is characterized by its history of past trips. This historical information associated with the ith connected vehicle is formalized in terms of a weighted directed graph, $\mathcal{G}^{(i)}$, and the associated connectivity matrices, $G^{(i)}$. The nodes and the edges of $\mathcal{G}^{(i)}$ are a subset of the nodes and edges of $\mathcal{G}_T$. In particular, a node/edge of $\mathcal{G}_T$ belongs to $\mathcal{G}^{(i)}$ if that node/edge has been used in the past by the ith vehicle. The weight associated with (l, m) is denoted by $w^{(i)}_{(l,m)}$ and is defined as the number of past trips for vehicle i that included link (l, m).

An obstruction is characterized by the road where the event occurred and by the corresponding maximum capacity for that road. That is, an obstruction is characterized by the tuple $[(l, m), c_{(l,m)}]$, where (l, m) is the link (i.e. the road) where the obstruction occurred and $c_{(l,m)}$ is the reduced capacity of that road due to the obstruction.

Finally, let $k \in \mathbb{Z}$, we denote by $x_{(l,m)}(k)$ the number of vehicles on link (l, m) at time k. We note that the variable $x_{(l,m)}(k)$ is associated with the current state of the road network. If an obstruction occurs within the road network, then we also associate a state variable with the ith vehicle. Specifically, $y_i(k)$ is a binary variable that takes value 1 if the ith vehicle is granted access to the obstructed link at time k and 0 otherwise. We also denote by $\bar{y}_i(k)$ the average access of the ith vehicle to the obstructed road up to time k. That is, $\bar{y}_i(k) := \frac{1}{k+1} \sum_{j=0}^{k} y_i(j)$.

3 The Service

The high level architecture of the service presented in this chapter is outlined in Fig. 1. As shown in such a figure, the architecture consists of a local component residing inside the connected vehicle (i.e. the in-car system), and of a global/remote component which has access to data regarding the geographic area of interest and which might be implemented via cloud technologies (i.e. the remote service). The modules composing the in-car system have the goal of predicting route and destination for the driver, and of detecting whether an obstruction is occurring along the predicted route. The remote service, instead, has the goal of computing the alternative routes. We now describe the key functions and algorithms of the modules in Fig. 1.

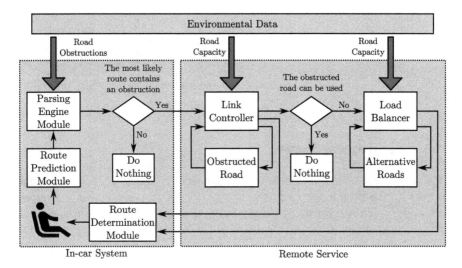

Fig. 1 Architecture of the service presented in this chapter. In the figure, both the key modules and their functional relations are shown

3.1 Route Prediction Module

The goal of the Route Prediction Module is to predict the next roads that will most likely be taken by the driver. That is, the module attempts to predict the route that the vehicle will take in the near future. This route is then used by the Route Parsing Engine described in Sect. 3.2 to detect whether an obstruction is going to affect the trip. This module gathers information from past trips of the host vehicle to create a picture of where the user typically goes along a journey, and by integrating information from the current trip the prediction is further refined.

From the functional viewpoint, the route prediction module can be broken down into three sub-modules: (1) the *Edge Ranking Routine*, (2) the *Edge Weighting Routine*, and (3) the *Prediction Routine*.

3.1.1 Edge Ranking Routine

The goal of this routine is to characterize the "importance" of a given route from a topological viewpoint. This is done by ranking the roads (and hence the directions) taken by a driver during a journey. Thus, the *Edge Ranking Routine* takes as input the topology of $\mathcal{G}_T$ and applies a modified version of the PageRank algorithm to rank its edges, rather than its nodes [6]. Specifically, let $r_{(l,m)}$ denote the ranking of the edge (l, m). Then, $r_{(l,m)}$ is defined by

$$r_{(l,m)} := \left((1-d) + d \sum_{(p,q) \in \mathcal{N}^{in}_{(l,m)}} r_{(p,q)} \right), \qquad (1)$$

where d is a damping factor which, for web applications, was originally set to 0.85 [1]. In our study, inspired by [9], we set this factor to 0.93.

3.1.2 Edge Weighting Routine

This routine is in charge of computing the frequency of use of a given link for the ith vehicle. The input to this routine is the set of weights associated with $\mathcal{G}^{(i)}$ (recall that such a graph is constructed starting from the travel history of the ith vehicle). The frequency of use of the edge (l, m) of $\mathcal{G}^{(i)}$ is denoted by $f^{(i)}_{(l,m)}$ and is computed as

$$f^{(i)}_{(l,m)} := \frac{w^{(i)}_{(l,m)}}{\sum_{(p,q) \in \mathcal{N}^{in}_{(l,m)}} w^{(i)}_{(p,q)}}. \qquad (2)$$

Note that, physically, $f^{(i)}_{(l,m)}$ is the ratio between the number of times the ith vehicle used the link (l, m) in the past and the number of past trips that passed through node l.

3.1.3 Route Prediction Routine

Finally, this routine takes the edge rankings, r's, and the frequencies, $f^{(i)}$'s, as inputs, together with the current position of the car. The output is a sequence of links that are most likely going to be visited by the vehicle. Let (l, m) be the link (i.e. the road segment) onto which the car is currently traveling. We define $tr^{(i)}(l, W)$ as the *tree* of width W (i.e. the length of the prediction horizon) generated from the node l of the graph $\mathcal{G}^{(i)}$ and we denote by $br^{(i)}(j)$ the jth branch of such a tree. A score is then associated with each of the branches of $tr^{(i)}(l, W)$. Namely, the score associated with the jth branch of $tr^{(i)}(l, W)$ for the ith vehicle is given by

$$R^{(i)}_j := \sum_{(m,n) \in br^{(i)}(j)} f^{(i)}_{(m,n)} r_{(m,n)}. \qquad (3)$$

The predicted route for the ith vehicle is then the one that corresponds to the branch having the maximum $R^{(i)}_j$. Note that the prediction horizon W is a design parameter which affects the computational complexity of the route prediction algorithm. In the validations of this chapter illustrated in Sect. 4, we set $W = 5$. The choice of such a value offered a good compromise between computational efficiency and prediction accuracy. The macro steps for route prediction are given in Algorithm 1.

Remark: Note that the use of (1) with $d \neq 1$ is chosen for reasons of numerical robustness and to guarantee that all the involved edges (p, q) have a ranking score $r_{(p,q)}$ different than zero so that sink edges are avoided. Note that the associated transition matrix is not used for the predictions of each vehicle i but rather the connectivity matrices $G^{(i)}$, and thus every analyzed branch (and in turn the final prediction) is always feasible.

Algorithm 1 Route Prediction Algorithm

1: $(l, m) \leftarrow$ current link on which the vehicle i is traveling
2: $tr^{(i)}(l, W) \leftarrow$ tree of width W rooted from node l
3: **for** each edge (p, q) in $tr^{(i)}(l, W)$ **do**
4:     Compute $r_{(p,q)}$ using (1)
5:     Compute $f_{(p,q)}^{(i)}$ using (2)
6: **end for**
7: **for** each branch $br^{(i)}(j)$ in $tr^{(i)}(l, W)$ **do**
8:     Compute $R_j^{(i)}$ using (3)
9: **end for**
10: **return** $S^{(i)} \leftarrow$ sequence of W links with largest $R_j^{(i)}$

3.2 Route Parsing Engine

The objective of the *Route Parsing Engine* is to parse information along the sequence of links, $S^{(i)}$, predicted by the Route Prediction Module of Sect. 3.1 (see Algorithm 1). Specifically, in the context of this chapter, the goal of this module is to check, in case an obstruction occurs, if the link affected by the obstruction belongs to the sequence $S^{(i)}$. Recall that an obstruction is characterized by the tuple $[(l, m), c_{(l,m)}]$. This module then checks whether the obstructed link (l, m) belongs to $S^{(i)}$. If this is the case, then a notification is sent to the *Link Controller* of Sect. 3.3, which, in turn, will regulate the flow of vehicles passing through the obstructed link (l, m). The key conceptual steps for this module are reported in Algorithm 2. Note that this algorithm returns a binary variable $C^{(i)}$ which can take value 0 if the obstruction does not belong to the sequence $S^{(i)}$ or it can take value 1 if the obstruction belongs to $S^{(i)}$. The variable $C^{(i)}$ is taken as input by the Link Controller. In particular, if $C^{(i)} = 1$, then the Link Controller is activated.

3.3 Link Controller

This module is invoked whenever an obstruction is detected, i.e. whenever the output variable, $C^{(i)}$, of the Route Parsing Engine is equal to 1. We assume that the link

Algorithm 2 Route Parsing Engine Algorithm

1: Compute $S^{(i)}$ from *Route Prediction Module* (Algorithm 1)
2: Gather environmental data
3: $C^{(i)} \leftarrow 0$
4: **if** obstruction $[(l, m), c_{(l,m)}]$ is detected **then**
5:    **if** $(l, m) \in S$ **then**
6:       $C^{(i)} \leftarrow 1$
7:    **end if**
8: **end if**
9: **return** $C^{(i)}$
---

affected by the obstruction is (l, m) and we build a mechanism around such a link to regulate access of vehicles to (l, m), so that $c_{(l,m)}$ is not exceeded. The output of this module is the probability for the vehicles demanding allocation to link (l, m). We propose here two Link Controllers: the first is for irregular obstructions, while the second is for regular obstructions. The use of two different controllers for such cases is motivated by the fact these two types of obstructions are fundamentally different in nature from each other. In particular, we can leverage the predictability of regular obstructions to design more tailored algorithms.

3.3.1 Allocation Through Irregular Obstructions

We start by considering the case of an irregular obstruction along the link (l, m). The goal is to regulate access to this link so that $x_{(l,m)}(k) \leq c_{(l,m)}$, $\forall k$. In order to do so, the link controller returns a signal, $P_{(l,m)}(k)$, to the *Route Determination Module* of the in-car system of the vehicles demanding allocation. This signal physically represents the probability for a vehicle to be allocated to (l, m). Such a probability is computed as

$$P_{(l,m)}(k) := \begin{cases} 0, & \text{if } e_{(l,m)}(k) \leq 0, \\ \frac{e_{(l,m)}(k)}{c_{(l,m)}}, & \text{if } 0 < e_{(l,m)}(k) \leq c_{(l,m)}. \end{cases} \quad (4)$$

where $e_{(l,m)}(k) = c_{(l,m)} - x_{(l,m)}(k)$.

3.3.2 Allocation Through Regular Obstructions

Now, we analyze the case of regular obstructions. As we describe below, the predictability of those obstructions can be leveraged to have a more tailored design, which e.g. includes balancing, over time, the average access of drivers to the obstructed road. In particular, the closed loop mechanism we design, which offers a generalization of the one proposed in [4], allows to regulate access of vehicles to the obstructed route in a way such that, on average in the long run, the marginal cost of each of the vehicles associated with not using the obstructed link is the same. Moreover, the marginal cost for each of the vehicles will be dependent on the available capacity along the obstructed link. In the case of regular obstructions, the link controller calculates a time-varying coefficient, $\gamma_{(l,m)}$, which is broadcast to the all in-car system of the vehicles that are likely going to pass through the obstructed link (l, m). The coefficient $\gamma_{(l,m)}$ is calculated as follows:

$$\gamma_{(l,m)}(k+1) = \alpha \gamma_{(l,m)}(k) + (1-\alpha) g\left(e_{(l,m)}(k)\right), \tag{5}$$

where, $0 < \alpha < 1$ is a design parameter, $e_{(l,m)}(k) = c_{(l,m)} - x_{(l,m)}(k)$, and the function $g : \mathbb{R} \to [0, 1]$ is a piece-wise continuous function defined similarly to Eq. 4, this is, $g\left(e_{(l,m)}(k)\right) = \max\left(0, \frac{e_{(l,m)}(k)}{c_{(l,m)}}\right)$. Clearly, the dynamical system in (5) is a stable discrete-time dynamical system and it converges towards the value $\bar{g} = g(e_{(l,m)})$.

The macro steps for the Link Controller are given in Algorithm 3.

3.4 Load Balancer

The load balancer is responsible for calculating a set of probabilities associated with links which are alternative to the obstructed link (l, m). We assume that, for each link of $\mathcal{G}_T$, this module has built-in a set of $N_{(l,m)}$ alternative links. Let $\mathcal{A}_{(l,m)} = \left\{(l, m)_1, \ldots, (l, m)_{N_{(l,m)}}\right\}$ be the set of links alternative to link (l, m). We denote the cardinality of $\mathcal{A}_{(l,m)}$ by $|\mathcal{A}_{(l,m)}|$, i.e. $|\mathcal{A}_{(l,m)}| = N_{(l,m)}$. We build a feedback loop which, based on the state of the links of $\mathcal{G}_T$, reallocates the vehicles not granted to use link (l, m) to one of the alternatives routes in $(l, m)_j$, $j = 1, \ldots, N_{(l,m)}$. Specifically, the output of this module is a set of time-varying probabilities, say $\left\{P_{(l,m)_j}(k)\right\}_{j=1,\ldots,N_{(l,m)}}$, associated with each of the alternative links in $\mathcal{A}_{(l,m)}$.

We start by considering the case when the number of vehicles along the alternative link $(l, m)_j$ is different from 0, $\forall j = 1, \ldots, N_{(l,m)}$. That is, we start considering the case where $x_{(l,m)_j}(k) > 0$, $\forall j = 1, \ldots, N_{(l,m)}$. In this case, the probabilities

---

Algorithm 3 Link Controller Algorithm

1: **if** $C^{(i)}$ is 1 **then**
2:     $c_{(l,m)} \leftarrow$ capacity of obstructed link
3:     $x_{(l,m)}(k) \leftarrow$ number of cars on the obstructed link
4:     **if** Irregular obstruction **then**
5:        Compute $P_{(l,m)}(k)$ using (4)
6:        **return** $P_{(l,m)}(k)$
7:     **else**
8:        Compute $\gamma_{(l,m)}(k)$ using (5)
9:        **return** $\gamma_{(l,m)}(k)$
10:    **end if**
11: **end if**

associated by the Load Balancer module for the possible $N(l, m)$ alternatives to the obstruction at the link (l, m) are given by

$$P_{(l,m)_j}(k) = \frac{1/h_j(k)}{\sum_{q=0}^{N_{(l,m)}} (1/h_q(k))},$$
$$h_j(k) = \frac{x_{(l,m)_j}(k)}{\sum_{q=0}^{N_{(l,m)}} x_{(l,m)_q}(k)},$$
(6)

where, physically, $h_j(k)$ can be seen as the time-varying portion of the total load of vehicles which is passing through the alternative links $(l, m)_j$. Note that, in essence, the vector $P_{(l,m)_j}(k)$ is the same for all the vehicles forced to reroute at instant k, and it only depends on the current state of link $(l, m)_j$.

Instead, in the case when there are some alternative links having no vehicles, i.e. when there exists some $(l, m)_j \in \mathcal{A}_{(l,m)}$ such that $x_{(l,m)_j}(k) = 0$, then the probabilities are computed as

$$P_{(l,m)_j}(k) = \begin{cases} \frac{1}{N^{(0)}_{(l,m)}(k)}, & \forall (l, m)_j : x_{(l,m)_j}(k) = 0, \\ 0, & \text{otherwise}, \end{cases}$$
(7)

where $N^{(0)}_{(l,m)}(k)$ is the number of links alternative to (l, m) having no vehicles at time k. That is, $N^{(0)}_{(l,m)}(k) := |(l, m)_j \in \mathcal{A}_{(l,m)} : x_{(l,m)_j}(k) = 0|$.

The value $P_{(l,m)_j}$ obtained either via (6) or (7) is finally broadcast to the ith vehicle, for which the route predicted by the Route Prediction Module is affected by the obstruction. Specifically, such values are received by the *Route Determination Module* described below. The macro steps for the Link Controller are given in Algorithm 4.

Algorithm 4 Load Balancer Algorithm

1: $(l, m) \leftarrow$ obstructed road link
2: $\{(l, m)_1, \ldots, (l, m)_{N_{(l,m)}}\} \leftarrow N_{(l,m)}$ alternative links to (l, m)
3: **for** $j \in 1, \ldots, N_{(l,m)}$ **do**
4:    Compute $x_{(l,m)_j}(k)$
5:    **if** $x_{(l,m)_j}(k) > 0$ for all alternative links **then**
6:       Compute $P_{(l,m)_j}(k)$ using (6)
7:    **else**
8:       Compute $P_{(l,m)_j}(k)$ using (7)
9:    **end if**
10: **end for**
11: **return** vector $\{P_{(l,m)_j}(k)\}_{N_{(l,m)}}^{j=1}$

3.5 Route Determination Module

The Route Determination Module is responsible to determine, given the outputs of the Link Controller and of the Load Balancer, which route will be taken by the vehicle that is predicted to be affected by the obstruction along the link (l, m). For the sake of clarity, we separately discuss the case of irregular obstructions and the case of regular obstructions.

3.5.1 Route Determination for an Irregular Obstruction

In this case, the Route Determination Module of the ith vehicle affected by the obstruction receives as input the $[N_{(l,m)} + 1]$-dimensional vector of probabilities computed by the Link Controller and by the Load Balancer

$$\left[P_{(l,m)}(k), \left\{ P_{(l,m)_j}(k) \right\}_{N_{(l,m)}}^{j=1} \right]. \tag{8}$$

Essentially, the road to which the ith vehicle is allocated is determined by flipping a coin against the probabilities in (8). Specifically:

- First, a coin is flipped against the probability $P_{allocation}^{(i)} = P_{(l,m)}$. If this coin toss is successful, then the ith vehicle is allocated to the obstructed link (l, m).
- If the coin toss is not successful, then the ith vehicle is not granted access to the link (l, m) and the vehicle is allocated to one of the alternative links. In order to allocate the vehicle to one alternative link, a coin is again flipped, this time against the probabilities $\left\{ P_{(l,m)_j}(k) \right\}_{N_{(l,m)}}^{j=1}$.

3.5.2 Route Determination for a Regular Obstruction

In case of a regular obstruction, the Route Determination Module receives as input the $[N_{(l,m)} + 1]$-dimensional vector

$$\left[\gamma_{(l,m)}(k), \left\{ P_{(l,m)_j}(k) \right\}_{N_{(l,m)}}^{j=1} \right]. \tag{9}$$

The first step is then to devise the probability $P_{allocation}^{(i)}$ associated with the ith vehicle to be allocated the link affected by the regular obstruction (as $\gamma_{(l,m)}$ is not indeed a probability). Thus, in order to do so, and inspired by [4], we let:

$$P_{allocation}^{(i)} = \gamma_{(l,m)}(k) H_i(k). \tag{10}$$

In the above equation, the dynamics $H_i(k)$ is specific of the ith vehicle and is given by

$$H_i(k) = \frac{\bar{y}_i(k)}{\phi_i(\bar{y}_i(k))}, \quad (11)$$

where (i) $\bar{y}_i$ is the average value of allocation of resources for user i, y_i, and (ii) $\phi_i(\cdot)$ is a strictly increasing function (on the domain of interest). Once $P_{allocation}^{(i)}$ is obtained from (10), then the coin toss mechanism described in the case of irregular obstructions is applied. The macro steps for the Route Determination module are given in Algorithm 5.

Algorithm 5 Route Determination Algorithm

1: Compute $\{P_{(l,m)_j}(k)\}_{N_{(l,m)}}^{j=1}$ from Load Balancer
2: **if** Regular Obstruction **then**
3:    Compute $\gamma_{(l,m)}(k)$ from Link Controller
4:    Compute $P_{allocation}^{(i)}$ using (10) and (11)
5: **else**
6:    Compute $P_{(l,m)}(k)$ from Link Controller
7:    Set $P_{allocation}^{(i)} = P_{(l,m)}(k)$
8: **end if**
9: Coin Toss against $P_{allocation}^{(i)}$
10: **if** Coin Toss successful **then**
11:    Grant access of the obstructed link (l, m) to vehicle i
12: **else**
13:    Coin toss against $\{P_{(l,m)_j}(k)\}_{N_{(l,m)}}^{j=1}$
14:    Grant access of vehicle i to the alternative link for which the coin toss is successful
15: **end if**
16: **return** link to be used by the ith vechicle

---

4 Experimental Validation

We now validate the service presented in Sect. 3 for both regular and irregular obstructions. For this, we use the open-source, microscopic, traffic simulator SUMO [5], and perform both numerical and HIL simulations. In particular, a number of Python scripts were developed to interact online with the SUMO environment (the Python scripts make use of the *Traffic Control Interface* (TraCI) package[3]). In order to set up the simulations, in SUMO the following two layers need to be defined: (i) the *Network Layer*, defining the geographic area of the simulation; (ii) the *Vehicular Layer*, defining the vehicles entering in the simulation.

In all the simulations, for the Network Layer, we imported, via OpenStreet Map, the area between Drimnagh and Kimmage, Dublin (see Fig. 2). For the Vehicular Layer, we defined a single vehicle type with the following parameters: (i) Length:

---

[3] http://sumo.dlr.de/wiki/TraCI

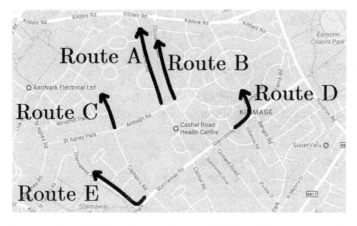

Fig. 2 Selected area between Drimnagh and Kimmage, Dublin, and main routes used to study the case of irregular obstructions

5 m; (ii) Acceleration: 0.8 m^2/s; (iii) Deceleration: 4.5 m^2/s; (iv) Car-following model: Krauss. Additionally, we generated 481 possible routes within the area of Fig. 2 and created a random number of vehicles using these routes.

4.1 Irregular Obstructions

Irregular obstructions were simulated in SUMO as a road link with a reduced speed limit. The reduced speed limit implies a reduction of the maximum link capacity. In order to validate the service for irregular obstructions, we designed an interactive method to generate, in real time, the obstruction on any arbitrary road link. Such a method is based on the data analysis (text mining) of tweets from Twitter® reporting about traffic incidents. This is possible via the Twitter® Streaming API accessed by *tweepy*,[4] a Python-based library. For this, we designed the dedicated Twitter® account "SmartTransport" (@smart_tran). The tweets posted by this account were of the form $< T1 > + < T2 > + < T3 > + < T4 > + < T5 >$, with: (i) T1: "New road incident: $< Location >$"; (ii) T2: "LatLon: $< latitude, longitude >$"; (ii) T3: "Maxcapacity: $< integer >$"; (iv) T4: "Maxspeed: $< speed >$ [km/h]"; (v) T5: "Time: $< timestamp >$". Similarly, the account was designed to post tweets to remove the obstruction in the simulation. Such tweets had the form: $< T6 > + < T7 >$, with: (i) T6: "Road incident closed: $< Location >$"; (ii) T7: "Time: $< timestamp >$".

With this method, the nearest road link in SUMO will be selected as the affected link once a new road incident is posted, and its maximum speed will be set as the provided "Maxcapacity" until such an incident is closed. In our tests, we considered:

---

[4]http://www.tweepy.org/.

location of the incident was selected at Cashel Rd North (latitude: 53.322340, longitude: −6.306612); maximum capacities in {3, 4, 5, 6}; and speed limit at 1.5 km/h.

Results from Numerical Simulations

We performed a number of simulations for 4 different scenarios corresponding to random realizations of the experiment for different values of the maximum capacity in {3, 4, 5, 6}. Each of these simulations evolved over a time span of 4 h, and the obstruction was included from the beginning of the simulation. The results of these experiments are presented in Figs. 3 and 4.

Figure 3 shows the vehicular flow in three different scenarios: (a) no obstruction, *baseline test*; (b) obstruction, *uncontrolled* case (i.e. the service is not active); (c) obstruction, *controlled* case (i.e. with active service). As it can be seen from the baseline simulation, Route A is used by a large amount of vehicular traffic in comparison with the other routes. As a result, the vehicular flow through Route A becomes saturated in the uncontrolled case when the obstruction occurs. Instead, if the service is active, then the vehicular traffic through Route A is regulated according to the maximum capacity of the obstructed link. At the same time, vehicles that would have entered in Route A are re-routed in a balanced manner by the service across a number of alternative routes.

In Fig. 4, the number of vehicles with access to the obstructed route is presented, showing that the service is effectively able to keep such a number below the corresponding maximum capacity.

Results from HIL Validation

We made use of a HIL platform (see e.g. [3]) to obtain a more realistic assessment of the proposed service. The simulations presented here involve the use of a real car (a 2015 Toyota Prius 1.8 VVT-I) equipped with the service designed in this chapter (also termed as *target car* in what follows). Within this set-up, the in-car system was deployed as a web service running on an Android tablet, while the remote service was implemented, in Python, on a server running at the University College Dublin campus. The real car was embedded into the traffic scenario described above (that is, the car was interacting, via the HIL platform, with simulated cars behaving in accordance to the scenario above). Re-routing indications were given to the driver via the web service hosted on the tablet.

This time, the obstruction was generated only after the simulation started. The obstruction was created, again, on Cashel Road North (latitude: 53.322334, longitude: −6.306720) using the Twitter®-based method described above (see also https://twitter.com/smart_tran for an illustration). Given this set-up, different experiments were performed. In most cases, the test car was predicted to use the obstructed route and allowed to use it (as a result of each particular traffic condition), as shown in Fig. 5a. In some cases, as shown in Fig. 5b, the target car was not allowed to use the obstructed route and an alternative was recommended to the driver, who accepted the recommendation. The recommendation was provided to the driver by highlighting

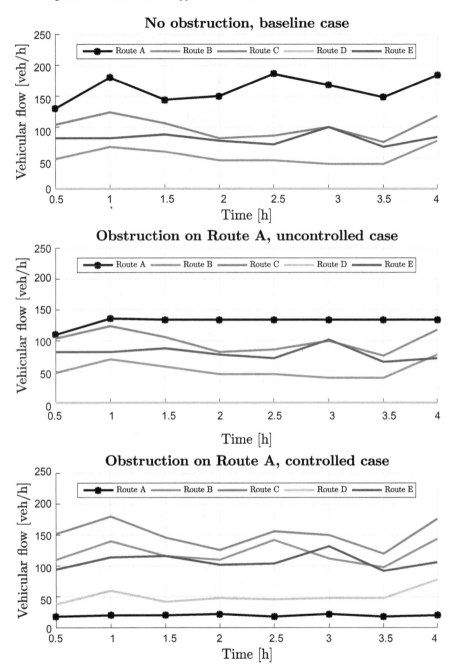

Fig. 3 Irregular obstruction: Vehicular flows observed for the selected routes. [Setup: incident on Cashel Rd North (included in Route A), maximum capacity equal to 3, speed limit at 2 km/h.]

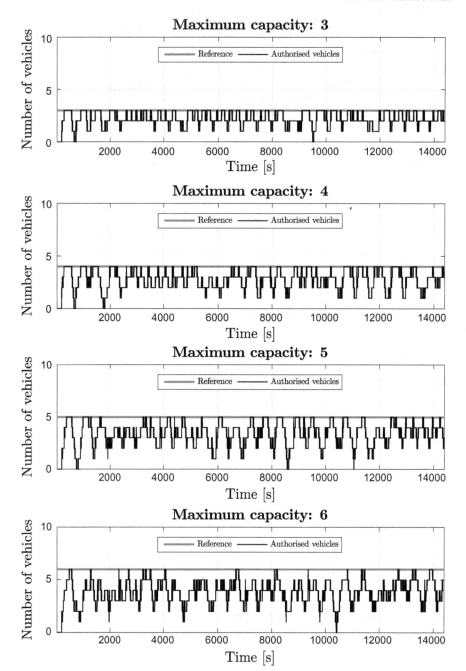

Fig. 4 Irregular obstruction: number of vehicles with access to the obstructed link, as the result of the link controller action. [Setup: incident on Cashel Rd North, maximum capacity in {3, 4, 5, 6}.]

the alternative route in yellow on the web service (as shown in Fig. 5b). Finally, in Fig. 6 it is shown how the service is, again, able to:

(i) successfully regulate the traffic along the obstructed route, by keeping the number of vehicles on Route A to be lower than 3 (this maximum capacity was specified in the test tweet); and
(ii) balance the traffic across the alternative (recommended) routes.

4.2 Regular Obstructions

Due to the regularity of these kinds of obstructions, these were studied on a longer time-span than the irregular obstructions: a period of 16 h per day, repeated over the course of a year (i.e. 365 times). Due to this fact, we do not use the traffic simulator SUMO but make use of a traditional numerical simulation in Python with the following setup:

- We define the simulation step k as 1 s, and emulate 2 access requests per 16-h period for the participating vehicles.
- We assume the error $e_{(l,m)}(k)$ used for the link controller as a 1-dimensional random walk between 0 and $c_{(l,m)}$, with $c_{(l,m)} \in \{3, 6\}$; we use a walking step given by $\lfloor \beta r(k) \rceil$ where $r(k)$ is a normally distributed random number, $\lfloor \bullet \rceil$ is the nearest integer function, and β is a design parameter to be adjusted. An empirical tuning process showed that $\beta = 0.35$ provides similar values of $e_{(l,m)}(k)$ as those obtained in the case of irregular obstructions.
- We analyze the performance of 10 cars, and use $\alpha = 0.1$ for Eq. 5 and $\phi_i(z) = 4z^3$ for the calculation of Eq. 11.

In the experiments, when a user i is required to compete for the affected link, his/her allocation history y_i is queried to calculate the average allocation $\bar{y}_i$ until the current simulation step so that Eqs. 10–11 can be solved. The allocation of the resource is then evaluated via a coin toss against $P_{allocation}^{(i)}$, after which y_i is updated in the following manner: if user i gets permission is allocated the resource, we append 1 to y_i; otherwise, we append 0.

Remark: Note that re-routing in the case of allocation denial does not affect (on average) the calculation of $P_{allocation}^{(i)}$, and thus it is not explicitly emulated in our experiments.

The numerical results from the above experiments are presented in Fig. 7. There, it can be noticed that all the average allocation values of the analyzed vehicles show convergence to a similar value (which is slightly higher in the case of a higher maximum capacity allowed for the regular obstruction), empirically showing the fairness of the proposed approach.

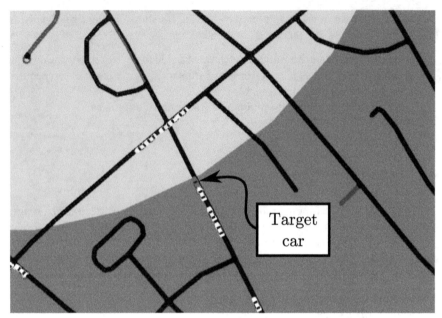

(a) Access to the limited resource is allowed.

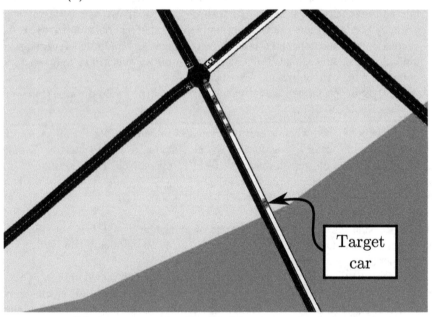

(b) Access to the limited resource is denied.

Fig. 5 Results from the HIL tests: **a** the target car is allowed to use the affected road link (associated car's avatar is coloured in blue), and **b** the target car must be re-routed (associated car's avatar is coloured in red), and thus new driving directions (yellow line) are generated

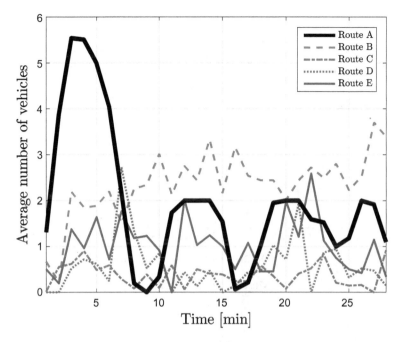

Fig. 6 Vehicular traffic information from HIL validation: Evolution of the average number of vehicles on the obstructed route (Route A) and the alternative routes (all the others). Obstruction was generated at around time 2 min

5 Conclusions and Future Work

In this chapter we presented the principled design of a parsing engine system for connected vehicles. This service we proposed is able to predict the likely route/ destination of a driver. If any obstruction is detected to occur along the predicted route/destination, then a set of recommendations is provided to the driver. The recommendations are computed via a closed-loop system and have the goal of: (i) regulating the number of vehicles gaining access to the obstructed route; (ii) and balancing traffic across a set of alternatives. Validation was performed via both a microscopic traffic simulator and via a HIL platform, which allowed the generation events via Twitter®feeds. Future work includes the design of more detailed route prediction algorithms which also take into account additional/contextual information from the driver and a traffic balancing algorithm that would take into account driver preferences.

Acknowledgements This work has been conducted within the ENABLE-S3 project that has received funding from the ECSEL Joint Undertaking under grant agreement No. 692455. This Joint Undertaking receives support from the European Union's Horizon 2020 research and innovation program and Austria, Denmark, Germany, Finland, Czech Republic, Italy, Spain, Portugal,

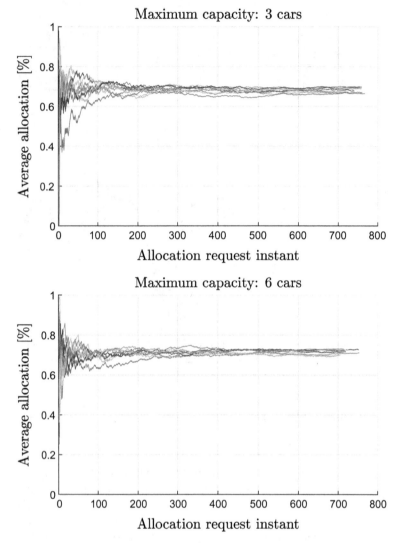

Fig. 7 Regular obstruction: Evolution of the average allocation (percentage) for the 10 simulated vehicles under study, using different values for the maximum capacity

Poland, Ireland, Belgium, France, Netherlands, United Kingdom, Slovakia and Norway. This work was also partially supported by SFI grant - 16/IA/4610.

References

1. Brin S, Page L (1998) The anatomy of a large-scale hypertextual web search engine. Comput Netw ISDN Syst 30:107–117
2. Godsil C, Royle GF (2001) Algebraic graph theory. Springer Science & Business Media
3. Griggs W, Ordóñez-Hurtado R, Crisostomi E, Häusler F, Massow K, Shorten R (2015) A large-scale SUMO-based emulation platform. IEEE Trans ITS 16(6):3050–3059. https://doi.org/10.1109/TITS.2015.2426056
4. Griggs W, Yu JY, Wirth F, Häusler F, Shorten R (2016) On the design of campus parking systems with qos guarantees. IEEE Trans Intell 17(5):1428–1437. https://doi.org/10.1109/TITS.2015.2503598
5. Krajzewicz D, Erdmann J, Behrisch M, Bieker L (2012) Recent development and applications of SUMO-simulation of urban mobility. Int J Adv Syst Meas 5(3&4)
6. Langville AN, Meyer CD (2012) Google's pagerank and beyond: the science of search engine rankings. Princeton University Press
7. McKinsey & Company (2014) What is driving the connected car. Technical report. http://goo.gl/8I4Lzx
8. Sinnott S, Ordóñez-Hurtado R, Russo G, Shorten R (2016) On the design of a route parsing engine for connected vehicles with applications to congestion management systems. In: 2016 IEEE 19th international conference on intelligent transportation systems (ITSC). pp 1586–1591. https://doi.org/10.1109/ITSC.2016.7795769
9. Song C, Qu Z, Blumm N, Barabási AL (2010) Limits of predictability in human mobility. Science 327(5968):1018–1021
10. Volkswagen ViaVision (2015) Shaping the future of mobility

Mobility on Demand in the United States

From Operational Concepts and Definitions to Early Pilot Projects and Future Automation

Susan Shaheen and Adam Cohen

Abstract The growth of shared mobility services and enabling technologies, such as smartphone apps, is contributing to the commodification and aggregation of transportation services. This chapter reviews terms and definitions related to Mobility on Demand (MOD) and Mobility as a Service (MaaS), the mobility marketplace, stakeholders, and enablers. This chapter also reviews the U.S. Department of Transportation's MOD Sandbox Program, including common opportunities and challenges, partnerships, and case studies for employing on-demand mobility pilots and programs. The chapter concludes with a discussion of vehicle automation and on-demand mobility including pilot projects and the potential transformative impacts of shared automated vehicles on parking, land use, and the built environment.

Keywords Mobility on demand · Mobility as a service · Shared mobility · Automation · Automated vehicles · Shared automated vehicles · Automated driving systems

1 Introduction

Technology is changing the way people travel, consume goods and services, and is reshaping cities and society. The integration of transportation modes, real-time information, and instant communication and dispatch all possible with the click of a mouse or a smartphone app is redefining traditional notions of auto mobility. The convergence of these trends coupled with the integration of innovative transportation

---

S. Shaheen (✉)
Department of Civil and Environmental Engineering and Transportation Sustainability Research Center, University of California, Berkeley, USA
e-mail: sshaheen@berkeley.edu

A. Cohen
Transportation Sustainability Research Center, University of California, Berkeley, USA
e-mail: apcohen@tsrc.berkeley.edu

© Springer Nature Switzerland AG 2020
E. Crisostomi et al. (eds.), *Analytics for the Sharing Economy: Mathematics, Engineering and Business Perspectives*,
https://doi.org/10.1007/978-3-030-35032-1_14

services and advanced technologies is reshaping traditional notions of public and private transportation.

In recent years, on-demand passenger and courier services known as Mobility on Demand (MOD)—have grown rapidly due to advancements in technology; changing consumer patterns (both mobility and retail consumption); and a combination of economic, environmental, and social forces. For example, there were 21 active carsharing programs in the United States (U.S.) with over 1.4 million members sharing more than 17,000 vehicles as of January 2017 [41]. Additionally, the U.S. had 261 bikesharing operators with more than 48,000 bicycles as of May 2018 (Russell Meddin, unpublished data). Moreover, as of December 2017 uberPOOL and Lyft Shared rides, a pooled version of transportation network companies (TNCs, also known as ridesourcing and ridehailing) known as ridesplitting, were available in 14 and 16 U.S. markets, respectively (Paige Tsai, personal communication; Peter Gigante, personal communication). Innovative carpooling apps, such as Scoop and Waze Carpool are enabling on-demand higher occupancy commuting. The growing popularity of on-demand mobility and delivery is contributing to a growing interest by the private sector. In the automotive sector, interest in MOD has taken a variety of forms including: acquisitions; investments; partnerships; internal development of technologies and services by original equipment manufacturers (OEMs), such as Fords acquisition of Chariot, Daimler and BMWs merger of car2go and Reach-Now; and numerous automotive interest in testing TNCs, shared automated vehicles (SAVs), and new business models [42]. In the logistics sector, companies are testing a variety of automated vehicle and drone delivery innovations. FedEx and UPS, for example, are developing delivery vans paired with drone systems that can make short-range aerial deliveries while a parcel van enroute makes another delivery [20, 44, 57]. Both Amazon and DHL are developing automated parcel stations, lockers, and delivery drones [20, 44, 57]. Across the U.S., startups such as Starship are developing automated delivery robots for e-commerce companies, such as DoorDash and Postmates [29, 51]. These trends require transportation practitioners to rethink both passenger and goods movement and foster innovative practices, strategies, and models for dynamically managing transportation supply and demand.

In this chapter, we briefly summarize the methodology used to research MOD for this paper. Then we explore the emerging concepts of MOD and explain how MOD differs from 2 Mobility as a Service (MaaS), including the MOD ecosystem: marketplace, stakeholders, and enablers. Next, we explore MOD opportunities and challenges; highlight case studies from the Federal Transit Administrations (FTA) MOD Sandbox demonstration program; and discuss the future of MOD and automation.

2 Methodology

As part of this research, we employed a multi-method qualitative approach to researching MOD and MaaS. First, we conducted a literature review of shared and on-demand mobility systems, including definitions and concepts. We supplemented published literature with an Internet-based review and targeted interviews and webinars with approximately 30 experts to categorize innovative and emerging technologies that facilitate MOD. Many of these sources filled gaps in the literature where existing publications have not kept pace with emerging MOD services and innovations. Additionally, in January 2017 and January 2018, we hosted a one-day workshop to engage MOD stakeholders at the Transportation Research Board Annual Meeting. Over 150 transportation researchers and practitioners representing the public and private sectors participated in each workshop comprised of plenary and breakout sessions, including moderated discussions and participant engagement [38, 39]. In particular, the breakout sessions included facilitated discussions on opportunities and challenges from public and private sector perspectives in four key areas: (1) managing and understanding pilot data; (2) equity and accessibility; (3) economic impacts and innovative business models; and (4) planning for MOD (e.g., land use and zoning). We also co-authored the U.S. Department of Transportation's (USDOT) MOD Operational Concept Report, a multi-modal effort initiated by the Intelligent Transportation Systems (ITS) Joint Programs Office (JPO) and the Federal Transit Administration (FTA) to study emerging mobility services; public transit operations; goods delivery services; real-time data services; and intelligent transportation systems that can enhance access to mobility, goods, and services for all. The purpose of the USDOTs MOD Operational Concept is to help guide MOD concept development, testing, demonstration projects, research, and public policy. For more background on this report, please see [42].

Between November 2017 and July 2018, we sponsored SAE International standard J3163^{TM} to develop definitions for terms related to shared mobility and enabling technologies. As part of this process, we engaged 12 experts as part of four expert panel meetings between March and August 2018. Between December 2017 and August 2018, we also engaged 30 experts as part of five task force meetings. Finally, we briefed the SAE Shared and Digital Mobility Committee, soliciting feedback from 30 voting members and approximately 100 participants on the committee through SAE's ballot and comment process. These engagements were intended to fill gaps in the literature and to validate our understanding. Participants included academic researchers, transportation professionals, policymakers, and service providers. Participants were selected by SAE based on their experience and knowledge of shared and on-demand mobility services. Each engagement averaged approximately one hour in length.

Finally, we are serving as members of the independent evaluation team for the USDOTs Mobility on Demand Sandbox demonstration, which has helped inform early lessons learned in the case studies. While our research approach documenting MOD concepts and definitions was extensive, it is important to note that the technology and concepts are rapidly evolving. Thus, it is possible that potential experts, literature, and case studies may not have been included in our review.

3 What Is Mobility on Demand (MOD)?

The USDOT's MOD Operational Concept Report defines MOD as an innovative transportation concept where consumers can access mobility, goods, and services on-demand by dispatching or using shared mobility, courier services, unmanned aerial vehicles, and public transportation strategies [42]. MOD is an emerging concept based on the principle that transportation is a commodity where modes have economic values that are distinguishable in terms of cost, journey time, wait time, number of connections, convenience, vehicle occupancy, and other attributes [42]. MOD passenger mobility can include bikesharing, carsharing, microtransit, ridesharing (i.e., carpooling and vanpooling), TNCs, scooter sharing, shuttle services, urban air mobility, and public transportation. MOD courier services can include app-based delivery services (known as courier network services (CNS)), robotic delivery, and aerial delivery (e.g., drones). Definitions for common and emerging MOD passenger and courier services are included in Table 1.

Reference [42] identify five key defining characteristics of MOD including:

- Commodifying transportation choices where modes have economic values based on cost, journey time, wait time, number of connections, convenience, and other attributes;
- Embracing the needs of all users (travelers and couriers), public and private market participants, and services across all modes including: motor vehicles, pedestrians, bicycles, public transit, for-hire vehicle services, carpooling/vanpooling, goods delivery, and other transportation services;
- Improving the efficiency of the transportation system and increasing the accessibility and mobility of all travelers;
- Enabling transportation system operators and their partners to monitor, predict, and influence conditions across an entire mobility ecosystem; and
- Maintaining the ability to receive data inputs from multiple sources and provide responsive strategies targeting an array of operational objectives.

The USDOT's MOD Operational Concept envisions MOD as a multimodal traveler and transportation management strategy that has the potential to enhance access, mobility, and goods delivery while simultaneously improving the operations and performance of the transportation network [42]. To make this happen, a number of stakeholders and enablers are important to MOD's success.

4 How Does MOD Differ from Mobility as a Service?

In Europe, another model of multimodal transportation known as MaaS is emerging. Although MOD and MaaS share a number of similarities, such as an emphasis on multimodal integration (physical co-location of services, fare payment, and digital integration), the concepts are fundamentally different. While MOD emphasizes the

Table 1 Definitions of common and emerging MOD passenger and courier modes. Source: Adapted from [7, 36, 40]

Mode	Definition
Bikesharing	Bikesharing provides users with on-demand access to bicycles at a variety of pick-up and drop-off locations for one-way (point-to-point) or roundtrip travel. Bikesharing fleets are commonly deployed in a network within a metropolitan region, city, neighborhood, employment center, and/or university campus [36, 40]
Carsharing	Carsharing offers members access to vehicles by joining an organization that provides and maintains a fleet of cars and/or light trucks. These vehicles may be located within neighborhoods, public transit stations, employment centers, universities, etc. Carsharing organizations typically provide insurance, gasoline, parking, and maintenance. Members who join a carsharing organization normally pay a fee each time they use a vehicle [40]
Courier Network Services (CNS)	Courier Network Services provide for-hire delivery services for monetary compensation using an online application or platform (such as a website or smartphone app) to connect couriers using their personal vehicles, bicycles, or scooters with freight (e.g., packages, food, etc.) [43]
Delivery drones	Delivery drones are unmanned aerial vehicles (UAVs) used to transport packages, food, or other goods
Microtransit	Microtransit is defined as a privately or publicly operated, technology-enabled transit service that typically uses multi-passenger/pooled shuttles or vans to provide on-demand or fixed-schedule services with either dynamic or fixed routing [36]
Ridesharing	Ridesharing (also known as carpooling and vanpooling) is the formal or informal sharing of rides between drivers and passengers with similar origin-destination pairings. Ridesharing includes vanpooling, which consists of 7 to 15 passengers who share the cost of a van and operating expenses and may share driving responsibility [40]
Scooter Sharing	Scooter sharing allows individuals access to scooters by joining an organization that maintains a fleet of scooters at various locations. Scooter sharing models can include a variety of motorized and non-motorized scooter types. The scooter service provider typically provides gasoline or power (in the case of motorized scooters), maintenance, and may include parking as part of the service. Users typically pay a fee each time they use a scooter [40]
Shuttles	Shuttles are shared vehicles (normally vans or buses) that connect passengers from a common origin or destination to public transit, retail, hospitality, or employment centers. Shuttles are typically operated by professional drivers, and many provide complimentary services to the passengers [7, 36]

(continued)

Table 1 (continued)

Mode	Definition
Taxis	Taxis provide prearranged and on-demand transportation services for compensation through a negotiated price, zone pricing, or taximeter (either traditional or GPS-based). Passengers can schedule trips in advance (booked through a phone dispatch, website, or smartphone app); street hail (by raising a hand on the street, standing at a taxi stand, or specified loading zone); or e-Hail (by dispatching a driver on-demand using a smartphone app) [7, 36]
TNC	TNCs (also known as ridesourcing and ridehailing) are prearranged and on-demand transportation services for compensation in which drivers and passengers connect via digital applications. Digital applications are typically used for booking, electronic payment, and ratings [36, 40]
Urban air mobility	A system for air passenger and car transportation within an urban area, inclusive of small package delivery and other urban Unmanned Aerial Systems (UAS) services, which supports a mix of onboard/ground-piloted and autonomous operations [30]

commodification of both passenger mobility and goods delivery and transportation systems management (e.g., supply and demand), MaaS focuses on mobility aggregation and subscription services, often facilitated through a smartphone application or website [8]. For example, the UbiGo pilot in Gothenburg, Sweden operated as a transportation brokerage service providing member households a mobility subscription in place of car ownership between November 2013 and April 2014 [50]. The monthly subscription allowed households to pre-purchase mobility access in a variety of increments on multiple modes, operating like a multimodal digital punch card for a number of transportation services (including public transportation, carsharing, rental cars, and taxis) [50]. Brokering travel with suppliers, repackaging, and reselling it as a bundled package is what distinguishes MaaS from MOD [50]. UbiGo was relaunched in March 2018 in partnership with Austrian IT supplier Fluidtime in Stockholm, Sweden. As part of the current project, households have access to public transportation, bikesharing, carsharing, rental cars, and taxis.

5 The MOD Ecosystem: Marketplace, Stakeholders, and Enablers

MOD enables an integrated and multimodal operations management approach that can influence the supply and demand sides of a broad mobility marketplace. The supply side of the MOD marketplace consists of the players, operators, and devices that provide transportation services for people or goods and service delivery. The demand side of the MOD marketplace is comprised of travelers and couriers, including their choices and preferences [42]. At the epicenter of the MOD marketplace is multimodal transportation operations management, which receives data from all portions of the system, assembles those data into an overall picture of current and predicted conditions, identifies problems considering a wide range of operational objectives applicable to the specific time period [42]. As such, the operational heart of the transportation network is able to draw upon pre-defined response strategies, identify interventions to be made by the transportation network manager(s) to address incidents, and ultimately generate and implement response and action plans dynamically [42]. Ideally and as the system evolves, the MOD marketplace will be able to dynamically generate and implement response and action plans optimized across a constantly changing array of outcomes from all areas of the transportation network, affecting a broad range of stakeholders that can vary in importance over time [42]. As such, MOD has the potential to support transportation demand management through strategies and policies to redistribute or reduce travel demand spatially (e.g., shifting demand to different routes or higher occupancy modes, etc.) and temporally (e.g., shifting demand to another time of day), making MOD a cost-effective method to manage and increase existing network capacity. Figure 1 depicts the supply-side, demand-side, operational management, stakeholder, and enabler components of the MOD marketplace.

In the sub-sections that follow, we explore: (1) the supply, demand, and operational management of the MOD marketplace; (2) four core MOD enablers; and (3) key stakeholders in greater detail.

5.1 The MOD Marketplace: Supply, Demand, and Operational Management

The supply side of the MOD marketplace consists of all the players, operators, and devices that provide transportation services for people or goods delivery including:

- Public transportation services (e.g., trains, buses, ferries, paratransit);
- Private-sector transportation services including: taxis, car rentals, microtransit (Chariot, Via, etc.); TNCs (Lyft, Uber, etc.), personal vehicles, volunteer drivers, other shared services (e-Hail, carsharing, ridesharing, bikesharing, scooter sharing, etc.);
- Goods delivery services including: first-and-last mile goods delivery, courier network services, drones, and robotic delivery;
- Transportation facilities including: parking, tolls, roadways, and highways;
- Vehicles of all types such as: public transit vehicles, private vehicles, goods delivery vehicles, and emergency vehicles, including connected and automated applications in the future;
- Transportation management and information systems such as: payment systems for parking, toll and public transit, signal systems, mobile applications for trip planning, booking, and payment (for all travelers), fleet management systems, and navigation systems; and

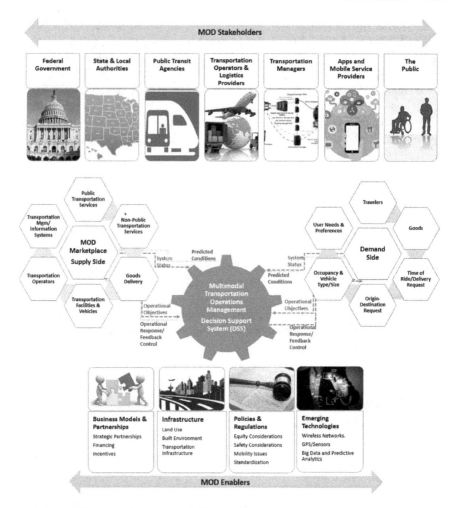

Fig. 1 The MOD marketplace. *Source* [42]

- Public and private transportation information services including: schedule information; 511; dynamic message signs; and mobile apps (i.e., Waze and Google Maps).

The demand side of the MOD marketplace consists of all the users and their travel choices and consumption preferences [42]. Examples of demand-side factors include:

- All travelers (e.g., pedestrians, riders, drivers, cyclists, older adults, people with disabilities, children, etc.);
- Goods and merchandise requiring physical delivery;
- Digital delivery of goods and services that impact traveler demand;

- Time of ride and/or delivery request that affects temporal choice and service availability;
- Origin-destination request that affects spatial demand and routing;
- Modal demand based on occupancy, size, or type of vehicle requested; and
- User needs and preferences.

Public agencies can leverage MOD to promote behavioral change for network efficiency, reduce congestion, and enhance traveler options. Operationally, MOD becomes a core component of multimodal transportation operations management strategies by implementing a proactive, anticipatory approach to identify problems ahead of time and intervene to manage demand and supply to meet the desired network performance. MOD paired with active multimodal operations management can help improve a transportation system's reliability, accessibility, and environment by evolving traffic management and operations paradigms from static and pre-set operations to a more dynamic, commoditized supply and demand management approach. Targeting activity-level decisions and providing travelers with travel choices, such as route choice, time of day choice, and mode choice, is a core component of the decision-support system.

5.2 Enablers

MOD is facilitated by four core enablers:

- **Business Models and Partnerships** include: financing structures, risk-sharing partnerships, incentive strategies, and strategic partnerships. Several MOD business models (e.g., business to consumer, business to government, business to business, and peer-to-peer (P2P)) have evolved to meet the diverse needs of consumers, service providers, and partners [42]. With different business models, there are also opportunities for a variety of financing structures, incentives, and partnerships that could be leveraged such as: (1) Non-Profit (owned and operated by an institution with the goal of covering operational costs), (2) Privately Owned and Operated (owned and operated by private entity), (3) Publicly Owned and Operated (operated and operated by a public agency), (4) Publicly Owned and Contractor Operated (owned by a public agency and operated by private vendor), and (5) vendor operated (owned and operated by the vendor that designs and/or manufacturers the MOD system equipment) [48]. Common partnership opportunities can include: user subsidies, discounts, tax incentives, risk-sharing partnerships, joint marketing, and other direct and indirect support [7, 37].
- **Infrastructure is** comprised of land use, the built environment, and transportation infrastructure (e.g., roads, sidewalks, bicycle paths, etc.) that can affect MOD use and operations. Urban density, walkability, the availability of active transportation infrastructure, and physical design are important MOD infrastructure enablers [42].

- **Policies and Regulations** enablers include: equity; safety; mobility; sustainability; accessibility considerations; and standardization (regulatory, data, legal definitions, etc.) efforts can help overcome challenges to existing laws and regulations and ensure accessibility to an array of user groups (e.g., people with disabilities, low-income households, digitally impoverished users, etc.) [7, 40]. The public sector has a major role as a stakeholder and enabler affecting different transportation modes by: defining legislative frameworks, ensuring fair market performance, establishing incentives, and initiating pilot programs [42].
- **Emerging Technology** enablers include: GPS, sensors, wireless systems, Internet of Things, mobile apps, automated aerial vehicles (AAVs), UAVs, robotic delivery, big data, data analytics and management systems, machine learning, artificial intelligence, virtual reality, inclusive information and communication technology, and universal design [42]. Technology is a key enabler of MOD and enables enhanced connectivity among travelers, goods, services, and infrastructure, which contributes to more efficient use of resources and emerging transportation and consumption choices.

5.3 Stakeholders

MOD can include an array of stakeholders and partners, such as public transit agencies, paratransit, MOD service providers, app developers, transportation and traffic managers, connected traveler services, metropolitan planning organizations, and local governments [42]. Common stakeholder roles include:

- **National Government** who establishes transportation strategies, policies, regulations, and legislation. The national government can also invest in pilot programs and develop national industry-wide standards.
- **State, Provincial, Regional, and Local Authorities** implement policy and regulations such as: issuing permits, managing public rights-of-way, and managing local and regional transportation planning and traffic management.
- **Public Transit Agencies** can play an important role fostering partnerships and implementing programs that bridge spatial and temporal gaps in the transportation network.
- **MOD Service Providers** are a critical supplier of on-demand mobility and delivery services.
- **Transportation/Traffic Managers** monitor the transportation system and can leverage MOD to manage overall supply and demand of the network.
- **Apps and Mobile Service Providers** enable the digital infrastructure of MOD by offering mobile ticketing, payment, navigation services, and other digital services.
- **Consumers** (including personal and business customers) are the end users who consume on-demand mobility and delivery services.

MOD stakeholders can play a variety of similar and differing roles such as: (1) commoditizing passenger mobility and goods delivery; (2) offering short-term, on-demand access to mobility and goods delivery strategies for users; (3) facilitating trip planning or delivery, payment, and other functions into a single interface; (4) offering on-demand mobility and delivery options; (5) providing transportation service to all users including people with special needs; and (6) increasing mobility and goods availability through specific partnerships or use cases (e.g., journeys previously inaccessible by a single mode, first-and-last mile connections, additional service offerings during off-peak or high-congestion travel times, and access to goods/services previously unavailable) [42].

6 MOD Opportunities and Challenges

Naturally, the benefits, opportunities, and challenges of MOD often vary depending on the stakeholder. Table 2 provides some examples of the diverse opportunities and challenges that can be confronted by the range of MOD stakeholders.

For the consumer, MOD can create opportunities to enhance access and equity by providing increased mobility options (e.g., fares, routes); increased travel speed and reliability; critical first-and-last-mile connectivity; and expanded coverage to historically underserved users or communities. However, the demographics of MOD users often differ from the general population. In general, MOD users tend to be younger, have higher levels of educational attainment and incomes, and are less diverse than the general population [47]. Older adults, low-income individuals, rural communities, and minority communities have historically been less likely to use MOD. Additionally, access to the Internet, smartphones, and banking services are a prerequisite for many MOD services, which tend to be lower among many of these groups [47]. MOD accessibility challenges can be generally categorized into four areas [45]. These include:

- **Access for People with Disabilities**: In East Asia, Europe, and the U.S., older adults are redefining longevity. By 2045, the number of Americans over the age of 65 will increase to 77%, and the number of people with disabilities will increase (an estimated 20% of the U.S. population has a disability today). Removing barriers to MOD services for people with visual, auditory, cognitive, mobility, and other disabilities is critical.
- **Un- and Under-Banked Households**: Many MOD services require debit/credit cards for payment and, in some cases, deposits or credit holds as collateral for vehicles or equipment. Providing alternative fare payment options for under-banked and unbanked users is key.
- **Low-Income Affordability and Service Equivalency**: Pay-as-you-go MOD pricing can be more expensive than walking, cycling, and public transportation. Equivalent level of service for low-income households and neighborhoods, including affordable mobility options, equivalent travel modes, comparable hours and frequency of service, and similar wait times, is important.

Table 2 Examples of potential opportunities and challenges for MOD stakeholders

Stakeholder	Opportunities	Challenges
Federal Government	– Potential to manage transportation supply and demand, mitigating the need for expensive capacity-enhancing capital projects	– Modes may lack clear and concise legal and regulatory definitions
		– Service providers may initiate service without the government's consent and/or exploit unclear legal or regulatory areas
		– It may be difficult for the government to keep up with dynamic, fast-changing developments
State and Local Authorities	– Potential to more effectively manage transportation supply and demand, while mitigating the need for expensive capacity-enhancing capital projects	– Modes may lack clear and concise legal and regulatory definitions
	– Opportunities to leverage MOD services to reduce vehicle miles traveled (VMT), greenhouse gas (GHG) emissions, and other public sector goals	– Providers may initiate service without the government's consent and/or exploit unclear legal or regulatory areas
	– Potential to expand service to underserved communities or user groups (e.g., people with disabilities, low-income neighborhoods, etc.)	– It may be difficult for the government to keep up with dynamic, fast-changing developments
		– The impacts of MOD services may be unclear or may have adverse impacts on travel behavior or the environment (i.e., increased congestion)
		– MOD could have unintended equity challenges (e.g., excluding digitally impoverished or underbanked households)
Public Transit Agencies	– Enhance public transit agency preparedness for MOD	– Potential competition from other transportation service providers
	– Bridge first-and-last mile gaps	– Private-sector service providers may not share data or be willing to work toward fare and digital integration
	– Reduce costs associated with low-rider/underperforming routes	– The future role of public transportation alongside MOD is evolving and unclear (i.e., bridging gaps vs. public transit replacement)
	– Potential for multimodal connections and mobility hubs	

(continued)

Mobility on Demand in the United States

Table 2 (continued)

Stakeholder	Opportunities	Challenges
Transportation Operators and Logistics Providers	– Opportunities to serve emerging markets and generate revenue	– MOD services may confront an uncertain or unfriendly regulatory environment (i.e., no regulation or over regulation)
	– Potential for public-private partnerships	– MOD services may have challenges meeting regulatory requirements (i.e., minimum service requirements or data sharing), while maintaining profitability and/or protecting consumer privacy
Transportation Managers	– Potential to dynamically manage transportation supply and demand near real time	– MOD services could create disruptions with other services or have unintended consequences
	– Opportunities to leverage MOD services to reduce VMT, GHG emissions, and address other public-sector goals	– The impacts of MOD services may be unclear or may have adverse impacts on travel behavior or the environment (i.e., increased congestion)
Apps and Mobile Service Providers	– Opportunities to provide digital services and/or integrate with or manage public sector fare payment, real-time information, and/or trip planning services	– The public sector may have complex requirements for data sharing or fare payment that limit or prohibit integration
	– Potential for public-private partnerships	
The Public	– Consumers can compare service options (i.e., cost, journey time, wait time, number of connections, convenience) and access to mobility and goods delivery services on-demand	– Services may not be available in all neighborhoods or to all users (i.e., unbanked users, people with disabilities, etc.)
		– Services may be less reliable or more expensive than existing service options

- **Digital Poverty**: MOD services typically require a smartphone and data packages to access services. This can be a barrier to low-income and rural households who may not be able to afford or lack data coverage to access MOD. Alternatives such as digital kiosks, telephone services, and non-tech access (such as street hail) can help overcome these challenges.

Ensuring equitable access should be a priority among public and private stakeholders. Legislation and regulation can play a notable role in safeguarding transportation equity by mitigating emerging MOD technological and access barriers, although more research and policy guidance is needed to clarify the applicability and scope of existing statutes.

In response to many of these equity challenges, a number of public agencies have implemented public policies and developed pilot programs in an attempt to address these equity challenges and test innovative approaches to enhancing MOD accessibility. One notable national initiative is the UDOT's MOD Sandbox, a US$8 million federal grant program that funds pilots that test innovative MOD business models that deliver high-quality, seamless, and equitable mobility options for all travelers [16].

7 Partnerships Between Public Transportation and MOD Service Providers

Six types of partnerships between public transportation and MOD service providers were identified through our literature review and 10 targeted expert interviews. Please note that all of the examples provided below are outside of the FTA MOD Sandbox demonstration projects. The partnership categories include:

- **Trip Planning and Fare Integration** partnerships that integrate traveler options and fare payment into a single user interface. A common goal of trip planning and fare integration for public transit agencies is to reduce common barriers associated with multimodal public transit trips. For example, Xerox's Go-LA app allows Angelinos to plan a trip using multiple public transit and MOD modes (e.g., Lyft, taxis, and Zipcar) [9].
- **First and Last Mile** partnerships where a public-sector partner subsidizes a MOD service operator to provide services to or from a public transit stop or station. In some cases, the public agency may geofence an eligible service area, geographically limiting the eligibility of the subsidy. For example, the City of Summit, New Jersey has partnered with Lyft and Uber to provide free rides to and from their station during weekday commute hours. A primary goal of this partnership was to increase station passenger throughput without having to build additional parking.
- **Low-Density Service and Public Transit Replacement** partnerships subsidize a MOD provider to offer service in a lower-density area. These types of partnerships can allow public transit agencies to reduce or replace low ridership transit service with a lower-cost alternative. For example, the City of Arlington, Texas replaced local bus service with the microtransit service, Via. Via operates a fleet of a 10 commuter vans in downtown Arlington and charges a fare of US$3 per ride [13].
- **Off-Peak Service** partnerships subsidize MOD services during late night or other public transit off-peak times. For public transit agencies, it can be expensive to run a bus or train during the middle of the night because there are not as many riders. In Pinellas County, Florida, the Pinellas Suncoast Transit Authority has funded US$300,000 to subsidize free, late-night rides for low-income residents

and workers. As part of the program, riders can request up to 23 rides per month between 9pm and 6am, if traveling to or from a residence or a workplace [33].
- **Paratransit** partnerships leverage MOD services to supplement or replace an existing paratransit service. In Boston, the Massachusetts Bay Transportation Authority (MBTA) has partnered with Lyft and Uber to provide MBTA's existing paratransit riders with US$1 uberPOOL rides, and US$2 uberX or Lyft rides. MBTA also pays any trip costs over US$15. Both Lyft and Uber have increased the number of wheelchair accessible vehicles, and Lyft offers a telephone service that allows people to schedule a ride. The program has reduced MBTA paratransit costs about 20%, while riders take 28% more trips and save an average of 6% on per-trip costs [28].
- **Guaranteed Ride Home (GRH)** partnerships involve a MOD provider subsidizing a service, a public agency providing GRH services for MOD users, or both. For example, in San Diego, the San Diego Association of Governments has partnered with Uber to provide a guaranteed ride home for commuters. Uber subsidizes this program up to US$20,000 annually (SANDAG, unpublished data, March 2018).

These partnerships exemplify the diverse ways public transit agencies are partnering and collaborating with MOD service providers to achieve a variety of public-sector goals.

8 FTA's MOD Sandbox Demonstration Program

In addition to a variety of pilots and ongoing partnerships across the country, FTA has also been researching innovative MOD and public transit partnerships. Recognizing the importance of multimodal transportation, the growth of MOD, and the commoditization of transportation services, FTA developed the MOD Sandbox Demonstration program in 2016. The aim of the MOD Sandbox demonstration is to explore opportunities and challenges for public transportation related to technology-enabled mobility services including: ways that public transit can learn from, build on, and interface with innovative transportation modes from a user, business model, technology, and policy perspective [17]. Key objectives of the sandbox include:

- Enhancing public transit industry preparedness for MOD;
- Assisting the public transit industry to develop the ability to integrate MOD practices with existing transit services;
- Validating the technical and institutional feasibility of innovative MOD business models and documenting MOD best practices that may emerge from the demonstrations;
- Measuring the impacts of MOD on travelers and transportation systems; and
- Examining relevant public sector and federal requirements, regulations, and policies that may support or impede transit sector adoption of MOD.

The MOD Sandbox demonstration includes 11 project demonstrations across the U.S. Each project demonstration pilots a variety of concepts such as: smartphone

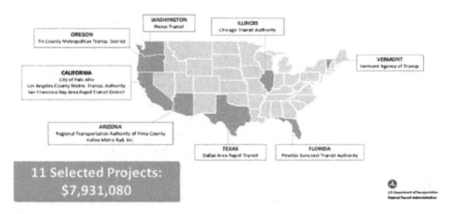

Fig. 2 MOD sandbox demonstration sites (FY16). Image Source: U.S. Department of Transportation

applications and trip planners, integrated fare payment, first-and-last mile connections to public transportation, and paratransit. A map of the pilots and description is included in Fig. 2 and Table 3. In the next sections, we review two MOD Sandbox case studies from the Bay Area Rapid Transit (BART) district and Pinellas Suncoast Transit Authority (PSTA).

9 FTA MOD Sandbox: Case Studies

In this section, we highlight two case studies from the 11 MOD Sandbox pilot projects: (1) BART and Scoop carpooling and (2) PTSA innovative paratransit.

9.1 Bay Area Rapid Transit (BART) and Scoop Carpooling Demonstration Project

BART owns and operates more than 47,000 parking spaces across the entire system. BART parking is very competitive with over 35,000 people on monthly reserved parking waitlists. BART's carpool program offers approximately 900 of BART's parking spaces to carpoolers at 12 of the systems stations [39]. BART's existing carpool parking program employs manual parking enforcement that requires that staff monitor single occupant driver parking in carpooling-only spaces [18]. BART's carpooling parking program coordinates with the regional metropolitan planning organization, the Metropolitan Transportation Commission to operate its permit program. When a user registers with the regional 511 Ridesharing program, they identify a specific station and are provided with a carpool permit that they print at home. To use a des-

Table 3 Overview of the MOD sandbox demonstration sites. Source: Gustave Cordahi and Susan Shaheen, unpublished data, 2018

State	Project sponsor	Description
IL	Chicago Transit Authority (CTA)	Incorporate the local bikesharing company, Divvy, a 580-station bikesharing service, into CTA's existing transit trip planning app (US$400,000)
TX	Dallas Area Rapid Transit (DART)	Integrate ridesharing services into its GoPass ticketing app to solve first-mile/last-mile issues (US$1,200,000)
CA and WA	Los Angeles County Metropolitan Transportation Authority	Two-region MOD partnership with the microtransit company, Via, in Los Angeles and Seattle to provide first- and last-mile strategies to public transit (US$1,350,000)
CA	City of Palo Alto	Proposed solutions seek to reduce Bay Area single occupancy vehicle (SOV) commute share from 75% to 50% through a Fair Value Commuting (FVC) solution ($1,080,000)
WA	Pierce County Public Transportation Benefit Area Corporation	The Utilize Limited Access Connections project is an initiative connecting Pierce Transit local service and Sound Transit/Sounder regional service with local microtransit companies to increase regional transit use (US$206,000)
AZ	Regional Transportation Authority (RTA) of Pima County	The Adaptive Mobility with Reliability and Efficiency (AMORE) project integrates fixed-route, subscription-based TNCs and social carpooling services into an existing data platform to provide affordable, convenient, and flexible service (US$670,000)
FL	Pinellas Suncoast Transit Authority (PSTA)	A set of partnerships with Lyft, United Taxi, CareRide, the Center for Urban Transportation Research (CUTR), and Goin Software to develop a model to provide more cost effective on-demand door-to-door paratransit service (US$500,000)
CA	San Francisco Bay Area Rapid Transit	Partnership between Scoop Technologies, Inc. (Scoop), the San Francisco Bay Area Rapid Transit (BART) District, and the Metropolitan Transportation Commission (MTC) to better integrate carpool access to public transit by matching passengers according to their destination and by providing a way to reserve and pay for parking spaces at BART stations (US$358,000)
OR	Tri-County Metropolitan Transportation District	Incorporate shared mobility options into the Open Trip Planner (OTP) project that will create a platform integrating public transit and shared mobility options (US$678,000)
AZ	Valley Metro Rail, Inc.	Smartphone mobility platform that integrates mobile ticketing and multimodal trip planning (US$1,000,000)
VT	Vermont Agency of Transportation	Statewide public transit trip planner that will enable flex-route, hail-a-ride, and other non-fixed route services to be incorporated in mobility apps (US$480,000)

ignated carpooling park space, all people in a vehicle must display their permits on the dashboard. This typically requires that users carpool together in both directions [18, 39]. Since the carpooling parking spaces are provided on a first-come, first-serve basis, there is no guarantee that users with a permit will find a parking space. Due to limited parking supply, long waitlists, and difficult enforcement, the carpooling program has a high percentage of parking violators [18, 39].

In response to these challenges, BART proposed a MOD Sandbox pilot in 2016 that partners with the technology company Scoop to facilitate carpooling matching to public transit and enforce parking [39]. Scoop is a company that focuses on increasing carpooling usage by overcoming challenges associated with carpooling, such as separating morning and evening trips and allowing a person to change their carpooling schedule day-to-day [18]. Part of BARTs motivation in partnering with Scoop was to create an innovative parking program that allows BART riders to dynamically find carpool matches (increasing ridership) and improve parking enforcement [39]. Partnership goals include: reducing single occupant vehicle travel, increasing ridership, and reducing the VMT associated with first- and last-mile connections to the BART network [18]. The program has launched at 12 BART stations. As part of the partnership, Scoop shares license plate data with BART carpool parking enforcement staff. While the pilot is still underway, BART anticipates the following stakeholder outcomes:

Commuters:

- Improved opportunities to carpool to BART stations;
- Ridematching assistance and the ability to share the cost of a carpooling trip (including parking fees); and
- The ability to arrive at the carpool's desired time (not at the time parking lots are anticipated to fill.

BART District:

- Better parking use and simplified, more effective enforcement of parking resources; and
- Increased ridership.

Metropolitan Transportation Commission:

- Increased use of existing carpooling infrastructure.

Scoop:

- Increased commuter exposure to carpooling apps, and
- Improved integration with BART parking payment systems and public transit schedules.

FTA:

- A public-private partnership that does not require an operational subsidy;
- Increased use of existing public transportation parking and public transit capacity; and

- The ability to test replicable carpool matching and parking enforcement approach that could be applied to other public transit agencies.

9.2 Pinellas Suncoast Transit Authority (PSTA) Innovative Paratransit Demonstration Project

The Pinellas Suncoast Transit Authority (PSTA) serves Pinellas County, the western county of the Tampa Bay metropolitan area, including the cities of St. Petersburg and Clearwater, Florida. PSTA operates fixed-route bus service, paratransit, shuttles, and tourist trolleys. Beginning in 2015, the PSTA began exploring ways the agency could pilot with MOD providers such as: first-and-last mile connections to public transportation, late-night service, and public transit replacement [49]. By July 2015, PSTA identified a number of ways to leverage MOD to consolidate and discontinue services based on low levels of farebox recovery, passengers per revenue hour, and passengers per revenue mile. PSTA opted to discontinue a few routes with an average of less than five passengers per revenue mile (compared to a system-wide average of 18.7 passengers per revenue mile [49]. In October 2015, PSTA launched the Transportation Alternatives Pilot Program to provide on-demand service from Uber, United Taxi, and Wheelchair Transport to a few bus stops in two service areas throughout the county from 7 am to 7 pm. PSTA provided a subsidy of US$3 per trip. Between February and August 2016, the pilot completed a total of 385 trips (averaging three trips per day) [49]. Limited program outreach and difficulty retaining former fixed-route public transit riders and transitioning them to the new flexible-route service may have contributed to limited ridership. To attract additional customers, rider subsidies were increased to US$5 per trip, and the service coverage area was expanded to include eight zones covering the entire county as part of a rebranded pilot in October 2016. Since launching the expanded service area, the pilot averages approximately 30 riders per day [49]. Leveraging the expertise from this initial partnership with MOD providers, PSTA applied for FTA's MOD Sandbox demonstration project and developed another pilot in an effort to reduce paratransit costs. PSTA currently spends an average of US$22.50/ride for the more than 275,000 annual paratransit trips for a total of approximately US$6.2M annually or 10% of the agencys operating budget [15]. With an aging population contributing to increasing paratransit demand and no new revenue sources, rising paratransit costs could result in the diversion of funding away from fixed route services. PSTAs current paratransit services are operated by one company that provides both ambulatory and wheelchair service that requires eligible riders to reserve trips by no later than 5:00 p.m. the day before a trip. The overall lack of flexibility is a common complaint with PSTA's paratransit operations [15]. PSTA's goal for the MOD Sandbox demonstration is to improve the mobility of paratransit customers and to provide more cost effective on-demand service offerings [15]. As part of this pilot, PSTA is experimenting with multiple service providers for paratransit trips. While the pilot is ongoing, PSTA expects to see notable cost savings compared to traditional paratransit services. Additionally, by offering on-demand options, PSTA anticipates providing riders with a higher quality service [15].

9.3 MOD Sandbox Early Lessons Learned

Lessons learned and best practices are still emerging from the MOD Sandbox with respect to public-private partnerships. While the independent evaluation of all MOD Sandbox sites is ongoing, early lessons learned from the MOD Sandbox demonstration include:

- Project stakeholders should focus (or remain focused) on project goals (and not get sidetracked overcoming technological and implementation challenges);
- Like any public-private partnership, it is important for all stakeholders to remain flexible and open to change (e.g., technologies, partners, etc.); and
- There is a need to overcome data sharing challenges and protect personal identifiable information and proprietary data from public records requests. As a recommended best practice, the ITS JPO of USDOT recommends three levels of data sharing and access: (1) MOD site level and partner data (for operations and internal reporting); (2) controlled access data for independent evaluation purposes; and (3) public research and access data releasable by the USDOT (Robert Sheehan and Ariel Gold, unpublished data, January 2018), [38, 39].

10 Automation and the Future of MOD

The convergence of MOD, automation, and electric drive technology have the potential to make the car more cost effective, efficient, and convenient especially when shared [46]. Potential benefits of an automated MOD future include: increased safety, more efficient road use, increased driver productivity, and energy savings [6, 12].

SAE International has defined five levels of vehicle automation; see Fig. 3. Level 1 describes vehicles that automate only one primary control function (e.g., self-parking or adaptive cruise control). Level 5 enables vehicles capable of driving in all environments without human control [54].

Today, the majority of SAV pilots are targeting Level 4 automation where a human operator does not need to control the vehicle as long as it is operating in suitable roadways and conditions (referred to as the operational design domain or ODD). The ODD describes the specific conditions under which a given automated feature is intended to function. The ODD is defined by key roadway characteristics (i.e., roadway types and speed limits) and conditions (i.e., weather conditions, time of day, etc.) an AV is designed to operate in [54]. Beginning in 2017, a number of SAV

SOCIETY OF AUTOMOTIVE ENGINEERS (SAE) AUTOMATION LEVELS

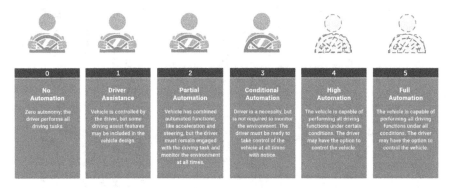

Fig. 3 Society of automotive engineers (SAE) automation levels. Image Source: National Highway Traffic Safety Administration (NHTSA)

pilots have emerged across the U.S. to test automated driving systems with a safety engineer.

Today, over a dozen SAV pilot projects are currently underway in the U.S. with more planned in the coming years. In many cases, AVs on private roads are not subject to state regulations. However, these classifications may evolve over time as services advance from testing to deploying services on public rights-of-way. The SAV pilots operating on private roads and in planned communities are typically low-speed (under 30 MPH) deployments operating in controlled environments. These pilots often focus on serving specific markets such as: office parks, housing developments, retirement communities, and universities. The other group of SAV pilots are operating on urban streets, typically using conventional vehicles equipped with AV technology to navigate their surroundings and traffic [52]. Table 4 provides a summary of SAV pilots currently underway in the U.S. as of July 2018. Temporary demonstrations or pilots that have ceased operations and those that are not carrying passengers (or very close to carrying passengers) are excluded.

Naturally, SAVs will change America's relationship with the automobile. For some households, automation may allow them to move closer to urban centers and sell private vehicles in favor of shared taxis, public transportation, and active modes (e.g., cycling and walking). While the potential impacts of automation on future ownership and modal choice patterns are unknown, the impacts will be notable. As SAVs begin to scale, policymakers will need to rethink traditional notions of access, mobility, and auto mobility [7]. Cities may have to reconsider parking minimums and consider replacing existing parking with infill development and affordable housing. Cities may be able to repurpose on-street parking for other uses (such as wider curbs, bicycle lanes, and loading zones for SAVs) [7]. What is clear is that automation will likely have a transformative impact on MOD, traveler mobility, and goods delivery [7].

Table 4 SAV pilots currently underway in U.S., as of July 2018

Operators	Location	Automation level	Vehicle type	Operational design domain (ODD)	Description
Aptiv and Lyft	Las Vegas, NV	Level 4 High Automation	Conventional	Public Roads and City Streets	A commercial pilot project accessed through the Lyft app, 20 SAVs are servicing popular destinations on the Las Vegas strip [22]
Cruise/GM	San Francisco, CA	Level 4 High Automation	Conventional	Public Roads and City Streets	In 2017, Cruise launched its pilot, Cruise Anywhere, a SAV service for its employees to use for pre-selected destinations in San Francisco. Cruise intends to launch a commercial SAV offering in 2019 [19, 26]
Drive.ai	Frisco, TX	Level 4 High Automation	Conventional	Public Roads and City Streets	After a test period that began in January 2018, a limited SAV pilot launched in July 2018. The vehicles feature LED screens that display messages to pedestrians and other road users [23]
Easymile/Contra Costa Transportation Authority	San Ramon, CA	Level 4 High Automation	Low-Speed Shuttle	Public Roads and City Streets	In March 2018, Easymile began a SAV shuttle service at Bishop Ranch, an office park of about 30,000 employees. The SAV shuttle service is one of the first to be granted approval by the California Department of Motor Vehicles to operate on public roads [3]
Easymile/Transdev	Babcock Ranch, FL	Level 4 High Automation	Low-Speed Shuttle	Private Roads/Planned Communities	A shuttle that operates as an amenity for homeowners in the Babcock Ranch development. Passengers can access the shuttles fixed-route services for free or request on-demand rides for a fee [11, 56]
Easymile/Transdev	Gainesville, FL	Level 4 High Automation	Low-Speed Shuttle	Public Roads and City Streets	A first-and-last mile downtown service planned to launch in Summer 2018 [5]
Local Motors Olli, IBM	National Harbor, MD	Level 4 High Automation	Low-Speed Shuttle	Public Roads and City Streets	With a 3-D printed, "crowd funded" design, Olli has had their shuttles in the DC area streets since 2016 [24, 55]

Table 4 (continued)

Operators	Location	Automation level	Vehicle type	Operational design domain (ODD)	Description
May mobility and Quicken loans	Detroit, MI	Level 4 High Automation	Low-Speed Shuttle	Public Roads and City Streets	A pilot launched in June 2018 in Detroit, with plans to remove the safety attendant from vehicles in early 2019. At present, May Mobility is responding to proposal requests from other municipalities and plans to expand its services [27, 31]
Navya and MCity Uni. of Michigan	Ann Arbor, MI	Level 4 High Automation	Low-Speed Shuttle	Private Roads/ Planned Communities	Launched in 2018, this SAV shuttle carries students and faculty from a campus research complex to a parking facility and bus stops. This shuttle is part of a research endeavor to understand how passengers react to SAVs and will gauge consumer acceptance of the technology [53]
Nuro and Kroger foods	Scottsdale, AZ	Level 4 High Automation	Conventional	Public Roads and City Streets	As of August 2018, Nuro is running its grocery delivery pilot using Toyota Pruises, but it intends to start delivery with its specialized R1 vehicle in Fall 2018. The R1 is designed to exclusively have space for delivery goods, without any passengers [32]
NuTonomy and Aptiv	Boston, MA	Level 4 High Automation	Conventional	Public Roads and City Streets	NuTonomy has been testing their vehicles in the Seaport neighborhood of Boston since 2017, and as of June 2018, they have been approved for testing city-wide. They are required to submit quarterly update reports to the City of Boston [25]
Optimus ride	Boston, MA; South Weymouth, MA	Level 4 High Automation	Low-Speed Shuttle	Public Roads and City Streets	They have been testing in Boston since 2017. Optimus Ride is now in agreement with the Union Point development in South Weymouth and is testing to provide SAV services for the "smart city" [14, 34]

Table 4 (continued)

Operators	Location	Automation level	Vehicle type	Operational design domain (ODD)	Description
Uber	Pittsburgh, PA; Tempe, AZ (ended)	Level 3 – Conditional Automation	Conventional	Public Roads and City Streets	In September 2016, Uber began a SAV pilot in Pittsburgh, and it was the first SAV service in the U.S. to serve passengers selected from the public. However, testing stopped in both cities after the high-profile crash and death in Tempe, Arizona in 2018. While Uber is banned from testing in Arizona, in Pittsburgh testing was briefly halted from March through July 2018 and has since resumed with a specialist in control at all times [2, 35]
Voyage	The Villages, San Jose, CA	Level 4 High Automation	Conventional	Private Roads/ Planned Communities	Voyage operates SAV pilots at The Villages retirement community in San Jose. It has operated in San Jose since 2017 [4]
Voyage	The Villages, FL	Level 4 High Automation	Conventional	Public Roads and City Streets	Voyage operates SAV pilots at The Villages retirement community in Central Florida. Service launched in Florida in 2018 [10]
Waymo	Phoenix, AZ	Level 4 High Automation	Conventional	Public Roads and City Streets	Waymo launched the Early Rider program in early 2017, allowing select Phoenix residents to request rides in their automated minivans. Waymo engineers have now moved to the backseat, as of November 2017 [1]

11 Conclusion

MOD is driving fundamental changes in the way people travel, consume goods and services. Defining characteristics of MOD include: (1) the commoditization of transportation choices, (2) embracing the needs of all users and stakeholders, (3) enhancing accessibility and improving efficiency of the transportation system, and (4) the ability to predict and respond to operational changes in the transportation network. MOD differs from MaaS in that MOD emphasizes the commoditization of both passenger mobility and goods delivery, whereas MaaS focuses on mobility aggregation and subscription services, typically through a smartphone or website. Because of MOD's role in managing the supply and demand of the transportation network, a number of marketplace stakeholders and enablers support the MOD ecosystem. Common stakeholders include: (1) the federal government, (2) state and local authorities, (3) public transit agencies, (4) transportation operators and logistics providers, (5) transportation managers, (6) apps and mobile service providers, and (7) the public. Business models and partnerships, infrastructure, public policy, and technology are key enablers of MOD. For the consumer, MOD can create opportunities to enhance access and equity by providing increased mobility options (e.g., fares, routes); increased travel speed and reliability; critical first-and-last-mile connectivity; and expanded coverage to historically underserved users or communities. However, the demographics of MOD users often differ from the general population raising concerns about equity challenges such as: serving people with disabilities, low-income communities, and unbanked households. The FTA MOD Sandbox demonstration program highlights the importance of pilots and demonstrations to enhance public transportation's preparedness for paradigm shifts in transportation including: MOD, the importance of public-private partnerships and mobility innovation, and the critical need to understand the impacts of MOD on the entire transportation network. More research and proactive public policies are needed to guide sustainable outcomes as MOD becomes automated in the future.

Today, more than a dozen SAV pilot projects are currently underway in the U.S. with more planned in the coming years. In the future, as more companies test AVs in urban settings, cities may be confronted with a variety of regulatory questions such as: (1) Who should regulate? (2) How does an agency regulate? and (3) What should be regulated (e.g., vehicles, consumer protections, pricing, etc.)? Additionally, in the future rights-of-way management could emerge as a more prominent issue as demand for a limited amount of curb space grows. With automation and big data, public agencies may be able to actively monitor curb space management. With better data, cities may be able to pursue curb space demand management programs with policies, such as dynamic pricing that lowers or raises with demand. Finally, real-time data sharing between SAVs and public transportation could allow for more efficient multimodal interactions, since each could share information on planned arrivals and departures to minimize conflicts and congestion. Although SAV impacts have yet to be seen, it is clear that data, research, and proactive policy will be necessary to maximize their potential benefits and mitigate any unforeseen impacts.

Acknowledgements The U.S. Department of Transportation generously funded this research. We would like to give special thanks to the academic researchers, transportation professionals, policymakers, and service providers who made this research possible. We also thank Gustave Cordahi, Sara Sarkhili, and Balaji Yelchuru of Booz Allen Hamilton and Adam Stocker, Emma Lucken, Marcel Moran, and Michael Randolph of the University of California, Berkeley's Transportation Sustainability Research Center for their support with this research. The contents of this chapter reflect the views of the authors and do not necessarily indicate sponsor acceptance.

References

1. Barr A, (2018) Waymo gets the O.K. for a commercial driverless ride-hailing service. https://www.bloomberg.com/news/articles/2018-02-16/waymo-gets-o-k-for-commercial-driverless-ride-hailing-service. Accessed 16 Feb 2018
2. Bliss L (2018) Uber just laid off its pittsburgh autonomous car drivers. https://www.citylab.com/transportation/2018/07/uber-just-fired-its-pittsburgh-av-drivers/564947/. Accessed 11 July 2018
3. Bloom J (2018) Californias first driverless bus hits the road in San Ramon. https://www.bishopranch.com/media-coverage/californias-first-driverless-bus-hits-the-road-in-san-ramon/. Accessed 28 Aug 2018
4. Cameron O (2017) Voyages first self-driving car deployment. https://news.voyage.auto/voyages-first-self-driving-car-deployment-29c7688c6a1. Accessed 4 Oct 2017
5. Caplan A (2018) Self-driving shuttle hits the streets. https://www.gainesville.com/news/20180503/self-driving-shuttle-hits-streets. Accessed 3 May 2018
6. Chen D, Kockelman K (2016) Management of a shared autonomous electric vehicle fleet: implications of pricing schemes. Transp Res Rec: J Transp Res Board 2572:37–46
7. Cohen A, Shaheen S (2016) Planning for shared mobility. American Planning Association, Chicago
8. Mobility as a Service (MaaS) and Mobility on Demand (MOD) via Blockchain (2018) https://medium.com/iomob/mobility-as-a-service-maas-and-mobility-on-demand-mod-via-blockchain-64e36a2f6676. Accessed 27 Aug 2018
9. Conduent Inc. n.d. Go-LA App (2018) https://itunes.apple.com/us/app/go-la/id1069725538?mt=8. Accessed 31 Aug 2018
10. Corder DR (2018) http://www.thevillagesdailysun.com/news/villages/driverless-taxi-service-coming-to-the-villages/article_9c59d096-f5c3-11e7-ad8c-d3c196b2ee33.html
11. Daniel D (2018) In Southwest Florida, taking a shine to the nations first solar-powered town. February 1. https://www.washingtonpost.com/lifestyle/travel/in-southwest-florida-taking-a-shine-to-the-nations-first-solar-powered-town/2018/02/01/cc7b710a-fd69-11e7-ad8c-ecbb62019393_story.html?noredirect=on&utm_term=.fbfa1013f6aa
12. Davidson P, Spinoulas A (2016) Driving alone versus riding together - How shared autonomous vehicles can change the way we drive. Road Transp Res: J Aust N Z Res Pract 25(3):51–66
13. Etherington D (2018) Texas replaces local bus service with Via on-demand ride-sharing, March 12. https://techcrunch.com/2018/03/12/arlington-texas-replaces-local-bus-service-with-via-on-demand-ride-sharing/. Accessed 31 Aug 2018
14. Etherington D (2017) Optimus ride will provide self-driving vehicles to Boston community residents. https://techcrunch.com/2017/11/28/optimus-ride-will-provide-self-driving-vehicles-to-boston-community-residents/. Accessed 28 Nov 2017
15. Federal Transit Administration (2018) Mobility on Demand (MOD) Sandbox Pinellas Suncoast Transit Authority (PSTA) Public-Private-Partnership for Paratransit Mobility on Demand Demonstration (P4-MOD). https://www.transit.dot.gov/sites/fta.dot.gov/files/FTA%20MOD%20Project%20Description%20-%20Pinellas%20Suncoast.pdf. Accessed 28 Aug 2018

16. Federal Transit Administration (2016) Mobility on Demand (MOD) Sandbox Program, May 3. https://www.transit.dot.gov/funding/applying/notices-funding/mobility-demand-mod-sandbox-program Accessed 27 Aug 2018
17. Federal Transit Administration (2018) Mobility on demand sandbox program, June 26. https://www.transit.dot.gov/research-innovation/mobility-demand-mod-sandbox-program.html. Accessed 27 Aug 2018
18. Federal Transit Administration (2018) Mobility on demand sandbox San Francisco bay area rapid transit district integrated carpool to transit access program. https://www.transit.dot.gov/sites/fta.dot.gov/files/FTA%20MOD%20Project%20Description%20-%20BART.pdf. Accessed 27 Aug 2018
19. Felton R (2018) GM cruise prepping launch of driverless car pilot in San Francisco: Emails. Retrieved from Jalopnik. https://jalopnik.com/gm-cruise-prepping-launch-of-driverless-car-pilot-in-sa-1826571157. Accessed 5 June 2018
20. Franco M (2016), DHL uses completely autonomous system to deliver consumer goods by drone, New Atlas, May 10, 2016. http://newatlas.com/dhl-drone-delivery/43248/. Accessed 17 May 2017
21. Greenblatt J, Shaheen S (2015) Automated vehicles, on-demand mobility, and environmental impacts. Curr Sustain/Renew Energy Rep 2:74–81
22. Hawkins AJ (2018) Lyft and Aptiv have completed 5,000 paid trips in their self-driving taxis. https://www.theverge.com/2018/8/21/17718018/lyft-aptiv-self-driving-cars-las-vegas-5000-trips. Accessed 21 Aug 2018
23. Hawkins AJ (2018) The self-driving cars hitting the road in Texas today are unlike any we've seen before. https://www.theverge.com/2018/7/30/17622540/drive-ai-self-driving-car-ride-share-texas. Accessed 30 July 2018
24. Local Motors (2018) https://localmotors.com/meet-olli/
25. Locklear M (2018) nuTonomy can test autonomous vehicles city-wide in Boston. https://www.engadget.com/2018/06/21/nutonomy-test-autonomous-vehicles-city-wide-boston/. Accessed 21 June 2018
26. Marshall A (2017) My Herky-Jerky ride in general motors' ultra-cautious self driving car. https://www.wired.com/story/ride-general-motors-self-driving-car/. Accessed 29 Nov 2017
27. Martinez M (2018) Detroit autonomous shuttle venture may mobility eyes more cities. http://www.autonews.com/article/20180801/OEM11/180809934/may-mobility-detroit-autonomous-shuttles-expansion. Accessed 1 Aug 2018
28. Massachusetts Governor's Office (2016) Governor Baker, MBTA launch innovative program to enlist Uber, Lyft to better serve paratransit customers, September 16. https://blog.mass.gov/governor/transportation/governor-baker-mbta-launch-innovative-program-to-enlist-uber-lyft-to-better-serve-paratransit-customers/. Accessed 31 Aug 2018
29. McFarland M (2017) Robot deliveries are about to hit U.S. streets, January 18. https://money.cnn.com/2017/01/18/technology/postmates-doordash-delivery-robots/index.html. Accessed 21 Aug 2018
30. NASA Embraces Urban Air Mobility (2017) Calls for market study, November 7. https://www.nasa.gov/aero/nasa-embraces-urban-air-mobility. Accessed 21 Aug 2018
31. Noble B (2018) Self-driving shuttles begin running in downtown Detroit. https://www.detroitnews.com/story/business/autos/mobility/2018/06/26/self-driving-shuttles-downtown-detroit-gilbert/719933002/. Accessed 26 June 2018
32. Scottsdale (2018) Meet Nuro. https://medium.com/nuro/az-pilot-launch-33cceb55c871. Accessed 16 Aug 2018
33. Pinellas Suncoast Transit Authority (2018) PSTA, Uber Offer Free, Late-Night Rides for Low-Income Residents. https://www.psta.net/about-psta/press-releases/2016/psta-uber-offer-free-late-night-rides-for-low-income-residents/. Accessed 31 Aug 2018
34. Building a Connected City From the Ground Up (2018) https://www.nytimes.com/2018/04/03/business/smart-city.html. Accessed 3 April 2018
35. Rogers S (2018) Uber puts self-driving cars back to work but with human drivers. https://interestingengineering.com/uber-puts-self-driving-cars-back-to-work-but-with-human-drivers. 26 July 2018

36. SAE International (2018) Taxonomy and definitions for terms related to shared. SAE International, Detroit
37. Shaheen S, Cohen A, Roberts D (2006) Carsharing in North America: market growth, current developments, and future potential. Transp Res Rec: J Transp Res Board 1986:116–124
38. Shaheen S, Cohen A, Martin E (2017) The U.S. department of transportation's smart city challenge and the federal transit administration's mobility on demand sandbox, in E-Circular. Transportation Research Board, Washington D.C
39. Shaheen S, Cohen A, Martin E (2018) US DOTs Mobility on Demand (MOD) initiative: moving the economy with innovation and understanding, in E-circular. Transportation Research Board, Washington D.C
40. Shaheen S, Cohen A, Zohdy I (2016) Shared mobility current practices and guiding principles. U.S. Department of Transportation, Washington D.C
41. Shaheen S, Cohen A, Jaffee M (2018) Innovative mobility carsharing outlook. University of California, Berkeley
42. Shaheen S, Cohen A, Yelchuru B, Sarkhili S (2017) Mobility on demand operational concept report, U.S. Department of Transportation, Washington D.C
43. Shaheen S, Cohen A, Zohdy I, Kock B (2016) Smartphone applications to influence traveler choices practices and policies. U.S. Department of Transportation, Washington D.C
44. Shaheen S, Cohen A (2017) Mobility innovations take flight: flying cars are on their way, in InMotion, March 31. https://www.inmotionventures.com/mobility-innovations-flying-cars/. Accessed 16 May 2017
45. Shaheen S, Cohen A (2018) Mobility On demand and transportation equity, March 15. https://www.move-forward.com/mobility-on-demand-and-transportation-equity/. Accessed 27 Aug 2018
46. Shaheen S, Cohen A (2018) The seismic shift in transportation. https://www.inmotionventures.com/seismic-shift-transportation/. Accessed 29 Aug 2018
47. Shaheen S, Bell C, Cohen A, Yelchuru B (2017) Travel behavior: shared mobility and transportation equity. U.S. Department of Transportation, Washington D.C
48. Shaheen S, Martin E, Chan N, Cohen A, Pogodzinski M (2014) Public bikesharing in North America during a period of rapid expansion: understanding business models, industry trends, and user impacts. Mineta Transportation Institute, San Jose
49. Shared-Use Mobility Center (2018) Direct connect case study. Shared-Use Mobility Center, Chicago
50. Sochor J, Stromberg H, Karisson MA (2015) Implementing mobility as a service: challenges in integrating user, commercial, and societal perspectives. Transp Res Rec: J Transp Res Board 2536:1–9
51. Starship (2018) Starship. https://www.starship.xyz/. Accessed 21 Aug 2018
52. Stocker A, Shaheen S (2018) Shared automated vehicle (SAV) pilots and automated vehicle policy in the U.S. Springer, New York
53. University of Michigan (2018) Mcity driverless shuttle launches on U-Ms North Campus. https://news.umich.edu/mcity-driverless-shuttle-launches-on-u-ms-north-campus/. 4 June 2018
54. USDOT (2017) USDOT, September. https://www.nhtsa.gov/sites/nhtsa.dot.gov/files/documents/13069a-ads2.0_090617_v9a_tag.pdf
55. Warren T (2016) This autonomous, 3D-printed bus starts giving rides in Washington, DC today. https://www.theverge.com/2016/6/16/11952072/local-motors-3d-printed-self-driving-bus-washington-dc-launch. 16 June 2016
56. WINK News (2018) New driverless shuttles to hit the roads in Babcock Ranch. http://www.winknews.com/2018/01/12/new-driverless-shuttles-hit-roads-babcock-ranch/. 12 Jan 2018
57. Yvkoff L (2017) FedEx sees robots, not drones, as the next big thing in logistics, in the drive, February 7. http://www.thedrive.com/tech/7430/fedex-sees-robots-not-drones-as-the-next-big-thing-in-logistics. Accessed 17 May 2017

Data-Driven Rebalancing Methods for Bike-Share Systems

Daniel Freund, Ashkan Norouzi-Fard, Alice Paul, Carter Wang, Shane G. Henderson and David B. Shmoys

Abstract As bike-share systems expand in urban areas, the wealth of publicly available data has drawn researchers to address the novel operational challenges these systems face. One key challenge is to meet user demand for available bikes and docks by rebalancing the system. This chapter reports on a collaborative effort with Citi Bike to develop and implement real data-driven optimization to guide their rebalancing efforts. In particular, we provide new models to guide truck routing for overnight rebalancing and new optimization problems for other non-motorized rebalancing efforts during the day. Finally, we evaluate how our practical methods have impacted rebalancing in New York City.

Keywords First keyword · Second keyword · More

This work was partially supported by National Science Foundation grants CCF-1526067, CMMI-1537394, CCF- 1522054, and CCF-1740822 as well as Army Research Office grant W911NF-17-1-0094.

D. Freund (✉)
MIT, Cambridge, USA
e-mail: dfreund@mit.edu

A. Norouzi-Fard
EPFL, Lausanne, Switzerland
e-mail: ashkan.norouzifard@epfl.ch

A. Paul
Brown University, Providence, USA
e-mail: alice_paul@brown.edu

C. Wang
Motivate International, New York, USA
e-mail: carterwang@motivateco.com

S. G. Henderson · D. B. Shmoys
Cornell University, Ithaca, USA
e-mail: sgh9@cornell.edu
e-mail: dbs10@cornell.edu

© Springer Nature Switzerland AG 2020
E. Crisostomi et al. (eds.), *Analytics for the Sharing Economy: Mathematics, Engineering and Business Perspectives*,
https://doi.org/10.1007/978-3-030-35032-1_15

1 Introduction

Bike-share systems have become a fixture of urban transportation landscapes, offering sustainable alternatives for both tourists and everyday commuters. In providing connections between existing transportation modes, these systems enable users to change their entire commute to a more sustainable one. In addition, they provide access to transportation in neighborhoods that historically had none. The wealth of data these systems collect and the high predictability of aggregate user traffic allow for in-depth analysis of customer demand and system operations.

More so than with other modes of transportation, users directly affect the state of bike-share systems. This presents novel operational challenges. In particular, asymmetric demand creates empty and full stations across the city; while empty stations prevent customers from accessing the system, full stations, where every dock is taken, prevent them from leaving the bike there. Thus, one of the operator's key challenges is to meet demand by relocating bikes and temporarily increasing station capacity. This is referred to as rebalancing.

In the past few years, bike-share systems have developed different approaches to rebalancing. In the most common approach, trucks are used to move bikes to high-demand areas. This is particularly effective overnight, when both traffic and demand are low. During the day, vehicular traffic impairs these efforts, and instead operators use *trikes* or *corrals*. A trike is a trailer that typically holds at most five bikes and is towed by a cyclist to relocate bikes between a station A, with high supply, and a station B, with low supply (cf. Fig. 1). A corral, on the other hand, artificially increases the capacity of a popular station by having an employee store bikes in between docks, thereby using all of the available space (cf. Fig. 1).

We provide data-driven methods to help New York City Bikeshare (NYCBS) improve overall utilization of the Citi Bike system and reduce customer dissatisfaction. This chapter formulates and attacks the underlying optimization problems that arise in truck-based rebalancing overnight, trike-based rebalancing during the day, and the placement of a limited number of seasonal corrals. For each of these settings, we describe the methods developed and their impact for NYCBS.

First, we consider optimally routing trucks to relocate bikes overnight. This is called the *overnight rebalancing problem*. Our objective in this problem is to minimize expected customer dissatisfaction over the next day. We present an integer program (IP) that constructs routes for a given number of trucks and show how to find good solutions when given a fixed amount of computation time in practice. Next, we consider the *mid-rush rebalancing problem*, studying how to optimally assign trikes to cycle between pairs of stations. Here, we use a maximum-weight k-edge matching to assign trike routes and maximize the impact on customer satisfaction. Finally, we consider how to optimally place corrals at stations where our goal is to minimize the number of customers who cannot find an open dock within a quarter mile of their preferred destination. We model this question as a maximum coverage problem that we solve within seconds using a simple integer programming formulation.

Fig. 1 Pictures of a corral and a trike being used in NYC

Each of the methods in this paper are in different development stages for NYCBS. In particular, we have completed trial runs using our overnight rebalancing schedules, routing three to four of their trucks over an eight-hour period. In the trials we ran, our routes were able to reduce expected customer dissatisfaction an additional 12% on average when compared to their usual efforts. NYCBS has also used our proposed placement of corrals in 2016. Further ahead, we have implemented our method to match trikes to stations and are in discussions with NYCBS to have our model inform their operational schedule in future seasons. We are also in discussions with bike-share systems in other major American cities (e.g., Washington D.C., Boston, Bay Area) about how these methods can make local transportation more sustainable and user-friendly.

This chapter thus summarizes our methodological contributions and their impact on day-to-day operations in New York City.

2 Related Work

Our objective in overnight rebalancing is to minimize the *expected number of dissatisfied customers*, or the expected number of customers who are not able to access or leave the system. This objective (cf. Sect. 3) function, often referred to as the *user dissatisfaction function* (UDF), has first been defined in [33]; different ways of computing it were suggested in [16, 30, 40].

The definition of the UDF triggered a line of work that used this objective to formulate a static rebalancing problem, wherein capacitated trucks are routed in a bike-sharing system over a finite time interval so as to minimize the subsequent number of dissatisfied users. Examples, solving larger and larger instances, include [12, 17, 44]. Our optimization model in Sect. 4 is similar to these, though it differs with respect to underlying constraints.

This approach is very much related to the so-called *Static Bicycle Rebalancing Problem*, which was first introduced in [1]. Here, each station has a target number of bikes, and the goal is to route finite capacity trucks to pick up and drop off bikes to meet these targets while minimizing the route length. Variations of this formulation have been studied in [2, 10, 11]. The last of these provides an IP formulation that allows vehicles to visit the same location repeatedly; [4] shows that conditions exist that guarantee that multiple visits to the same location are unnecessary.

Whereas most of the previously mentioned approaches are based on integer programming techniques, [32] defines greedy heuristics to solve a similar formulation. In fact, this and subsequent works like [9] as well as [23, 29] combine minimizing travel time and maximizing impact in their objectives. Reference [23] consider what they refer to as the *dynamic case*, where demand occurs both during and after the interval in which rebalancing happens. Fundamentally different methodological approaches to rebalancing have been taken by [25] whose rebalancing approach is based on a clustering of stations and [14, 15, 26] who develop a robust optimization framework.

Beyond rebalancing, researchers have also looked at the question of what the optimal allocation of bikes should look like at any given time. This can be thought of as solving a rebalancing problem in which the time taken for rebalancing is neither bounded through constraints nor penalized in the objective. Methodologically, this has been approached through the UDFs and simulation optimization frameworks. [13, 16] apply UDFs; [19, 20] develop a simulation optimization that aims to minimize the number of dissatisfied users, but takes into account (more explicitly) the interdependencies between stations. Reference [7] also takes into account interdependencies between stations and additionally even considers customers' behavioral responses to empty/full stations with respect to transportation alternatives. Rather than minimizing the number of users that do not get served, [7] then minimize the total travel-times of all users, including ones that had to find an alternative to the bike-sharing system.

There are also flow formulations to capture the optimal allocations of bicycles rather than the routing problem: examples of such approaches include [6, 41, 45]. These do not explicitly take into account routing, though they do contain some costs to account for rebalancing between stations.

Work on the optimal allocation of bikes at a given time has been complemented by studies of the system design itself. Such studies so far have mostly focused on the (re-)allocation of dock capacity within the system. Examples include the aforementioned [19] and, closely related, [39] (simulation), [13, 16] (UDF). A somewhat orthogonal approach, using econometric techniques to infer the impact on demand from reallocated capacity, is found in [21, 47].

In addition to the widespread study of optimization problems in the design and operation of bike-sharing systems, there are also a variety of forecasting-related problems that have been explored. In particular, there are several examples of demand forecasting, including [24, 28, 36–38]—in this chapter we apply the decensoring techniques suggested in [28]. Reference [22] identified unusable bikes, given usage data, which has relevance for routing problems in maintenance as studied by [31]. Researchers have also focused on more specific demand problems: [18] examines customer choices at empty or full stations, [46] studies the likely destination of each customer conditioned on the customer's attributes (e.g., gender, age, origin), and [42] tries to predict demand in neighborhoods based on census data.

Although there has been extensive work on overnight, motorized rebalancing, comparably little research has been conducted on non-motorized rebalancing efforts, which form a crucial part of NYCBS's operations. One such work is [16], which studies the routing of trikes. This paper partitions the set of stations into *producers*, which are stations likely to fill up, and *consumers*, likely to empty out. In investigating the problem of setting trike routes, the paper aims to minimize the distance of any consumer (producer) to another consumer (producer) that is rebalanced by one of the trikes. While our work on trikes is driven by the same application, our objective (cf. Sect. 5) is again to minimize the expected number of dissatisfied customers. A different kind of non-motorized rebalancing has been investigated by [5, 43]: they each investigate operational questions related to an incentive program like NYCBS's *Bike Angels*, which crowdsources some of the rebalancing efforts to the user base.

Finally, our work distinguishes itself in that it was conducted in close cooperation with NYCBS and has had an impact on their day-to-day operations. This stands in contrast to the findings of [8], which conclude that very little of the existing work on rebalancing has had an impact in practice.

3 User Dissatisfaction Function

Prior work by [27, 33, 40] defines a *user dissatisfaction function* that uses demand information to map the number of bikes at a station at the beginning of a time interval to the expected number of customers that are unable to access/leave the system at that station over the course of the interval.

The user dissatisfaction function is based on an $M/M/1/\kappa$ queue at each station s, in which the state of the queue $X_s(t)$ at time t corresponds to the number of bikes in the station at that time. More precisely, there are two Poisson processes with rates μ and λ. Both rates can be functions of time, but they are assumed to be exogenous and independent of the operator's actions. An arrival of the Poisson process with rate μ moves the state of the queue from i to $i-1$ if $i > 0$. If $i = 0$, the state of the queue does not change. This represents the arrival of a user who wants to pick up a bike but cannot due to a lack of available bikes. We assume that such a customer leaves the system dissatisfied. Similarly, an arrival of the Poisson process with rate λ moves the state of the queue from i to $i+1$ if $i < \kappa$. If $i = \kappa$, the state of the queue does not change; the user is also dissatisfied in this case as she cannot drop off her bike due to a lack of available docks. The UDF F_s uses these Poisson processes to map an initial number of bikes to the expected number of such dissatisfied users over a given finite time horizon (cf. Fig. 2). In the prior work mentioned above it was shown that $F_s(\cdot)$ is convex and can be efficiently computed for time-invariant λ and μ. In later work (cf. [30]) a dynamic program was derived to extend those methods to the time-varying rates we have used in our work. To obtain values for λ and μ we use the decensoring method introduced in [28].

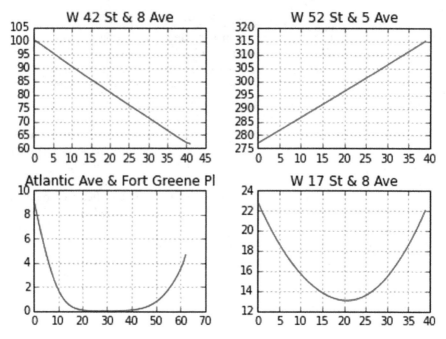

Fig. 2 User dissatisfaction function for four stations, with different demand patterns, in NYC for 7AM–10PM

4 Overnight Rebalancing

New York City Bikeshare deploys between three and five box trucks every night to redistribute bikes. Based on an analysis of past usage data (cf. [28]), NYCBS aims to ensure that both bikes and open docks are available where needed. This is critical because customers' most cited complaint about bike sharing is the lack of available bikes or docks (cf. [3]). Given the latent demand, perfect customer satisfaction, i.e. always having at least one bike and one dock available at every station, seems unattainable: our estimates for the user dissatisfaction function suggest that with the optimal distribution of bikes across the system and no rebalancing thereafter, there would still be more than 12000 dissatisfied over the course of the day (cf. Fig. 2). Thus, obtaining perfect availability would require the relocation of an estimated 6000 bikes (each relocated bike can help satisfy demand for 2 customers: one searching a dock at the station the bike is taken from, and one searching a bike at the station it is placed at), which is beyond the current scope of NYCBS's efforts. However, rebalancing can nevertheless yield improvements for the system when it decreases the (expected) number of dissatisfied customers. In this section we discuss our work

Fig. 3 Truck routes for three trucks on August 8, 2016. Each circle corresponds to a station at which at least one of the trucks stops. A white outer circle corresponds to a pick-up, a black outer circle to a drop-off. Initially, all trucks start at a NYCBS depot in the East Village. (Map data: Google)

with NYCBS to increase the efficiency of their overnight rebalancing towards that goal (see Fig. 3 for a visualization of resulting routes).

As was done in the earlier work on overnight rebalancing, we use an IP formulation to model this optimization problem. The main distinguishing feature of our approach is that we greedily partition the problem into subproblems, thereby controlling the size of the IP. In particular, we find routes for subintervals of the entire time horizon, subsets of trucks, and subsets of stations. Our choice of stations is based on carefully weighing the potential benefits of rebalancing at each station and the distance of the station from the location of the truck at the beginning of the interval.

4.1 Integer Programming Formulation

We begin by introducing our integer program. The formulation is time-indexed, and we assume the given time for overnight rebalancing has been broken into T identical time steps. In each time step, a truck will either pick up/drop off bikes or move to an adjacent station. In that way, the edges between the stations are unweighted and traversing any edge takes one time step. In order to make this assumption reasonable, we add dummy stations to break long distances into individual time steps. More precisely, if the distance between two stations s_1 and s_2 is ℓ time steps, we add a path of $\ell - 1$ stations between them, that allows us to move between such two stations in exactly ℓ time steps by traversing one edge in each step. This technique significantly reduces the dimension of the IP and distinguishes our formulation from previous ones.

Notation.

After adding all dummy stations, let S be the set of stations, T be the number of time steps, and K be the number of trucks. For $s \in S$, $t \in [T]$, and $k \in [K]$, the variable x_{stk} represents whether or not truck k is at station s at time t. Similarly, the variable y_{stk} represents the number of bikes at station s at time t to which truck k has access. This prevents multiple trucks from moving the same bikes. Lastly, the variable b_{tk} represents the number of bikes in truck k at time t.

We use the following notation:

- $N(s)$ denotes the neighborhood of s, that is, the stations to which a truck can move from s in a single time step.
- γ is the number of bikes that can be picked up or dropped off in one time step.
- start(s) is the number of bikes in station s at time $t = 1$.
- min(s) is the minimizer of the user dissatisfaction function at station s. That is, the number of bikes at station s that minimizes the expected number of dissatisfied customers.
- $c_s = \frac{F_s(\text{start}(s)) - F_s(\text{min}(s))}{|\text{start}(s) - \text{min}(s)|}$ is a linear approximation of the slope of F_s and gives the improvement per bike moved at s (see Fig. 4).
- S_+ is the set of stations s for which start$(s) > $ min(s). For example, the station in Fig. 4 is in S_+.
- S_- is the set of stations s for which start$(s) \leq $ min(s).

For ease of presentation, we state here only the main constraints of the IP before we explain the effect of each. The reader is advised to read the IP in parallel with the explanations below.

$$\text{maximize}_{x,y,b} \sum_{s \in S, k \in [K]} (y_{s1k} - y_{sTk}) c_s$$

subject to:

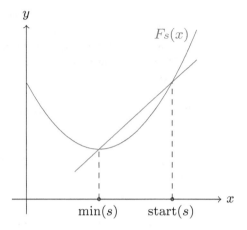

Fig. 4 Linearization of F_s

$$x_{stk} \leq x_{s(t-1)k} + \sum_{s':s\in N(s')} x_{s'(t-1)k}, \qquad \forall s,t,k; \qquad (1)$$

$$\sum_{s\in S} x_{stk} = 1, \qquad \forall t,k; \qquad (2)$$

$$\sum_{k\in[K]} y_{s1k} = \text{start}(s), \qquad \forall s; \qquad (3)$$

$$\text{start}(s) \leq \sum_{k\in[K]} y_{stk} \leq \min(s), \qquad \forall s \in S_-, t; \qquad (4)$$

$$\min(s) \leq \sum_{k\in[K]} y_{stk} \leq \text{start}(s), \qquad \forall s \in S_+, t; \qquad (5)$$

$$\sum_{s\in S} y_{stk} + b_{tk} = \sum_{s\in S} y_{s1k} + b_{1k}, \qquad \forall t,k; \qquad (6)$$

$$|y_{stk} - y_{s(t-1)k}| \leq \gamma x_{stk}, \qquad \forall s,t,k; \qquad (7)$$

$$|y_{stk} - y_{s(t-1)k}| + \gamma |x_{stk} - x_{s(t-1)k}| \leq \gamma, \qquad \forall s,t,k. \qquad (8)$$

Below we explain the function of each part of the IP.

- The objective function is the summation of changes in the linearized user dissatisfaction functions at each station, i.e. the reduction in expected number of dissatisfied customers due to the relocation of bikes.
- **Constraint (1)** allows each truck to only move to a station adjacent to the one it is currently at.
- **Constraint (2)** indicates that at each time step, each truck must be in exactly one station.

- **Constraint (3)** initiates the number of bikes at every station. Notice that at this point already the bikes are distributed among the K trucks.
- **Constraints (4)** and **(5)** guarantee that the number of bikes in each station s remains between start(s) and the minimizer min(s). In other words, we enforce that moving a bike only improves the setup (cf. *Pareto Constraints and Optimal Fleet Size*).
- **Constraint (6)** enforces that the total number of bikes in the system does not change over time.
- **Constraint (7)** makes sure that we pick/drop bikes at a station from a truck only if the truck is at that station and that the number of bikes moved is bounded by γ, the number of bikes rebalancers are able to move within one period.
- **Constraint (8)** ensures the truck either moves or picks/drops bikes in one time step but not both. In most of the previous works, researchers have omitted this constraint. This constraint makes the IP significantly harder to solve but makes the resulting path viable in practice. Notice that the absolute values in the constraints can be linearized.

Moreover, we add capacities to the truck by bounding b_{tk}. In practice, we extend this IP to fix the starting/finishing stations for each truck, as well as the number of bikes in each truck at the beginning of the night.

Pareto Constraints and Optimal Fleet Size

Notice that the extension of the linearization of F_s beyond the point min(s) does not capture the actual behavior of the UDF. In particular, at that point the UDF sees more dissatisfied customers while the linearization sees fewer. This, however, is not the reason we impose the *Pareto constraints* (4) and (5), since optimizing over the linear envelope of F_s is not significantly harder than over the linearization; instead, the constraints are imposed to ensure that service for customers at one station is not sacrificed for improved service for customers elsewhere. This is particularly true when, as is the case in NYC, $\sum_s \text{start}(s) << \sum_s \text{min}(s)$ (at the time we started the pilots, the bike fleet size was about 7.000 whereas the minimizers summed to about 10.000). In such a setting, not having fairness constraints can yield undesirable outcomes. For instance, without fairness constraints we might find solutions in which bikes can be picked up from a station far away that has more bikes than needed but are instead picked up from a station nearby that has fewer than needed. While this may lead to fewer dissatisfied customers in total, NYCBS aims for rebalancing to be a Pareto improvement to the system (i.e., no station is worse off, some are improved); thus, we include constraints (4) and (5).

4.2 Solution Methods

As presented, state-of-the-art IP solvers are too slow to solve this formulation within the limited time window between when the operator receives the data and when the trucks must start their routes. The following heuristic methods help decrease the computation time to solve this IP and improve the quality of the solution returned.

Reducing the number of edges. To avoid adding too many dummy stations and inflating the size of the IP, we choose a threshold d on the distance between two stations and only add a path between stations s_1 and s_2 if they are at most d time steps apart.

Dividing T into smaller time intervals. At the start of the rebalancing period, there is little time (typically about 20 min) between the time all the data (state of the system, number of bikes on each truck, etc.) becomes available and the time when trucks are meant to begin their routes. To gain computation time, we break T into smaller intervals and only route trucks for the first few hours of the route. While this part of the route is being executed, we then use the time to solve for the next interval. This segmentation of the computation time greatly improved the quality of our overall routes.

Greedily selecting stations. For each time interval, we further reduce the size of the IP by removing stations where the room for improvement is low. We rank stations based on a combination of c_s, the potential benefit of each bike picked or dropped, and $|\min(s) - \text{start}(s)|$, the number of available/required bikes, and run the IP with roughly 40 stations, further refined by excluding stations too far away from the starting point of the truck.

Splitting trucks. Instead of solving one IP for all K trucks, we break the computation time into K equal pieces and solve for the route of the first truck, then the second truck, and so forth. For example, if we have two trucks and two hours of computation time, we would solve for the route of each truck in one hour.

To decrease the size of the IP, we only add stations in reasonable proximity to the current position of the truck. We then compute the path over one time interval and update the set of stations. On the one hand, our IP can find very good solutions for the smaller instances on which we solve. On the other hand, by picking stations in the described greedy fashion, we ensure that the combination of the solutions to the small instances has objective close to the global optimum. The routes we construct for each truck and time interval are compatible in that they can be pieced together to form one coherent route. We show in our results that these heuristics still yield good solutions to the original IP, that is, we compare our objectives to the LP relaxation of the full IP.

Fig. 5 A posteriori optimization for overnight truck rebalancing in Manhattan

360 Stations		
# of Trucks	Avg Objective Function	Avg Gap
1	148	28.1%
2	232	22.1%
3	301	12.3%
4	330	9.5%

4.3 Results

In this section, we summarize our results. First, we report the gap between the solution we return using the techniques above and an optimal solution of a valid LP relaxation. Second, we compare our solutions to the current routes employed by NYCBS, as guided by tools based on our earlier work. On average, our solutions reduce customer dissatisfaction by 20% compared to this previous computationally informed approach. All results were produced using Gurobi v6.5 on a machine with 8GB RAM and an Intel i7-2600 processor with 3.4 GHZ.

IP gap

When solving the IP, we assume the number of time steps is 60, each time step corresponds to 6 min, and that workers can load/unload up to 7 bikes in each time step. These numbers are based on discussions with NYCBS. To evaluate the performance of our IP, we used our IP to route various numbers of trucks in Manhattan, which had around 360 stations at the time of our study. As inputs for start(s), we used the number of bikes at each station at midnight over the course of a week. In Fig. 5, we report the average impact on the number of dissatisfied customers and the average integrality gap over the week. The worst integrality gap occurred Sunday night when the distribution of bikes was furthest from the commuter demand during the week. We emphasize that this gap is computed with respect to the LP-relaxation on all 360 stations, all trucks, and all time-steps, i.e. the LP-relaxation of the provided integer program; hence, it includes both the integrality gap we find for the smaller problem instances (on 30–40 stations) and the gap between the optima of these smaller instances and the global optimum.

The average integrality gap decreases as the number of trucks grows. This seems to be a consequence of the Pareto constraints, since there are fewer bikes available to be moved (per truck). With more trucks, it is easier to achieve the best possible solution. We remark that a much smaller integrality gap can be obtained without constraints (7) and (8) as has been done in [12, 17, 34]. The solutions obtained this way, however, allow for as many stops (with loading/unloading of few bikes at each) as there are time periods; in discussions with NYCBS we found that in many cases the time required to find a place to park the truck is not sufficiently dominated by driving/(un)loading time that it could be ignored. We thus had to add these constraints, even though it makes the IP much harder to solve.

Practical Results

Dispatchers at NYCBS currently use a myopic decision aid to route trucks. This aid is based on user dissatisfaction functions and was developed in close cooperation with our group. While this decision aid shows dispatchers the optimal fill level at stations and indicates stations where rebalancing could yield large improvements, it does not provide optimized routes. In contrast, our IP solutions look to globally optimize routes. We formulated our model with feedback from NYCBS; their expertise led, for example, to the refinement that in each time step a truck can either move or pick/drop bikes but not both. Over the course of a week in July 2016, we then compared our proposed routes with the manual routes executed by the dispatchers at NYCBS. On average our results showed objectives about 20% higher. Together with the NYCBS management, we reviewed our proposed routes before running pilots to route between 2 and 5 of their trucks overnight.

Pilot Experiences

Despite the improvements, the pilots conducted with NYCBS proved to be more difficult than expected, as we quickly describe here.

Unknown constraints. Our first attempt to pilot our method had to be aborted because we routed a truck to a station that was in a narrow street. Indeed, the street was so narrow that the driver did not feel comfortable driving the truck to the station. Since we did not know how to handle the missed stop (and the resulting difference in number of bikes aboard the truck) at subsequent stops, we aborted that first attempt as a pilot. While events like this are rare, they do happen and when they do, they significantly complicate the routing problem. As an heuristic to overcome such problems, our subsequent pilots included alternative stations (greedily selected after solving the IP) at which each truck could stop without interfering with the routes of the other trucks.

Cost of solving offline. Solving the system offline can have detrimental consequences when unexpected demand occurs late at night. While the system significantly slows down at night, we have encountered cases where our route dictated picking up some number of bikes from a station and by the time the truck had arrived there, fewer than that were left. Similarly, it may occur that the Pareto constraint is violated in the execution of the route because the number of bikes at a station changes between the time we solve and the time the truck arrives. To avoid this occurring regularly, we excluded certain areas that tend to slow down later than others (e.g., East Village) in the first part of the route.

Evaluating results. In evaluating our results, we found that certain system conditions greatly influence the efficiency of rebalancing. On nights when the system is in great imbalance, it is not uncommon for each truck to move enough bikes to reduce the expected number of dissatisfied customers by >100 users. In contrast, on nights were the system is already reasonably balanced, it is often not feasible for a truck to move bikes to reduce the objective by >50.

While the final point indicates that evaluating results is difficult, the improvements through rebalancing on the nights during which we ran this pilot were on average 12% higher than in the rest of that month.

5 Trikes

In this section, we study the impact of trikes on the expected number of dissatisfied users and solve a corresponding optimization problem. Recall that a trike is a trailer towed by a cyclist that holds at most 5 bikes at a time (cf. Fig. 1). NYCBS uses trikes between fixed pairs of stations to move bikes from high-demand, low-supply stations to low-demand, high-supply stations. NYCBS considers trikes a preferred choice of rebalancing during rush hour when trucks are slowed down by traffic. We first assume that every station may have only one trike route incident to it. In this regime, the problem of finding the optimal m trike routes can be formulated as a bipartite maximum m-edge matching, where the weight of an edge between two stations corresponds to the reduction in user dissatisfaction with a trike added between them. Next, we generalize these ideas to the case where a station can utilize multiple trikes. With a slight variation, we show that this as well as can be formulated as a matching problem and efficiently solved.

5.1 Model

Similar to the user dissatisfaction function defined in Sect. 3, we model the *user dissatisfaction function with a trike* between stations A and B as follows: Poisson processes with rates $\lambda_A, \mu_A, \lambda_B, \mu_B$ correspond to arrivals and departures of users at station A and B, respectively. We use the random variable $X_s(t) \in \{0, \ldots, k_s\}$ to represent the number of bikes at station $s \in \{A, B\}$ at time t, where k_s is the capacity of station $s \in \{A, B\}$. As before, a dissatisfied customer at s corresponds to an arrival (resp. departure) when s is full (resp. empty), i.e., $X_s(t) = k_s$ (resp. $X_s(t) = 0$).

In addition to arrivals and departures of users, we also have a trike with capacity k_R that moves as many bikes as possible from A to B. We assume that at times $t_1, \ldots, t_r$ the trike stops at one of the stations. Without loss of generality, the stop at t_i is at A if i is odd and at B if i is even. In other words, the trike cycles back and forth between A and B. When the trike arrives at station A it picks up as many bikes as possible given the number of bikes at A and the number of bikes already in the trike; similarly, when the trike arrives at station B it drops off as many bikes as possible given the number of available docks at B and the number of bikes already in the trike. We use the random variable $X_R(t) \in \{0, \ldots, k_R\}$ to represent the number of bikes in the trike at time t.

We are interested in the expected number of dissatisfied users at A and B over the time horizon from t_0 to t_{r+1}. We can write the expected number of dissatisfied users at stations A and B as follows:

$$F_{A \to B} = \sum_{i=0}^{r} \sum_{j=0}^{k_A} \Pr(X_A(t_i^+) = j) F_A^i(j)$$
$$+ \sum_{i=0}^{r} \sum_{j=0}^{k_B} \Pr(X_B(t_i^+) = j) F_B^i(j)$$

The two terms correspond to the expected number of dissatisfied users at each station; to obtain them, notice that at the beginning of time-interval i with probability $\Pr(X_s(t_i^+) = j)$ there are j bikes at station s and in that case an expected $F_s^i(j)$ users will be dissatisfied in that interval. Thus, we need to find $\Pr(X_s(t_i^+) = j) \, \forall s, i, j$. For ease of notation, we denote $\Pr(X_A(t_i^+) = \alpha, X_B(t_i^+) = \beta, X_R(t_i^+) = \rho)$ as $\pi^i(\alpha, \beta, \rho)$ and remark that by setting $X_R(t_0^+) = 0$ with probability one and assuming that $X_A(t_0), X_B(t_0)$ are independent we obtain:

$$\pi^0(\alpha, \beta, \rho) = \begin{cases} 0, \text{ if } \rho > 0 \\ \Pr(X_A(t_0^+) = \alpha) \cdot \Pr(X_B(t_0^+) = \beta), \text{ else.} \end{cases}$$

This will represent the base case to recursively compute $F_{A \to B}$.

To calculate $\pi^i(\alpha, \beta, \rho)$ in general we need to analyze the system changes in each time interval. Specifically, given the Poisson process rates and the current number of bikes x at a station $s \in \{A, B\}$, we can compute the expected number of dissatisfied customers $F_s^i(\cdot)$ in an interval (t_i, t_{i+1}). In doing so, we also obtain the probability that there are y bikes in station s at the start of the next time interval. More precisely, $\forall s \in \{A, B\}, x, y \in \{0, \ldots, k_s\}$ we let

$$P_{x,y}^{s,i} := \Pr(X_s(t_{i+1}^-) = y | X_s(t_i^+) = x),$$

where $X_s(t^+) := \lim_{\epsilon \to 0^+} X_s(t + \epsilon)$, i.e., $X_s(t_i^+)$ is the number of bikes at s just after the trailer has stopped at the station and $X_s(t^-) := \lim_{\epsilon \to 0^+} X_s(t - \epsilon)$ is the number of bikes at s just before the trailer has stopped at the station.

Thus, for even i, we have $\pi^{i+1}(\alpha, \beta, \rho) = 0$ if $\alpha > 0$ and $\rho < k_R$, as otherwise the trike would pick up more bikes. Otherwise, we obtain

$$\pi^{i+1}(\alpha, \beta, \rho) = \sum_{x=0}^{k_A} \sum_{y=0}^{k_B} \sum_{z=0}^{\rho} \pi^i(x, y, z) P_{x, \alpha+\rho-z}^{A,i} P_{y,\beta}^{B,i},$$

where we define $P_{x,y}^{s,i} = 0$ for $y \notin \{0, \ldots, k_s\}$. This is because, with $X_R(t_i^+) = z$, the event that $\{X_A(t_{i+1}^+) = \alpha$ and $X_R(t_{i+1}^+) = \rho\}$ happens if and only if the number of bikes at A before the pick-up at t_{i+1} is $\alpha + \rho - z$.

Similarly, for odd i, we have $\pi^{i+1}(\alpha, \beta, \rho) = 0$ if $\beta < k_B$ and $\rho > 0$. Otherwise,

$$\pi^{i+1}(\alpha, \beta, \rho) = \sum_{x=0}^{k_A} \sum_{y=0}^{k_B} \sum_{z=\rho}^{k_R} \pi^i(x, y, z) P_{x,\alpha}^{A,i} P_{y,\beta+\rho-z}^{B,i}.$$

Recognizing that

$$\Pr(X_A(t_i^+) = j) = \sum_{y=0}^{k_B} \sum_{z=0}^{k_R} \pi^i(j, y, z) \quad \text{and}$$
$$\Pr(X_B(t_i^+) = j) = \sum_{x=0}^{k_A} \sum_{z=0}^{k_R} \pi^i(x, j, z),$$

we can now compute all $\Pr(X_s(t_i^+) = j)$ and $\pi^i(\alpha, \beta, \rho)$ recursively starting with the base cases for $i = 0$ and incrementing i. Thus, we can efficiently compute $F_{A \to B}$.

Given $F_{A \to B}$ for each pair of stations, we use these values to assign the trike routes. To formulate the problem as a maximum m-edge matching problem, we create a bipartite graph $G = (X \cup Y, E)$ where $X = Y$ is the set of stations. We set the weight of edge (A, B) equal to the difference of the expected number of dissatisfied users at stations A and B without the trike, given by

$$\sum_{j=0}^{k_A} \Pr(X_A(t_0) = j) F_A(j) + \sum_{j=0}^{k_B} \Pr(X_B(t_0) = j) F_B(j),$$

and the expected number with the trike, $F_{A \to B}$. Thus, the weight of any matching is equal to the reduction in the expected number of dissatisfied customers from the corresponding trike routes and vice versa. Further, there is a well-known and efficient algorithm for finding the maximum m-edge matching for any bipartite graph.

Relaxed Assumptions

If we allowed more than one trike incident to a particular station, the dimensionality of the dynamic program explodes, yielding it infeasible. A slight variation of the model, however, allows us to model this problem as a maximum-weight matching as well. Instead of coupling the random variables of numerous stations and trikes, we estimate for each station how many bikes trikes can pick up/drop off bikes over the time horizon in which they operate. To find a solution for k trikes we then create for each station s nodes $s_1, \ldots s_k$ and create an edge (A_i, B_j) with weight set to be the improvement at station A through the ith additional trike and at B through the jth additional trike. Notice that, in contrast to the earlier formulation, this does not incorporate the chance that a trike incident to station A does not have an effect because there are no bikes to pick up at the adjacent station B.

Fig. 6 Trike routes identified by the maximum-weight matching formulation. Red lines indicate trikes that pick up bikes at white circles and drop them off at black ones. (Map data: Google)

5.2 Results

Setting $k_R = 5$ and eight trike stops between 7:30 AM and 9:15 AM (thus moving a total of at most 20 bikes per station pair), we use the above dynamic program to compute the weight of the maximum matchings of various sizes; in Fig. 6 we display the maximum matching of size 8, in Fig. 7 the weight of all maximum matchings of size 1 through 20. Notice that for up to 5 trailers, there is an improvement of about 20 (fewer dissatisfied customers). While these numbers can be viewed in comparison to the numbers in Fig. 5, the costs of operating a truck are significantly higher than those of a trailer. Decisions about which to operate thus also depend on the relative costs of trucks and trailers.

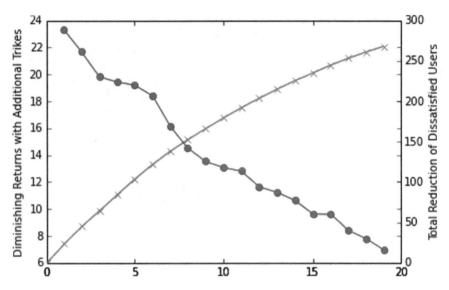

Fig. 7 Total improvement (red) of trikes and corresponding diminishing returns (blue)

6 Corrals

Our work on corrals is motivated by studies on the correlation between distance to transportation modes and willingness to use these modes. A study in [35], for example, claims that commuters are much more likely to use public transportation when living within a quarter mile of a station than when further away. Furthermore, it was shown in [21] that shorter distances to available stations correlates with increased demand for bike-share systems.

6.1 Model

Based on the findings described before, we say a station has a *shortage* if no station within a quarter mile has at least 15% of its docks available. In other words, a station is in shortage if a user intending to end their trip at that station will likely have to search more than a quarter mile away to find an available dock. Thus, stations in shortage significantly impact customer utilization.

We define a shortage measure for the system given by the total time stations are in shortage. Formally, let $N(s) \subseteq S$ be the set of neighboring stations within a quarter mile of s, including s itself, and let $A_{s,t}$ the indicator of the event that station s has at least 15% of its docks available at time t. For a set of stations S and a set of points in time T, we define the shortage measure as

Data-Driven Rebalancing Methods for Bike-Share Systems

$$w(S, T) = \sum_{s \in S} \sum_{t \in T} \max\{0, 1 - \sum_{j \in N(s)} A_{j,t}\}.$$

It is possible to extend this measure to a weighted version that associates a time-dependent coefficient based on dock demand to each station. Since the unweighted version was the one that informed decision-making for NYCBS, we restrict ourselves to that.

By placing a corral at a station s, the amount of time that stations in $N(s)$ are in shortage is significantly reduced. Given a budget B, the goal is to place at most B corrals to minimize the shortage measure. For a set of past time points in T', let

$$w_s = \sum_{t \in T'} \max\{0, 1 - \sum_{j \in N(s)} A_{j,t}\}.$$

This represents the amount of time that station s is in shortage and is equivalent to the reduction in the shortage measure for s if a station in $N(s)$ is assigned a corral. It is then natural to model corral placement as a maximum coverage problem via the following IP:

$$\text{maximize}_{x,y} \sum_{s \in S} w_s x_s$$

$$\text{subject to: } x_s \leq \sum_{s' \in N(s)} y_{s'} \quad \forall s$$

$$\sum_{s \in S} y_s \leq B$$

$$x_s, y_s \in \{0, 1\} \quad \forall s.$$

In the given IP, the variable y_s represents whether or not a corral is placed at station s and the variable x_s represents whether or not there is a corral placed within a quarter mile from station s. The first constraint sets x_s to be 0 if no corral is assigned in $N(s)$, and the second constraint corresponds to the constraint that we can assign at most B corrals.

6.2 Results

The evaluation of our work in this section is two-fold. Given the real implementation of our suggested corrals by NYCBS, we are able to observe the change in shortages from 2015 (without corrals) to 2016 (with corrals). However, we cannot be certain this corresponds 1-to-1 with a decrease in unsatisfied customers. Since the presence/absence of these is in general difficult to measure, we also use a discrete-event simulation to estimate the reduction in dissatisfied customers through our choice of corrals.

Our discrete-event simulation is based on Poisson arrivals with fixed destinations at each station. In contrast to the user dissatisfaction function described in Sect. 3, the discrete-event simulation captures interdependencies between stations. For instance, customers who do not find a dock at a particular station, roam around to nearby stations until eventually they do. For details of the implementation, we refer the reader to [19, 27]; we use the same simulation with the addition of allowing for some stations to have temporarily increased capacity between 7.30AM and 5.30PM.

Simulated Impact on Unsatisfied Customers

To compare the performance of our coverage formulation, we consider several allocations of corrals (cf. Fig. 8):

1. *coverage '15/'16* are the allocations of corrals obtained by solving the coverage problem using data from 10 days in June 2015, and June 2016 respectively.
2. *k-median* is based only on the biking distance between separate stations. Using the Google Maps API, we obtained pairwise distances and then solved the resulting k-median instance.
3. *max minutes* relies only on the number of minutes a station had no docks available in the month of May 2016.
4. *None* allocates no corrals at all.

Our results are visualized in Fig. 8. Based on data from June 2016, we simulated 40 realizations of demand and computed, using common random numbers, for each realization and each allocation of corrals, the number of unsatisfied customers. Notice first that adding corrals, usually, only ever improves the objective (though corrals may also have second-order effects that increase the expected number of dissatisfied customers, which the plot shows). Moreover, the solutions obtained from the coverage formulation (run both with data from 2015 and with more current data from 2016) outperforms the solutions that are based only on geography (*k-median*) or only on number of full minutes (*max minutes*): on average the two coverage solutions observe 304 and 162 fewer unsatisfied customers, compared to just 4 fewer for the k-median solution; the max minutes solutions ranks in the middle with an average reduction of 77. The results thus show the necessity of taking both usage data and geography into account.

Observed Real Impact

Six corrals identified by the maximum coverage formulation (see Fig. 9) with $B = 6$ have been in place for most of Summer 2016. To evaluate the benefit of these corrals, we first consider the value of the shortage measure in July 2016 (with the corrals) and in July 2015 (without). As the size of the system has expanded significantly, we

Fig. 8 Results of 40 simulated days with different sets of corrals; × denotes the average performance

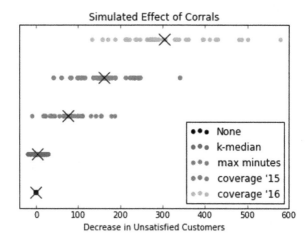

restricted the shortage measure to only include stations S in Manhattan that were in use for most of July 2015 and July 2016. For reference, $|S| = 277$. We find a 31.6% reduction in the time that stations were in shortage. Further, to show that the majority of the improvement is indeed due to the effects of the corrals, we also calculate the shortage measure in July 2015 and July 2016 for $\bar{S}$ which excludes the stations with corrals from S. This should give an idea as to how much the shortage measure has decreased through other rebalancing efforts. Here, we only find a 14.6% improvement which shows that the majority of the improvement was due to the added corrals. Results of these analyses are summarized in Fig. 10.

7 Conclusion

We provide models and solutions for three different optimization problems that arise in rebalancing bike-share systems. For overnight rebalancing and trailer routing, we focus on minimizing the expected number of dissatisfied users. First, we present our IP for the overnight rebalancing problem. This formulation, along with our preprocessing techniques, allows us to route trucks for New York City's large and complicated bike-share system. Here, we are able to improve on current rebalancing efforts by 20% on average. Next, we consider how to optimally add trikes between stations. We formulate this problem as a maximum-weight matching. In focusing on the same objective for these two problems, the expected number of dissatisfied users, we provide the operator with a quantitative comparison between two rebalancing methods. Lastly, we consider where to add corrals. Our analysis shows that a small number of corrals significantly improves user access to the system. While our work considered the effect of these rebalancing efforts separately, future work could consider how to combine these efforts. For example, it would be interesting to consider how adding an additional trike affects overnight rebalancing and analyzing

Fig. 9 Corral stations in NYC with $\frac{1}{4}$-mile radius. (Map data: Google)

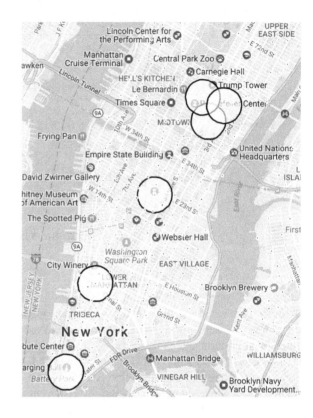

Fig. 10 Shortage measure for July 2015 and July 2016

	S	$\tilde{S}$
July 2015	49435	56989
July 2016	33804	48657
Percent Reduction	31.6%	14.6%

the cost-effectiveness of these efforts. Since combining these efforts would likely lead to intractable optimization problems, a promising direction is to incorporate rebalancing into simulation frameworks.

References

1. Benchimol M, Benchimol P, Chappert B, De La Taille A, Laroche F, Meunier F, Robinet L (2011) Balancing the stations of a self service bike hire system. RAIRO-Oper Res 45(1):37–61
2. Bulhoes T, Subramanian A, Erdoğan G, Laporte G (2018) The static bike relocation problem with multiple vehicles and visits. Eur J Oper Res 264:508–523
3. Bikeshare C (2014) capital bikeshare member survey report
4. Casazza M, Ceselli A, Calvo RW (2017) Inventory rebalancing in bike-sharing systems. In: 15th cologne-twente workshop on graphs and combinatorial optimization. p 35
5. Chung H, Freund D, Shmoys DB (2018) Bike angels: an analysis of citi bike's incentive program. In: Proceedings of the 1st ACM SIGCAS conference on computing and sustainable

societies. ACM, p 5
6. Contardo C, Morency C, Rousseau LM (2012) Balancing a dynamic public bike-sharing system, vol 4. Cirrelt, Montreal
7. Datner S, Raviv T, Tzur M, Chemla D (2017) Setting inventory levels in a bike sharing network. Transp Sci 0 (0):null https://doi.org/10.1287/trsc.2017.0790
8. de Chardon CM, Caruso G, Thomas I (2016) Bike-share rebalancing strategies, patterns, and purpose. J Transp Geogr 55:22–39
9. Di Gaspero L, Rendl A, Urli T (2013) Constraint-based approaches for balancing bike sharing systems. In: International conference on principles and practice of constraint programming. Springer, pp 758–773
10. Erdoğan G, Battara M, Calvo R (2015) An exact algorithm for the static rebalancing problem arising in bicycle sharing systems. Eur J Oper Res 245:667–679
11. Erdoğan G, Laporte G, Calvo R (2014) The static bicycle relocation problem with demand intervals. Eur J Oper Res 238:451–457
12. Forma IA, Raviv T, Tzur M (2015) A 3-step math heuristic for the static repositioning problem in bike-sharing systems. Transp Res Part B: Methodol 71:230–247
13. Freund D, Henderson SG, Shmoys DB (2017) Minimizing multimodular functions and allocating capacity in bike-sharing systems. In: International conference on integer programming and combinatorial optimization. Springer, pp 186–198
14. Ghosh S, Trick M, Varakantham P (2016) Robust repositioning to counter unpredictable demand in bike sharing systems. In: Proceedings of the twenty-fifth international joint conference on artificial intelligence. AAAI Press, pp 3096–3102
15. Ghosh S, Varakantham P, Adulyasak Y, Jaillet P (2017) Dynamic repositioning to reduce lost demand in bike sharing systems. J Artif Intell Res 58:387–430
16. Henderson SG, O'Mahony E, Shmoys DB (2016) (citi)bike sharing. Submitted
17. Ho SC, Szeto W (2014) Solving a static repositioning problem in bike-sharing systems using iterated tabu search. Transp Res Part E: Logist Transp Rev 69:180–198
18. Hsu YT, Kang L, Wu YH (2016) User behavior of bikesharing systems under demand–supply imbalance. Transp Res Rec: J Transp Res Board 117–124
19. Jian N, Freund D, Wiberg HM, Henderson SG (2016) Simulation optimization for a large-scale bike-sharing system. In: Proceedings of the 2016 winter simulation conference. IEEE Press, pp 602–613
20. Jian N, Henderson SG (2015) An introduction to simulation optimization. In: Proceedings of the 2015 winter simulation conference. IEEE Press, pp 1780–1794
21. Kabra A, Belavina E, Girotra K (2016) Bike-share systems: accessibility and availability
22. Kaspi M, Raviv T, Tzur M (2016) Detection of unusable bicycles in bike-sharing systems. Omega 65:10–16
23. Kloimullner C, Papazek P, Hu B, Raidl GR (2014) Balancing bicycle sharing systems: an approach for the dynamic case. European conference on evolutionary computation in combinatorial optimization. pp 73–64
24. Li Y, Zheng Y, Zhang H, Chen L (2015) Traffic prediction in a bike-sharing system. In: Proceedings of the 23rd SIGSPATIAL international conference on advances in geographic information systems. ACM, p 33
25. Liu J, Sun L, Chen W, Xiong H (2016) Rebalancing bike sharing systems: a multi-source data smart optimization. In: Proceedings of the 22nd ACM SIGKDD international conference on knowledge discovery and data mining. ACM, pp 1005–1014
26. Lowalekar M, Varakantham P, Ghosh S, Jena SD, Jaillet P (2017) Online repositioning in bike sharing systems. In: Proceedings of the international conference on automated planning and scheduling (ICAPS)
27. O'Mahony E (2015) Smarter tools For (Citi) bike sharing. PhD thesis, Cornell University
28. O'Mahony E, Shmoys DB (2015) Data analysis and optimization for (citi) bike sharing. In: Twenty-Ninth AAAI conference on artificial intelligence
29. Papazek P, Kloimüllner C, Hu B, Raidl GR (2014) Balancing bicycle sharing systems: an analysis of path relinking and recombination within a grasp hybrid. In: International conference on parallel problem solving from nature. Springer, pp 792–801

30. Parikh P, Ukkusuri SV (2014) Estimation of optimal inventory levels at stations of a bicycle sharing system
31. Paul A, Freund D, Ferber A, Shmoys DB, Williamson DP (2017) Prize-collecting tsp with a budget constraint. In: LIPIcs-Leibniz international proceedings in informatics, vol 87. Schloss Dagstuhl-Leibniz-Zentrum fuer Informatik
32. Rainer-Harbach M, Papazek P, Hu B, Raidl GR (2013) Balancing bicycle sharing systems: a variable neighborhood search approach. In: European conference on evolutionary computation in combinatorial optimization. Springer pp 121–132
33. Raviv T, Kolka O (2013) Optimal inventory management of a bike-sharing station. IIE Trans 45(10):1077–1093
34. Raviv T, Tzur M, Forma IA (2013) Static repositioning in a bike-sharing system: models and solution approaches. EURO J Transp Logist 2(3):187–229
35. Regional Plan Association (1997) Building transit-friendly communities a design and development strategy for the tri-state metropolitan region
36. Riquelme C, Johari R, Zhang B (2017) Online active linear regression via thresholding. In: AAAI pp 2506–2512
37. Rudloff C, Lackner B (2014) Modeling demand for bikesharing systems: neighboring stations as source for demand and reason for structural breaks. Transp Res Rec: J Transp Res Board 1–11
38. Salaken SM, Hosen MA, Khosravi A, Nahavandi S (2015) Forecasting bike sharing demand using fuzzy inference mechanism. In: ICONIP 2015: Proceedings of the 22nd international conference on neural information processing. Springer, pp 567–574
39. Saltzman RM, Bradford RM (2016) Simulating a more efficient bike sharing system. J Supply Chain Oper Manag 14(2):36
40. Schuijbroek J, Hampshire R, van Hoeve WJ Inventory rebalancing and vehicle routing in bike sharing systems. Euorpean J Oper Res to appear
41. Shu J, Chou MC, Liu Q, Teo C-P, Wang I-L (2013) Models for effective deployment and redistribution of bicycles within public bicycle-sharing systems. Oper Res 61(6):1346–1359
42. Singhvi D, Singhvi S, Frazier PI, Henderson SG, O'Mahony E, Shmoys DB, Woodard DB (2015) Predicting Bike usage for New York city's bike sharing system. Computational Sustainability. In: AAAI Workshop
43. Singla A, Santoni M, Bartók G, Mukerji P, Meenen M, Krause A (2015) Incentivizing users for balancing bike sharing systems. In: AAAI. pp 723–729
44. Szeto W, Liu Y, Ho SC (2016) Chemical reaction optimization for solving a static bike repositioning problem. Transp Res Part D: Transp Environ 47:104–135
45. Vogel P, Saavedra BA, Mattfeld DC (2014) A hybrid metaheuristic to solve the resource allocation problem in bike sharing systems. International workshop on hybrid metaheuristics. pp 16–29
46. Zhang J, Pan X, Li M, Philip SY (2016) Bicycle-sharing system analysis and trip prediction. In: Mobile data management (MDM), 2016 17th IEEE international conference on, vol 1. IEEE, pp 174–179
47. Zheng F, He P, Belavina E, Girotra K (2018) Customer preference and station network in the london bike share system

Peer-to-Peer Energy Trading

Thomas Morstyn and Malcolm D. McCulloch

Abstract This chapter presents peer-to-peer energy trading as an emerging case study for the sharing economy. Power systems are undergoing a fundamental transition due to the emergence of distribution network prosumers; proactive-consumers with renewable sources and flexible energy resources that actively manage their consumption, production and storage of energy. Peer-to-peer energy trading has been proposed as a new strategy for integrating prosumers into power system operations. The chapter presents an overview of different business models peer-to-peer energy trading can support, and reviews the mathematical frameworks that have been proposed for designing peer-to-peer and other prosumer-centric energy markets. Case studies are presented, showing how a peer-to-peer energy trading platform can create value for prosumers with a mix of PV generation sources, home battery systems and electric vehicles. Networked matching market theory is used to design a peer-to-peer negotiation mechanism through which the prosumers can agree on a set of mutually beneficial energy transactions none of them wish to renegotiate. Local demand constraints can be enforced by including the prosumers' supplier in the peer-to-peer market.

Keywords Bilateral contract · Blockchain · Energy management · Matching market · Market design · Peer-to-peer · Prosumer · Trading network

1 Introduction

Power systems are undergoing a fundamental transition due to the emergence of distribution network prosumers; proactive-consumers with renewable sources and flexible energy resources that can actively manage their consumption, production and storage of energy [1]. Drawing from sharing economy ideas successfully applied in other sectors, peer-to-peer (P2P) energy trading has been proposed as a new strategy

---

T. Morstyn (✉) · M. D. McCulloch
Department of Engineering Science, the University of Oxford, Parks Road, Oxford OX1 3PJ, UK
e-mail: thomas.morstyn@eng.ox.ac.uk

© Springer Nature Switzerland AG 2020
E. Crisostomi et al. (eds.), *Analytics for the Sharing Economy:
Mathematics, Engineering and Business Perspectives*,
https://doi.org/10.1007/978-3-030-35032-1_16

to incentivise coordination between prosumers and integrate them into power system operations [2].

Electricity markets rely on a shared physical power system, and therefore require design and regulation [3]. Early power systems were operated by vertically integrated utilities, but their monopoly structure raised concerns that electricity prices did not reflect an efficient economic cost of supply [4]. This motivated electricity market liberalisation, which has been a global trend for the last thirty years [5]. A central feature of the liberalised model is a wholesale market allowing generators to compete to fulfil the demand of large industrial consumers and retail suppliers. The wholesale market is overseen by a transmission system operator responsible for managing aggregate balancing and transmission constraints.

Small-scale distribution network consumers are not integrated directly into wholesale electricity markets. Instead, they sign long-term contracts with large suppliers in a retail market, and their supplier obtains energy on their behalf [6]. This arrangement is based on the assumption that small-scale consumers have limited flexibility, and therefore the cost and complexity of integrating them into the wholesale market would not be justified. This assumption also informs the arrangements of distribution system operators responsible for managing local power flows [7]. Rather than pursuing active management like transmission system operators, distribution system operators have relied primarily on long-term investments in physical network reinforcement.

The assumption that distribution networks are primarily made up of passive consumers is being challenged by the adoption of distributed renewable generation sources and flexible energy resources, including electric vehicles, home battery systems and heat-pumps [8]. These resources, along with smart meters and energy management systems for communications and control, have allowed previously passive consumers to become active prosumers.

Coordinating prosumer distributed energy resources could offer significant value in a range of areas. Matching supply and demand on a local level could alleviate the need for investments in generation and transmission infrastructure, and reduce transmission losses [9]. Distributed energy resources could also help manage local power flow and voltage constraints [10]. This is expected to be particularly important for the low-cost electrification of heat and transport, which are key steps towards decarbonisation [11].

However, existing market arrangements are not set up to incentivise or facilitate localised coordination between distributed energy resources owned by individual prosumers. Retail suppliers meter prosumers' individual energy demands, which only incentivises prosumers to consider their own energy use when scheduling flexible resources [12]. Time-of-use energy prices have been proposed to incentivise prosumers to shift their energy use to periods with low expected demand [13]. However, if prosumers all take advantage of the same low price periods, this can result in worse demand peaks [14].

Another strategy is to organise virtual power plants—aggregated groups of distributed energy resources that can operate like a traditional generator in the wholesale market, and provide ancillary services to the transmission system operator [15].

Virtual power plant plant formation can be accomplished through direct control if distributed energy resources have a single owner/operator, but is more challenging for resources that are owned by individual prosumers. The virtual power plant aggregator needs to achieve controllability, while providing sufficient flexibility and end-use functionality to appeal to prosumers [16].

P2P energy trading offers a new strategy for integrating prosumers into electricity markets, while providing transparency, autonomy and scalability. The extension of communications and automation to the individual consumer level opens the possibility for P2P platforms that enable prosumers to negotiate energy transactions directly with one another, and with upstream electricity market participants. Three factors make a market amenable to P2P trading: (i) low economies of scale, (ii) demand variability and diversity and (iii) the ability to efficiently match small-scale buyers and sellers [17]. These factors are increasingly relevant for electricity. Local renewable generation sources are challenging the assumption that large centralised power plants have a definitive economies of scale advantage for energy production [18]. Also, with variable renewable sources and flexible loads embedded in distribution networks, it is increasingly important for transactions to account for local conditions and constraints [19].

This chapter presents P2P energy trading as an emerging case study for the sharing economy. The rest of the chapter is organised as follows. Section 2 presents an overview of four different business models that can be supported by P2P energy trading, and Sect. 3 reviews different mathematical frameworks that have been proposed for designing P2P and other prosumer-centric energy markets. Section 4 presents a new framework for P2P energy market design based on networked matching market theory. Within this framework, prosumers use P2P negotiation and autonomous decision making to reach agreement on a set of mutually beneficial energy transactions none of them wish to renegotiate. Section 5 presents case studies, showing how this framework can be used to design a P2P trading platform for distribution network prosumers with a mix of PV generation sources, home battery systems and electric vehicles. It is also shown how local demand constraints can be enforced by including the prosumers' supplier in the P2P market. Section 6 concludes the chapter.

2 Business Models

Depending on how prosumer owned distributed energy resources are coordinated, they can create value (and costs) for different electricity market participants. This section presents an overview of four business models that can be supported by P2P energy trading: (Sect. 2.1) behind-the-meter trading, (Sect. 2.2) local flexibility, (Sect. 2.3) multi-class energy trading and (Sect. 2.4) federated power plant formation. Figure 1 presents high-level diagrams of these business models.

Although research into P2P energy trading is at an early stage, a large number of pilot projects and trials are already underway. Detailed reviews of the business

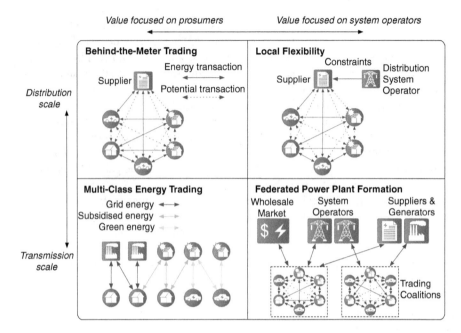

Fig. 1 High-level digrams for four different business models (described in Sects. 2.1–2.4) that can be supported by P2P energy trading. The business models vary in terms of whether they are targeted at the distribution or transmission network scale, and whether the value created is focused on prosumers or on transmission/distribution system operators

models for particular P2P energy trading projects are presented in [20–22]. In [23], a business model is defined as having four components:

1. Value proposition: 'What value is being created for stakeholders?'
2. Products (or services): 'What is being traded?'
3. Value architecture: 'Who are the value partners and what are the underlying firm resources and coordination mechanism?'
4. Revenue model: 'How is revenue generated?'

Within this framework, P2P energy trading platforms can be seen as a value architecture, consisting of value partners (prosumers, suppliers, system operators), firm resources (software, data, subscriber relationships) and a coordination mechanism (the P2P market mechanism). Revenue models from other P2P market platforms can be applied to P2P energy trading platforms, including subscription and transaction fees, advertising, selling data products and sponsorship by an external beneficiary. A key consideration are cross-side network effects between different types of subscribers [24]. For example, consider a P2P platform allowing system operators to obtain flexibility from small-scale prosumers. For the system operators, the attractiveness of the platform is primarily driven by the number of prosumers offering flexibility (a cross-network effect). Therefore, it makes sense for the platform owner to make the platform attractive to prosumers by subsidising their participation, and to recover costs from the system operators.

2.1 Behind-the-Meter Trading

Virtual net metering has been proposed as a new arrangement for retail electricity markets to improve the utilisation of local renewable generation and flexible energy resources [25]. Under this arrangement, groups of prosumers are charged according to their combined net energy demand, rather than their individually metered demand. Under a retail energy contract, prosumers are generally paid less for excess supply than they are charged for demand, since their generation is not considered to be dispatchable in the wholesale energy market [12]. Therefore, when metered as a group, prosumers have an incentive to sell excess renewable generation to other prosumers with flexible loads and storage systems, rather than exporting it upstream through their supplier. P2P energy trading platforms have a key role to play in this, by providing a negotiation mechanism to arrange prosumer-to-prosumer transactions [26].

Virtual net metering can be implemented in buildings and developments with privately owned distribution infrastructure. It is being extended to distribution networks in France by the collective self-consumption rules [27], and Ofgem is trialling similar virtual private wire network arrangements in the UK [28]. Value is created for the participating prosumers, which benefit through lower energy costs and increased utilisation of their energy resources. System benefits include reduced losses and pollution due to increased local consumption of renewable energy.

2.2 Local Flexibility

P2P energy trading can also be used to help manage local demand constraints. This is relevant for distribution networks with thermal and voltage limits, and islanded microgrids where supply must match demand to maintain stability. In the case where energy import/export is managed by a single supplier (or microgrid operator), the supplier can organise a local P2P energy market, and buy/sell energy subject to a supply constraint [29, 30]. With a constraint on net energy imports, local generation and energy storage become more valuable. If local flexibility markets can alleviate the need for investments in network reinforcement, prosumers can also benefit through lower network charges. For off-grid applications, such as rural microgrids in the developing world, sharing flexible resources can significantly improve security of supply [31].

2.3 Multi-class Energy Trading

P2P trading offers the potential for prosumers to express energy preferences, beyond purely-financial ones, related to where their energy comes from and how it is generated. Suppliers already use renewable energy certificate (REC) schemes to offer green retail supply options [32]. RECs are also used by corporations to meet social responsibility obligations. However, existing REC markets suffer from a lack of transparency, and they may not incentivise efficient investments for offsetting carbon emissions, since RECs are not differentiated geographically or by time of generation [33, 34].

P2P REC trading could be done without significant regulatory changes, and could expand RECs from the MWh level to the kWh level. Also, P2P RECs could vary in multiple dimensions, such as generation technology, location and certifying agency. Vandebron in the Netherlands and Piclo Match in the UK are P2P platforms set up by retail suppliers, offering customers the ability to select preferred sources of energy as a value-adding service. P2P REC trading is also being targeted as an application for blockchain platforms [35].

In [36], an energy trading platform is proposed that allows prosumers to trade different energy classes, rather than treating energy as a homogeneous product. Environmentally conscious prosumers can buy renewable energy, while philanthropic prosumers can supply subsidised energy to low-income households. Since energy transactions are time-specific, owners of energy storage systems are rewarded for their role helping to match supply and demand for each product.

2.4 Federated Power Plant Formation

P2P energy trading offers a new mechanism for incentivising coordination between prosumers that may be willing to provide some flexibility, but are not willing to give up completed control to a virtual power plant aggregator. Facilitating the formation of cooperative federated power plants offers another business model for P2P energy trading platforms [37].

P2P energy trading provides a transparent mechanism supporting negotiation between system operators at different levels of the power system, and groups of prosumers which could operate together to provide flexibility services. By cooperatively coordinating their distributed energy resources, prosumers could provide frequency regulation and reserve services to help the transmission system operator manage aggregate power system balancing. They could also provide value to distribution system operators by helping to manage local thermal and voltage constraints. In addition, groups of prosumers with complementary resources could share risk and trade energy directly in the wholesale market, or enter into bilateral contracts with large suppliers and generators.

3 Prosumer-Centric Market Design Frameworks

An important component for P2P energy trading is a coordination mechanism allowing prosumers to negotiate mutually beneficial transactions. Market designs for P2P energy trading fit within a broader group of prosumer-centric energy market designs, which may rely on some degree of centralised communication and control, rather than being strictly based on P2P negotiation and autonomous decision making. Prosumer-centric market design frameworks include (Sect. 3.1) price-based distributed optimisation, (Sect. 3.2) mean-field game theory, (Sect. 3.3) Stackelberg leader-follower games, (Sect. 3.4) double-sided auctions, (Sect. 3.5) cooperative energy sharing and (Sect. 3.6) networked matching markets.

3.1 Price-Based Distributed Optimisation

Distributed energy resources that are controlled by a single owner/operator can be coordinated using centralised optimisation based on load, generation and upstream energy price predictions [10]. Distributed optimisation addresses scalability and data privacy, by dividing the optimisation problem into local agent sub-problems, with coupling constraints mediated by coordinator nodes or through distributed consensus [38–40].

Distributed optimisation strategies that are based on dual price variable adjustments operate like auction mechanisms, with the price of energy increasing in the case of excess demand, and decreasing in the case of excess supply, until equilibrium is achieved [41]. Distributed price-based optimisation strategies for prosumer coordination are presented in [30, 36, 42]. In [43], a private Ethereum Homestead blockchain is used to implement the coordination step of a distributed optimisation strategy in a transparent fashion, and to securely store the resulting transactions. A limitation of distributed optimisation approaches is that agents must participate cooperatively to achieve convergence by implementing correctly chosen penalty terms and step-sizes when solving their local sub-problems, rather than trying to maximise their individual utilities.

3.2 Mean-Field Game Theory

Mean-field game theory has been proposed as an alternative framework for price-based coordination between fleets of flexible loads (e.g. electric vehicles with unidirectional chargers) [44–46]. Within this framework, it is assumed that a central coordinator faces increasing marginal supply costs when the aggregate demand of the fleet is higher, and sets prices to recover these costs. Assuming that each agent has a negligible individual impact on overall demand, and a suitable iterative price-adjustment mechanism is implemented, it can be shown that the fleet will reach a Nash Equilibrium which flattens aggregate demand.

3.3 Stackelberg Leader-Follower Games

Stackelberg leader-follower games can be used to model the strategic behaviour of a supplier with monopoly power which sets energy prices for prosumers within a distribution network or microgrid [47–49]. Assuming the supplier does not face external competition and knows how the prosumers will respond to different prices, it can maximise its profits using bilevel optimisation. In [50], the Stackelberg game model is extended to consider net metered prosumers with an internal trading price based on the local supply and demand ratio to incentivise cooperative flexible load scheduling. However, it is noted that convergence to a mutually acceptable internal trading price is not guaranteed.

3.4 Double-Sided Auctions

Double-sided auctions allow direct energy trading between net metered generators and consumers [51–54]. Consumers submit bids to an auctioneer, specifying their energy demand given a maximum energy price, and generators submit offers specifying their capacity given a minimum energy price. The auctioneer iteratively adjusts the energy price to match supply and demand. The auctioneer can use different price-adjustment mechanisms to trade-off between competing objectives, such as economic efficiency and revenue maximisation. Double-sided auctions strictly divide participants into buyers and sellers, and only consider energy trading at a single time interval. This limits their applicability when prosumers have energy storage systems, and may buy or sell energy depending on the price, or may only be willing to buy energy at a particular time interval if they can arrange a future sale.

3.5 Cooperative Energy Sharing

Heuristic approaches for dividing costs between net-metered prosumers with PV sources include bill-sharing and mid-market rate allocation [55]. Bill-sharing involves dividing the prosumers' collective costs in proportion to their net consumption, while under a mid-market rate allocation scheme, local energy transactions are priced at the mid-point between the (higher) retail energy price and the (lower) retail feed-in tariff. In [56], cooperative game theory is used to derive a profit sharing scheme that ensures all prosumers are made better off by participating in virtual net metering. However, these strategies do not address cooperative scheduling and profit-sharing between prosumers with energy storage systems and flexible loads, who may need to be compensated for the use of their resources (e.g. due to battery degradation, reduced comfort).

In [57], cooperative game theory is used to find profit-sharing allocations for net-metered prosumers with energy storage systems which incentivises them to schedule their storage systems cooperatively, considering the possibility for prosumers to engage in strategic price-setting behaviour. However, finding allocations that are robust to strategic behaviour is exponentially complex, limiting scalability [58]. In [59], cooperative game theory is used for prosumers with flexible loads that provide demand response, with a focus on allocation fairness rather than robustness to strategic behaviour.

3.6 Networked Matching Markets

In [10], networked matching market theory is proposed as a framework for P2P energy market design. This framework addresses autonomous agents in a trading network that negotiate to find mutually beneficial transactions for non-homogeneous goods [60]. Given certain restrictions on agents' individual preferences, computationally scalable agent-to-agent negotiation mechanisms allow the agents to reach stable outcomes—sets of agreed transactions no agent, or group of agents, wish to deviate from [61]. The framework extends to hierarchical supply chains, and networks with cyclical trading relationships between agents [62]. In an energy trading context, this allows agents with energy storage systems to simultaneously negotiate multiple energy transactions for different time intervals, acting as a buyer in some and a seller in others. Networked matching market theory has been proposed for P2P energy trading in microgrids [10], electric vehicle smart charging markets [63] and markets for distribution system flexibility with multiple competing aggregators [64].

4 Networked Matching Markets for Behind-the-Meter and Local Flexibility P2P Energy Trading

In this section, a P2P energy market design is proposed, supporting the behind-the-meter trading and local flexibility business models from Sects. 2.1 and 2.2. Networked matching market theory is used to design a distributed price-negotiation mechanism that allows the prosumers to agree on a stable set of energy transactions.

Consider a distribution network with a retail supplier and a group of prosumers, which may have PV generation sources, battery energy storage systems and electric vehicles. The proposed P2P market design allows the prosumers to negotiate local energy transactions with other prosumers and their supplier during a future 24 h period, based on their individual preferences and energy requirements. Within the proposed market design, the potential trading relationships between the agents are described by a network of contracts. Each contract specifies a price for a discrete quantity of energy offered by one agent to anther for a particular time interval. Each contract has the same power level and duration, and thus the same energy-unit. For example, contracts could specify 0.5 kW supplied during 30 min intervals, giving a contract energy-unit of 0.25 kWh.

Energy which the prosumers do not obtain from peers must be obtained from their supplier. It is assumed that the supplier is willing to sell energy at a particular price during each time interval and to buy energy at a lower feed-in tariff. These prices may also be time-of-use dependent. Under the behind-the-meter trading business model, the supplier does not impose a demand constraint, whereas under the local flexibility business model, the supplier can only sell energy up to a maximum limit.

In practice, arrangements for real-time settlement would be required to deal with inaccurate prosumer load and renewable generation predictions. Also, only a single-round of ahead trading is considered here, but additional trading rounds could be introduced closer to real-time using a receding time-horizon approach.

4.1 Bilateral Contract Framework

Consider a trading network made up of agents $i \in \mathcal{A}$ connected by potential bilateral trades $\omega \in \Omega$. Each trade specifies a unit of energy offered by a seller agent $s(\omega) \in \mathcal{A}$ to a buyer agent $b(\omega) \in \mathcal{A}$, during a particular time interval $t_\omega \in \mathcal{T}$. $\mathcal{T} = \{1, \ldots, T\}$ is the set of time intervals of the P2P energy market.

Let X be the set of bilateral contracts offered between the agents, each specifying a price for a potential trade. A contract $x \in X$ is a tuple $x = (\omega, t_\omega, p_\omega)$ specifying a price p_ω for an underlying trade $\omega \in \Omega$. Each contract has a buyer $b(x) = b(\omega)$ and a seller $s(x) = s(\omega)$.

Agent $i \in \mathcal{A}$ is associated with contracts $X_i = \{x \in X | i \in b(x) \cup s(x)\}$. Of these, $Y_i = \{y \in X | b(y) = i\}$ are its upstream contracts and $Z_i = \{z \in X | s(z) = i\}$ are its downstream contracts. These can be further divided into the upstream and downstream contracts associated with each time interval, $Y_{it} = \{(\phi, t_\phi, p_\phi) \in X | b(\phi) = i, t_\phi = t\}$ and $Z_{it} = \{(\omega, t_\omega, p_\omega) \in X | s(\omega) = i, t_\omega = t\}$.

Each agent $i \in \mathcal{A}$ has a utility function $U_i(Y, Z)$ over sets of available upstream and downstream contracts, which specifies their individual energy preferences. The contracts agent i selects to maximise their utility is given by their choice correspondence,

$$C_i(Y, Z) = \underset{Y_i' \subseteq Y_i, Z_i' \subseteq Z_i}{\operatorname{argmax}} U_i(Y_i', Z_i') \tag{1}$$

The upstream and downstream contracts rejected by agent i are given by $R_i^B(Y|Z)$ and $R_i^S(Z|Y)$ where,

$$R_i^B(Y|Z) = \{y \in Y | b(y) = i, y \notin C_i(Y, Z)\} \quad (2)$$
$$R_i^S(Z|Y) = \{z \in Z | s(z) = i, z \notin C_i(Y, Z)\}. \quad (3)$$

4.2 Prosumer Preferences

Prosumers with inflexible loads, renewable generation sources, battery energy storage systems and electric vehicles are considered. Let $\mathcal{P} \subseteq \mathcal{A}$ be the set of prosumers.

Prosumers with only inflexible loads and renewable generation sources must trade with their supplier or other prosumers to balance their net inflexible demand (load less renewable generation). For prosumer i, let d_{it} be the net inflexible demand for time interval t. The utility function for prosumers with only inflexible loads and variable renewable generation sources is given by,

$$U_i(Y_i, Z_i) = \begin{cases} \sum_{(\omega, t_\omega, p_\omega) \in Z_i} p_\omega - \sum_{(\phi, t_\phi, p_\phi) \in Y_i} p_\phi, & |Y_{it}| - |Z_{it}| = d_{it} \; \forall t \in \mathcal{T} \\ -\infty, & otherwise. \end{cases} \quad (4)$$

$|X|$ gives the number of elements in set X. Note that d_{it} and the other energy quantities in this section have the same energy-units as the bilateral contracts. $\sum_{(\omega, t_\omega, p_\omega) \in Z_i} p_\omega$ is the total revenue from selling downstream energy contracts and $\sum_{(\phi, t_\phi, p_\phi) \in Y_i} p_\phi$ is the total cost of upstream energy contracts.

For prosumers with battery energy storage systems, let E_{i0} be the initial energy stored in the battery, let $\underline{E}_i, \bar{E}_i$ be the minimum and maximum energy capacities and let $\underline{P}_i, \bar{P}_i$ be the minimum and maximum power capacities. It is assumed that the batteries should be returned to E_{i0} by the end of the final time interval. Using a linear battery energy storage model, a suitable prosumer utility function is given by,

$$U_i(Y_i, Z_i) = \begin{cases} \sum_{(\omega, t_\omega, p_\omega) \in Z_i} p_\omega - \sum_{(\phi, t_\phi, p_\phi) \in Y_i} p_\phi - c_{bi}[d_{it} - |Y_{it}| + |Z_{it}|]^+, \\ \underline{P}_i \leq d_{it} - |Y_{it}| + |Z_{it}| \leq \bar{P}_i \; \forall t \in \mathcal{T} \\ \underline{E}_i \leq E_{i0} + \sum_{t=1}^{\tau}(|Y_{it}| - |Z_{it}| - d_{it}) \leq \bar{E}_i \; \forall \tau \in \mathcal{T} \\ \sum_{t \in \mathcal{T}}(d_{it} - |Y_{it}| + |Z_{it}|) = 0 \\ -\infty, \quad otherwise. \end{cases} \quad (5)$$

$[\cdot]^+ = \max\{\cdot, 0\}$. The battery output power at time interval t is given by $(d_{it} - |Y_{it}| + |Z_{it}|)$, and c_{bi} models a linear degradation cost associated with discharging the battery.

Finally, electric vehicle owners with unidirectional chargers are considered. It is assumed that the electric vehicle owners have an arrival time $t_{ia} \in \mathcal{T}$ and a departure deadline $t_{id} \in \mathcal{T}$, with $t_{ia} \leq t_d$, between which their vehicle must be fully charged. However, electric vehicle owners may also prefer their vehicle is charged earlier than their departure deadline, since this gives them additional flexibility. This is modelled with a utility coefficient γ_i for early charging. Let E_{ia} be the energy stored in the vehicle when it arrives, and let $\bar{E}_{id}$ be the energy required at departure. The electric vehicle also has a maximum charging power $\bar{P}_i$. The electric vehicle owner utility function is given by,

$$U_i(Y_i, Z_i) = \begin{cases} \sum_{(\omega, t_\omega, p_\omega) \in Z_i} (p_\omega + \gamma_i t_\omega) - \sum_{(\phi, t_\phi, p_\phi) \in Y_i} (p_\phi + \gamma_i t_\phi), \\ \quad 0 \leq |Y_{it}| - |Z_{it}| - d_{it} \leq \bar{P}_i \ \forall t \in \mathcal{T} \\ \quad E_{ia} + \sum_{t=t_{ia}}^{t_{id}} (|Y_{it}| - |Z_{it}| - d_{it}) = \bar{E}_{id} \\ -\infty, \quad otherwise. \end{cases} \quad (6)$$

4.3 Supplier Preferences

For the behind-the-meter trading business model, the prosumers operate under virtual net metering. The prosumers buy energy from their supplier which cannot be obtained from other prosumers, and sell excess local generation. It is assumed that the prosumers have a single supplier $\mathcal{S} \in \mathcal{A}$, which is willing to supply energy at time-of-use dependent retail energy prices $r_{\mathcal{S}t}^b$, and is willing to buy excess energy at lower feed-in tariffs $r_{\mathcal{S}t}^s \leq r_{\mathcal{S}t}^b, t \in \mathcal{T}$. The supplier's utility function is given by,

$$U_{\mathcal{S}}(Y_{\mathcal{S}}, Z_{\mathcal{S}}) = \sum_{(\omega, t_\omega, p_\omega) \in Z_{\mathcal{S}}} p_\omega - \sum_{(\phi, t_\phi, p_\phi) \in Y_{\mathcal{S}}} p_\phi + \sum_{t \in \mathcal{T}} (r_{\mathcal{S}t}^s |Y_{\mathcal{S}t}| - r_{\mathcal{S}t}^b |Z_{\mathcal{S}t}|) \quad (7)$$

Within the P2P market, the supplier selects contracts to maximise its utility, like the prosumers. However, in Sect. 4.5, the P2P price-negotiation mechanism is designed so that the prosumers are always made better off than they would be under individual metering.

For the local flexibility business model, the supplier must enforce minimum and maximum demand constraints $\underline{D}_{\mathcal{S}t}, \bar{D}_{\mathcal{S}t}$ for each time interval $t \in \mathcal{T}$. In this case, the supplier's utility function is given by,

$$U_{\mathcal{S}}(Y_{\mathcal{S}}, Z_{\mathcal{S}}) = \begin{cases} \sum_{(\omega, t_\omega, p_\omega) \in Z_{\mathcal{S}}} p_\omega - \sum_{(\phi, t_\phi, p_\phi) \in Y_{\mathcal{S}}} p_\phi \\ \quad + \sum_{t \in \mathcal{T}} (r_{\mathcal{S}t}^s |Y_{\mathcal{S}t}| - r_{\mathcal{S}t}^b |Z_{\mathcal{S}t}|), \\ \quad \underline{D}_{\mathcal{S}} \leq |Z_{\mathcal{S}t}| - |Y_{\mathcal{S}t}| \leq \bar{D}_{\mathcal{S}} \ \forall t \in \mathcal{T} \\ -\infty, \quad otherwise. \end{cases} \quad (8)$$

Note that there are several straightforward extensions. Multiple competing retail suppliers can be included (each with a portion of the total demand constraint for the local flexibility business model). Also, rather than assuming the supplier has predetermined prices at which it is willing to buy and sell energy, competing upstream generators with different marginal production costs could be included, with the suppliers acting as an intermediary between the generators and the prosumers (see e.g. [10]).

4.4 Stability and Competitive Equilibrium

A key market design objective is to provide a mechanism allowing agents to find stable market outcomes. Let X be the full set of contracts offered between agents within the trading network.

Definition 1 $A \in X$ is a *stable outcome* if it is individually rational for all agents $i \in \mathcal{A}$ and there is no blocking set of contracts $B \in X$. A is individually rational for agent i if $A \in C_i(A)$. B is a blocking set if $B \cap A = \varnothing$ and $B_i \subseteq C_i(B \cup A) \; \forall i \in \{\cup_{x \in B} b(x), s(x)\}$.

Stable outcomes are a desirable market solution concept since they ensure no individual agent prefers to drop a contract they have agreed to buy or sell, and that there are no alternative sets of contracts which agents would prefer to mutually deviate to. Depending on the preferences of individual agents, stable outcomes may not exist. In [62], it is shown that stable outcomes are guaranteed to exist when each agent's preferences satisfy full substitutability.

Definition 2 Agent $i \in \mathcal{A}$ has *fully substitutable* preferences if its preferences are same-side substitutable and cross-side complementary. Let $Y' \subseteq Y$ be upstream contracts and let $Z' \subseteq Z$ be downstream contracts. Agent i's preferences are same-side substitutable if,

$$R_i^B(Y'|Z) \subseteq R_i^B(Y|Z) \text{ and } R_i^S(Z'|Y) \subseteq R_i^S(Z|Y). \tag{9}$$

Agent i's preferences are cross-side complementary if,

$$R_i^B(Y|Z) \subseteq R_i^B(Y|Z') \text{ and } R_i^S(Z|Y) \subseteq R_i^S(Z|Y'). \tag{10}$$

Same-side substitutable preferences mean that being offered an additional contract will not cause the agent to accept a contract it previously rejected on the same-side of the market (upstream–upstream or downstream–downstream). Cross-side complementary preferences mean that being offered an additional contract will not cause the agent to reject a previously accepted contract on the other side of the market. In [64], full substitutability is shown for preferences defined by a general utility function, for which the agent utility functions in (4)–(8) are special cases. Note that certain types of relevant agent preferences do not satisfy full substitutability within the proposed setting. In particular, agents with decreasing marginal costs of generation would not satisfy same-side substitutability.

Another important market solution concept is competitive equilibrium, which occurs when all trades in the market have prices that balance supply and demand. When agents have fully substitutable preferences, stable outcomes coincide with competitive equilibria [62]. Also, when the agent utility functions are linear with respect to payments, as in (4)–(8), stable outcomes are economically efficient (they maximise the agents' collective utility).

4.5 Price Negotiation

The proof in [62] that stable outcomes exist for trading networks between agents with fully substitutable preferences is constructive. This provides the basis for an iterative P2P price negotiation algorithm, which is based on the concept of deferred acceptance.

Agents start by specifying the bilateral trades Ω they may be willing to make with other agents in the trading network. Each trade $\omega \in \Omega$ has a buyer price p_ω^b and a seller price p_ω^s. The market is designed with trades starting at a price of zero, and buyers making progressively higher offers to sellers. An exception is made for trades offered by prosumers to the supplier, which start at the feed-in tariff r_{St}^s.

At each iteration, agents (in parallel) select the subset of trades that maximise their individual utility, given the current buyer and seller prices. Then, trades demanded by the buyer, but not accepted by the seller, have their buyer or seller price increased by an increment Δp. The negotiation ends when no price changes occur during at iteration. This indicates a stable outcome has been found. The price-adjustment process is presented in Algorithm 1.

Algorithm 1 P2P Price-Negotiation

1: **Initialize:**
    $\Omega_i^B \leftarrow \{\omega \in \Omega | b(\omega) = i\}$, for each agent $i \in \mathcal{A}$
    $\Omega_i^S \leftarrow \{\omega \in \Omega | s(\omega) = i\}$, for each agent $i \in \mathcal{A}$
    $p_\omega^b \leftarrow 0$, $p_\omega^s \leftarrow 0$ for each trade $\{\omega \in \Omega | b(\omega) \notin \mathcal{S}\}$
    $p_\omega^b \leftarrow r_{St_\omega}^s$, $p_\omega^s \leftarrow r_{St_\omega}^s$ for each trade $\{\omega \in \Omega | b(\omega) \in \mathcal{S}\}$
    done $\leftarrow false$
2: **while** done is $false$ **do**
3:     done $\leftarrow true$
4:     **for** each agent $i \in \mathcal{A}$ (in parallel) **do**
5:         $Y_i \leftarrow \{\cup_{\omega \in \Omega^B}(\omega, t_\omega, p_\omega^b)\}$
6:         $Z_i \leftarrow \{\cup_{\omega \in \Omega^S}(\omega, t_\omega, p_\omega^s)\}$
7:         $Y_i^* \leftarrow C_i(Y_i, Z_i) \cap Y_i$
8:         $Z_i^* \leftarrow C_i(Y_i, Z_i) \cap Z_i$
9:     **for** each trade $\omega \in \Omega$ (in parallel) **do**
10:        **if** $\omega \in Y_{b(\omega)}^*$ and $\omega \notin Z_{s(\omega)}^*$ **then**
11:            done $\leftarrow false$
12:            **if** $p_\omega^b > p_\omega^s$ **then**
13:                $p_\omega^s \leftarrow p_\omega^s + \Delta p$
14:            **else**
15:                $p_\omega^b \leftarrow p_\omega^b + \Delta p$
16: The set of contracts $\cup_{i \in \mathcal{A}} Y_i^*$ is a stable outcome

---

Having contracts start at low prices with buyers making progressively higher offers is a design decision, which results in a stable outcome that favours energy consumers. An alternative arrangement would be for contracts to begin at high prices, with sellers making progressively lower offers to buyers. This would still result in a stable outcome, but one favouring energy suppliers.

For the behind-the-meter trading business model, the supplier does not have a demand constraint and the prosumers will always do as well or better within the P2P market as they would under individual metering. This is guaranteed, since prosumers with excess supply will at worst be able to sell it to the supplier at the feed-in tariff rates r^s_{St}, and prosumers with excess demand will at worst be able to buy energy from the supplier at the retail energy prices r^b_{St}.

Under the local flexibility business model, the price of energy contracts may go above r^b_{St} for periods with high demand due to the demand constraint. In this case, the benefit to consumers would be in the form of reduced distribution fees (external to the P2P negotiation mechanism).

4.6 Contract Selection

At each iteration of the price-negotiation process (Algorithm 1) each agent needs to select utility maximising sets of upstream and downstream contracts Y_i^*, Z_i^* from those they are offered Y_i, Z_i. This can be done in a computationally scalable manner since the agent preferences satisfy full substitutability. In [65], it is shown that agent preferences are fully substitutable if and only if they satisfy the single improvement property.

Definition 3 Let Y_i, Z_i be the upstream and downstream contracts available to agent i, with each contract associated with a unique trade. Let $Y_i' \subseteq Y_i, Z_i' \subseteq Z_i$ such that $U_i(Y_i', Z_i') > -\infty$ and $U_i(Y_i', Z_i') < \max_{Y_i^* \in Y_i, Z_i^* \in Z_i} U_i(Y_i^*, Z_i^*)$. Then, the preferences of agent i satisfy the *single improvement property* if $U_i(Y_i', Z_i')$ can always be increased by either (i) adding/removing a single contract to/from Y_i' or Z_i'; or (ii) simultaneously adding/removing an upstream and downstream contract to/from Y_i' and Z_i'.

The single improvement property suggests a two step process for agents to select utility maximising sets of contracts. First, agents need to select contracts until their constraints are satisfied, so that $U_i > -\infty$. Then, agents should select additional upstream and/or downstream contracts until their utility cannot be further improved.

The prosumers with utility functions (4)–(6) and the supplier under the local flexibility business model (8) have maximum and minimum power limits for each time interval. For each $t \in \mathcal{T}$ when an upper power limit is violated, agents should select upstream contracts (in ascending price order) until it is cleared. To clear lower power limit violations, downstream contracts should be selected (in descending price order).

The prosumers with battery energy storage systems (5) and electric vehicles (6) also need to ensure their energy level constraints are not violated. First, the time intervals when the energy limits are violated should be identified. Then working forward in time, each violation should be cleared by accepting contracts for preceding intervals which do not introduce new constraint violations. Contracts should be compared to one another and selected in descending order of the marginal utility which would be provided assuming the utility function was unconstrained. Finally, any additional upstream, downstream or upstream–downstream contract pairs which provide positive marginal utility should be selected.

5 Case Studies

Simulation case studies are presented demonstrating the operation of the P2P energy market design for a distribution network with 30 prosumers and a retail supplier. Prosumers 1–10 have PV generation sources (4 kWp) and battery energy storage systems (2 kW, 4 kWh), prosumers 11–20 own electric vehicles and prosumers 21–30 are consumers with only inflexible loads. Three case studies are considered: (i) individual metering, (ii) behind-the-meter P2P trading and (iii) local flexibility P2P trading with the supplier enforcing a 25 kW demand constraint.

The retail supplier applies time-of-use energy prices, with a higher price of £0.15/kWh from 6 am to 11 pm and a lower price of £0.07/kWh from 11 pm to 6 am. Under local flexibility P2P trading, the energy price demanded by the supplier may go above these values during periods of constrained supply. The retail feed-in tariff is £0.04/kWh.

The P2P energy market operates over a 24 h period from 8 am to 8 am. Contracts specify 0.25 kWh of energy delivered during a particular 30-min interval. The price negotiation increment is $\Delta p = £0.01$/kWh. Smart meter data from the UK Customer-Led Network Revolution project is used for the prosumer load and PV generation profiles [66]. For prosumers 1–10, a battery degradation coefficient of $c_{bi} = £0.01$/kWh is used. It is assumed that the batteries start at their minimum energy level at 8 am and must be returned to this energy level by 8 am the next day.

Prosumers 11–20 have 3.5 kW unidirectional chargers, and must fully re-charge their electric vehicles between their arrival and departure times. A dataset made available by NREL was used for the vehicle energy requirements [67]. Vehicle arrival and departure times were generated randomly, assuming normal distributions with mean times of 6 pm and 7 am respectively, and one hour standard deviations. The extra utility received for early charging was modelled by a utility coefficient $\gamma_i = £0.02$/kWh/h.

Figure 2 shows the total prosumer demand for the three case studies. Figure 3 shows the total output power of the battery energy storage systems owned by prosumers 1–10 and Fig. 4 shows the total output power of the electric vehicles owned by prosumers 11–20.

Individual metering results in larger energy export during the middle of the day, since the prosumers with PV sources and batteries are only incentivised to store energy to meet their individual energy needs. Under behind-the-meter P2P trading, there is a larger battery charge-cycle at 11 pm when the low-price period starts. The energy stored during this period is then sold to other prosumers when the higher energy prices return after 6 am. There is no difference in terms of electric vehicle

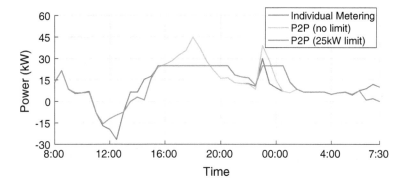

Fig. 2 Net demand of the prosumers under individual metering, behind-the-meter P2P trading and local flexibility P2P trading with a 25 kW demand limit

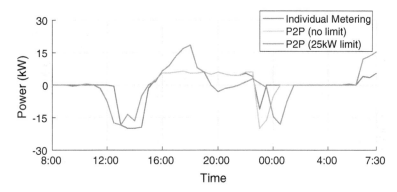

Fig. 3 Total output power of the battery energy storage systems owned by prosumers 1–10 under individual metering, behind-the-meter P2P trading and local flexibility P2P trading with a 25 kW demand limit

charging between the individual metering and behind-the-meter P2P trading case studies.

During the local flexibility P2P trading case study, the supplier enforces a 25 kW demand limit, reducing the peak demand by 15 kW. The battery energy storage systems adjust their discharge cycles to flatten the demand peak around 6 pm, and part of the electric vehicle demand is shifted later to the start of the low retail price period. The battery charge-cycle that occurs during the low retail price period is shifted to follow this new period of electric vehicle charging, since the electric vehicle owners have a preference for early charging.

Figure 5 shows the total utility accumulated by the different groups of prosumers under the three case studies. Figure 6 compares the utility accumulated by each of the prosumers for each case study.

Under behind-the-meter P2P trading, the total utility of the prosumers is increased by 15% compared with individual metering. In addition, Fig. 6 shows that every

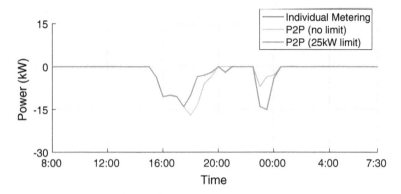

Fig. 4 Total output power of the electric vehicles owned by prosumers 11–20 under individual metering, behind-the-meter P2P trading and local flexibility P2P trading with a 25 kW demand limit

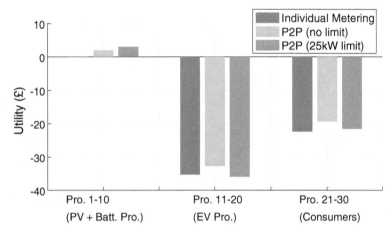

Fig. 5 Total utility accumulated by different groups of prosumers under individual metering, behind-the-meter P2P trading and local flexibility P2P trading with a 25 kW demand limit

individual prosumer does as well or better, including prosumers 21–30 that only have inflexible loads.

With the demand limit imposed, the prosumers with PV sources and battery energy storage systems achieve higher utility, since local energy generation becomes more valuable. In total, the utility received by the prosumers is higher than under individual metering by 5.6%, but some electric vehicle owners and consumers receive less utility than they would under individual metering. Depending on the alleviated infrastructure cost, the prosumers could be separately compensated through lower distribution charges. For all of the prosumers to do better under the local flexibility P2P trading business model than under individual metering, they would need to receive £0.016 per kW of peak demand reduction, and £0.028 per kW of peak demand reduction to do better than under the behind-the-meter P2P trading business model.

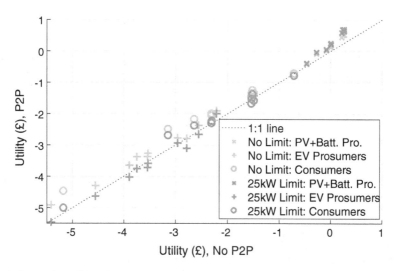

Fig. 6 Comparison of the utility accumulated by each prosumer under individual metering, behind-the-meter P2P trading and local flexibility P2P trading with a 25 kW demand limit

6 Conclusion

P2P energy trading creates the potential for new business models to be built around coordinating prosumer owned distributed energy resources. These business models vary in terms of the product being transacted, their physical scale and the value offered to different electricity market participants. Focusing on two distribution network scale business models, a new P2P energy market design has been presented which allows groups of prosumers operating under virtual net metering to negotiate mutually beneficial energy transactions with one another. Integrating the prosumers' supplier into the P2P market enables local demand constraints to be enforced. Networked matching market theory provides a mathematical framework for designing a distributed P2P market mechanism that does not require a coordinator or auctioneer. The design is based on individual decision making, parallel computation and agent-to-agent communication, offering advantages in terms of transparency, prosumer autonomy and scalability.

A number of important areas for future research remain. New mechanisms are needed to enable negotiation between prosumers and system operators at both the distribution and transmission levels of the power system, which may have conflicting flexibility requirements. Also, P2P energy trading introduces challenges for managing uncertainty, losses and transmission constraints, which have traditionally required aggregation and centralised coordination. The chapter has focused on P2P energy trading in electricity networks, but future applications could include integrated multi-energy systems in smart cities (combining electricity, heat and transport), and the formation of microgrids for energy access in the developing world.

References

1. Dimeas BA, Drenkard S, Hatziargyriou N, Karnouskos S, Kok K, Ringelstein J (2014) Smart houses in the smart grid. IEEE Electrification Mag 81–93
2. Parag Y, Sovacool BK (2016) Electricity market design for the prosumer era. Nat Energy 1(4):16032
3. Cramton P (2017) Electricity market design. Oxf Rev Econ Policy 33(4):589–612
4. Joskow PL (2008) Lessons learned from electricity market liberalization. Energy J, Int Assoc Energy Econ (Special I) 9–42
5. Wilson R (2002) Architecture of power markets. Econometrica 70(4):1299–1340
6. Defeuilley C (2009) Retail competition in electricity markets. Energy Policy 37(2):377–386
7. Ochoa LN, Pilo F, Keane A, Cuffe P, Pisano G (2016) Embracing an adaptable, flexible posture: ensuring that future european distribution networks are ready for more active roles. IEEE Power Energy Mag 14(5):16–28
8. Guerrero J, Blaabjerg F, Zhelev T, Hemmes K, Monmasson E, Jemeï S, Comech MP, Granadino R, Frau JI (2010) Distributed generation: toward a new energy paradigm. IEEE Ind Electron Mag 4(1):52–64
9. Pudjianto D, Gan CK, Stanojevic V, Aunedi M, Djapic P, Strbac G (2010) Value of integrating distributed energy resources in the UK electricity system. In: IEEE PES general meeting, pp 1–6
10. Morstyn T, Hredzak B, Agelidis VG (2018) Control strategies for microgrids with distributed energy storage systems: an overview. IEEE Trans Smart Grid 9(4):3652–3666
11. Baruah P, Eyre N, Qadrdan M, Chaudry M, Blainey S, Hall JW, Jenkins N, Tran M (2014) Energy system impacts from heat and transport electrification. Proc ICE - Energy 167(3):139–151
12. Ossenbrink J (2017) How feed-in remuneration design shapes residential PV prosumer paradigms. Energy Policy 108:239–255
13. Grünewald P, McKenna E, Thomson M (2015) Keep it simple: time-of-use tariffs in high-wind scenarios. IET Renew Power Gener 9(2):176–183
14. Papadaskalopoulos D, Strbac G (2013) Decentralized participation of flexible demand in electricity markets - Part I: market mechanism. IEEE Trans Power Syst 28(4):3658–3666
15. Pudjianto D, Ramsay C, Strbac G (2007) Virtual power plant and system integration of distributed energy resources. Renew Power Gener IET 1(1):10–16
16. Good N, Ellis KA, Mancarella P (2017) Review and classification of barriers and enablers of demand response in the smart grid. Renew Sustain Energy Rev 72:57–72
17. Einav L, Farronato C, Levin J (2016) Peer-to-peer markets. Annu Rev Econ 8(1):615–635
18. Allan G, Eromenko I, Gilmartin M, Kockar I, McGregor P (2015) The economics of distributed energy generation: a literature review. Renew Sustain Energy Rev 42:543–556
19. Villar J, Bessa R, Matos M (2018) Flexibility products and markets: literature review. Electr Power Syst Res 154:329–340
20. Park C, Yong T (2017) Comparative review and discussion on P2P electricity trading. Energy Procedia 128:3–9
21. Zhang C, Wu J, Long C, Cheng M (2017) Review of existing peer-to-peer energy trading projects. Energy Procedia 105:2563–2568
22. Orlov A (2017) Blockchain in the electricity market: identification and analysis of business models. Norwegian School of Economics & HEC Paris, pp 1–107
23. Stähler P (2002) Business models as an unit of analysis for strategizing. In: Proceedings of 1st international workshop on business models, vol 2001, pp 1–11
24. Kim J (2016) The platform business model and business ecosystem: quality management and revenue structures*. Eur Plan Stud 24(12):2113–2132
25. Heavner B, Patey A, Vogel C, Real JD, Morton K, Ninow S, McConnell ES, Passera L, Auck SB (2015) Virtual net metering policy background and tariff summary report. Center for sustainable energy, pp 1–66

26. Long C, Wu J, Zhang C, Thomas L, Cheng M, Jenkins N (2017) Peer-to-peer energy trading in a community microgrid. In: IEEE PES general meeting
27. Gaiddon B, Joos M (2016) Report on collective self-consumption of photovoltaic. Smarter together: smart and inclusive solutions for a better life in urban districts, pp 1–26. http://smarter-together.eu/file-download/download/public/429%0A%0A
28. Ofgem (2017) Local energy in a transforming energy system. Ofgem's future insights series. http://ofgem.gov.uk/system/files/docs/2017/01/ofgem_future_insights_series_3_local_energy_final_300117.pdf
29. Morstyn T, Teytelboym A, McCulloch MD (2019) Bilateral contract networks for peer-to-peer energy trading. IEEE Trans Smart Grid 10(2):2026–2035
30. Moret F, Pinson P (2018) Energy collectives: a community and fairness based approach to future electricity markets. IEEE Trans Power Syst PP(99)
31. Inam W, Strawser D, Afridi KK, Ram RJ, Perreault DJ (2015) Architecture and system analysis of microgrids with peer-to-peer electricity sharing to create a marketplace which enables energy access. In: 9th international conference on power electronics - ECCE Asia: "green world with power electronics", ICPE 2015-ECCE Asia, pp 464–469
32. Dagher L, Bird L, Heeter J (2017) Residential green power demand in the United States. Renew Energy 114:1062–1068
33. The European consumer organisation (BEUC) (2016) Current practices in consumer-driven renewable electricity markets. The European Consumer Organisation (BEUC), Technical Report
34. Will C, Jochem P, Fichtner W (2017) Defining a day-ahead spot market for unbundled time-specific renewable energy certificates. In: International conference on the European energy market, EEM
35. Castellanos JAF, Coll-Mayor D, Notholt JA (2017) Cryptocurrency as guarantees of origin: simulating a green certificate market with the ethereum blockchain. In: 2017 5th IEEE international conference on smart energy grid engineering, SEGE 2017, pp 367–372
36. Morstyn T, McCulloch MD (2019) Multi-class energy management for peer-to-peer energy trading driven by prosumer preferences. IEEE Trans Power Syst 34(5):4005–4014
37. Morstyn T, Farrell N, Darby SJ, McCulloch MD (2018) Using peer-to-peer energy-trading platforms to incentivize prosumers to form federated power plants. Nat Energy 3(2):94–101
38. Hug G, Kar S (2013) Consensus + innovations approach for distributed multi-agent coordination in a microgrid. IEEE Signal Process Mag 30(3):99–109
39. Kraning M, Chu E, Lavaei J, Boyd S (2014) Dynamic network energy management via proximal message passing. Found Trends Optim 1(2):70–122
40. Morstyn T, Hredzak B, Agelidis VG (2018) Network topology independent multi-agent dynamic optimal power flow for microgrids with distributed energy storage systems. IEEE Trans Smart Grid 9(4):3419–3429
41. Karlsson M, Ygge F, Andersson A (2007) Market-based approaches to optimization. Comput Intell 23(1):92–109
42. Sorin E, Bobo L, Pinson P (2018) Consensus-based approach to peer-to-peer electricity markets with product differentiation. CoRR. arXiv:abs/1804.03521, http://arxiv.org/abs/1804.03521
43. Munsing E, Mather J, Moura S (2017) Blockchains for decentralized optimization of energy resources in microgrid networks. In: 2017 IEEE conference on control technology and applications (CCTA), pp 2164–2171
44. Ma Z, Callaway DS, Hiskens IA (2013) Decentralized charging control of large populations of plug-in electric vehicles. IEEE Trans Control Syst Technol 21(1)
45. Deori L, Margellos K, Prandini M (2016) On the connection between Nash equilibria and social optima in electric vehicle charging control games. In: 20th world congress of the international federation of automatic control, pp 1–6
46. De Paola A, Angeli D, Strbac G (2017) Price-based schemes for distributed coordination of flexible demand in the electricity market. IEEE Trans Smart Grid 8(6):3104–3116
47. Carrion M, Conejo AJ, Arroyo JM (2007) Forward contracting and selling price determination for a retailer. IEEE Trans Power Syst 22(4):2105–2114

48. Wei W, Liu F, Mei S (2015) Energy pricing and dispatch for smart grid retailers under demand response and market price uncertainty. IEEE Trans Smart Grid 6(3):1364–1374
49. Jia L, Tong L (2016) Renewables and storage in distribution systems: centralized vs decentralized integration. IEEE J Sel Areas Commun 34(3):665–674
50. Liu N, Yu X, Wang C, Li C, Ma L, Lei J (2017) An energy-sharing model with price-based demand response for microgrids of peer-to-peer prosumers. IEEE Trans Power Syst 32(5):3569–3583
51. Lee J, Guo J, Choi JK, Zukerman M (2015) Distributed energy trading in microgrids: a game theoretic model and its equilibrium analysis. IEEE Trans Ind Electron 62(6)
52. Khorasany M, Mishra Y, Ledwich G (2017) Auction based energy trading in transactive energy market with active participation of prosumers and consumers. In: 2017 Australasian universities power engineering conference (AUPEC)
53. Faqiry MN, Das S (2016) Double-sided energy auction in microgrid: equilibrium under price anticipation. IEEE Access 4:3794–3805
54. Guerrero J, Chapman A, Verbic G (2017) A study of energy trading in a low-voltage network: centralised and distributed approaches. In: Australasian universities power engineering conference (AUPEC), pp 1–6
55. Zhou Y, Wu J, Long C (2018) Evaluation of peer-to-peer energy sharing mechanisms based on a multiagent simulation framework. Appl Energy 222:993–1022
56. Lee W, Xiang L, Schober R, Wong VWS (2014) Direct electricity trading in smart grid: a coalitional game analysis. IEEE J Sel Areas Commun 32(7):1398–1411
57. Han L, Morstyn T, McCulloch M (2019) Incentivizing prosumer coalitions with energy management using cooperative game theory. IEEE Trans Power Syst 34(1):303–313
58. Han L, Morstyn T, McCulloch M (2018) Constructing prosumer coalitions for energy cost savings using cooperative game theory. In: Power systems computation conference (PSCC), pp 1–7
59. Chapman AC, Mhanna S, Verbič G (2017) Cooperative game theory for non-linear pricing of load-side distribution network support. In: The 3rd IJCAI algorithmic game theory workshop
60. Fleiner T, Jankó Z, Tamura A, Teytelboym A (2015) Trading networks with bilateral contracts. In: Proceedings of the 3rd conference on auctions, market mechanisms and their applications. ACM, pp 1–39
61. Fleiner T, Jankó Z, Schlotter I, Teytelboym A (2018) Complexity of stability in trading networks. ArXiv e-prints, pp 1–18. arXiv:abs/1805.08758
62. Hatfield JW, Kominers SD, Nichifor A, Ostrovsky M, Westkamp A (2013) Stability and competitive equilibrium in trading networks. J Polit Econ 121(5):966–1005
63. Morstyn T, Teytelboym A, Mcculloch MD (2018) Matching markets with contracts for electric vehicle smart charging. In: IEEE power and energy society general meeting (PESGM), pp 1–5
64. Morstyn T, Teytelboym A, McCulloch MD (2019) Designing decentralized markets for distribution system flexibility. IEEE Trans Power Syst 34(3):2128–2139
65. Hatfield JW, Kominers SD, Nichifor A, Ostrovsky M, Westkamp A (2015) Full substitutability in trading networks. In: Proceedings of the 16th ACM conference on economics and computation - EC'15. ACM Press, New York, New York, USA, pp 39–40
66. Customer-Led network revolution, enhanced profiling of domestic customers with solar photovoltaics. http://networkrevolution.co.uk
67. Muratori M (2017) Impact of uncoordinated plug-in electric vehicle charging on residential power demand - supplementary data. http://data.nrel.gov/submissions/69

Healthcare and the Sharing Economy

Jad Bitar

Abstract The sharing economy has expanded rapidly in the last decade, with startups such as Uber and Airbnb providing consumers access to shared goods and services, in convenient and low-cost ways. This article highlights how sharing economy principles are turning ownership models of assets in healthcare upside down, and how a sharing mindset can provide answers to some of the major challenges facing the industry today.

Keywords Healthcare · Primary care · Medical recruitment · Medical equipment · Sharing economy

1 The Relevance of the Sharing Economy to Healthcare

While there has been tremendous progress in developing life-saving treatments and an increase in life expectancy in recent decades, the costs required to deliver high-quality care to an older and frailer population has become increasingly unsustainable within traditional healthcare structures.

New ways of thinking about how healthcare is organized and delivered are needed. Sharing economy could provide some of the answers to challenges such as an increase in demand, a burden on existing resources, unsustainable costs and a need for greater efficiency.

While the sharing economy proliferated by the likes of AirBnB and Uber has disrupted a wide range of industries to meet consumer demand, its influence on healthcare has been modest to date. However, several healthcare organizations are beginning to reap the benefits of embracing a sharing mindset to improve access to services and increase efficiencies. Below, we highlight some of the ways providers of healthcare services are using sharing principles to re-imagine how healthcare can

J. Bitar (✉)
Partner and Managing Director, Boston Consulting Group, Boston, USA
e-mail: bitar.jad@BCG.com

be organized and delivered, which in turn is increasing access and creating value for a range of stakeholders, from patients, providers, and industry.

2 The Economics of Sharing

The sharing economy is an economic model most commonly defined as a peer-to-peer (P2P) activity of acquiring, providing, or sharing access to goods and services, usually facilitated by an online platform [1]. The best-known sharing assets are cars and homes. Cars go unused for 95% of their lifetimes, according to the Brookings Institute. So why purchase and maintain a rarely used vehicle when Uber or Lyft will pick up and drive passengers to their destinations for a one-time fee? Homeowners use Airbnb to rent their spare bedrooms for weeks at a time, with Airbnb rates reported to be between 30 and 60% cheaper than hotel rates around the world. Its a win-win as homeowners make money and customers save money. Based on the success of Uber and Airbnb, Juniper Research predicts the sharing economy will grow from $18.6 billion in 2017 to a forecasted $40.2 billion by 2022. While sharing in the physical world can be time consuming and inefficient, sharing in the digital world can be more time and cost efficient [2]. Many successful examples of the sharing economy use technology to create a two-sided peer-to-peer market where one has never existed. In a sharing market, one party is either a buyer or a seller, and together they re-allocate under-utilized resources toward more efficient use. This is where the real economic value lies and is at the heart of the sharing economy. Typically, sharing platforms provide access to new markets and customer segments by lowering costs for consumers who want to use an asset but cannot afford to own it or who want to use it for only a short period of time. The majority of medical equipment sits idle for 42% of the time [3] and sharing platforms can increase the purchase price of shareable assets, since buyers are often willing to pay a premium for items that can generate revenue by being shared [3]. Applying sharing economy in healthcare also presents an opportunity to address deep-seated structural disconnects that obscure the relationship between the consumers of care (patients) and the providers. At present, the majority of healthcare is negotiated and commissioned via payors (providers, governments, insurance companies), rather than directly by patients. This disconnect also often leaves patients unable to transparently understand the relevance and quality of services they use. By establishing a more direct relationship between consumer and provider of a service, the sharing economy presents an opportunity in healthcare to give patients a greater say over the services they value and are willing to pay for.

3 Case Studies: Shaping the Healthcare Sharing Economy

Below are examples of how owners, users, and intermediaries in healthcare are benefiting from the sharing of assets and resources to improve access and enhance value-both in new revenue streams and market opportunities. By applying the economic and behavioral rationale for sharing, startups are already shaping the healthcare industry to meet the needs of healthcare consumers.

3.1 Heal: Improving Access to Primary Care

Problem:
In the U.S., the Association of American Medical Colleges estimates there will be a need for as many as 32,000 additional primary care physicians by 2025 [4], due to aging and population growth. And already, there is a problem with access to primary care, with available appointments to see a primary care physician sometimes exceeding 20 days [5] and doctors typically spending 55% of their time on administrative tasks, rather than with patients.

Solution:
In 2015, startup Heal launched an app to help patients in California book on-demand home visits from primary care physicians. Users download the app and provide their address and the reason for the visit and a physician, with a medical assistant, will arrive within 20–60 min for a flat fee. Heal doctors provide basic services on a house call, such as diagnosis and treatment of moderate ailments like bronchitis, give flu shots, stitch up a nasty cut, or write a prescription. Each visit costs patients a $99 flat fee, which the patient can pay for directly or through their insurance company. Heal receives a revenue share of the fee received by the doctor and assistant who make the house visits.

Impact:
Heal is now available to over 70% of California residents and most residents in Washington D.C. and Northern Virginia. By enabling patients to book home visits on-demand, Heal is providing patients with improved access to primary care in their own homes, which is ultimately delivering better outcomes more quickly, and in convenient locations such as their home or office.

Heal also helps physicians by taking care of some administrative tasks, such as preparing patients paperwork and billing, so that they can be freed up to spend more time with patients. As a result, Heal claims that physicians also are able to spend up to 28 min with each patient they visit: double the duration of the national average of 13 min.

3.2 Nomad Health: Shaking Up Medical Recruitment

Issue:
60% of healthcare providers costs are spent on staff, which means it is critical to get the right team in place quickly. However, recruitment can be a challenge with

traditional recruitment agencies proving costly and not always able to quickly fill labor gaps, which can lead to a delay in meeting patient demand and stretched resources.

Solution:
Startups such as Nomad Health have built an online platform to that connects healthcare systems directly with healthcare workers, who can search and sign up for short term and per-diem jobs, that suit their skillsets and individual circumstances. "Doctors can search for gigs [jobs] by specifying which state they want to work in, or which electronic medical-records system they know best, while healthcare providers describe the job in hopes of finding someone whos interested", explains Dr. Alexi Nazem, CEO of Nomad Health [6].

Nomad Health also offers services to physicians, such as negotiating contracts and salaries, to next-day payment processing and, in some cases, malpractice insurance.

Impact:
By matching up healthcare professionals looking for short-term jobs with providers, these startups are able to increase the productivity of the existing medical workforce.

These traditional recruiting agencies are gradually being supplanted by sharing startups. Nomad takes a 15% cut of physicians salaries, a rate much lower than the typical 30–40% commission brokers and agencies take.

Jill Schneider, Executive Director at Clinica, has already found great success with Nomad. "Nomad has changed how I hire. As an Executive Director, I don't have much time to spend recruiting and we've found other hiring methods to be too cumbersome. On Nomad, it is incredibly simple to post jobs and the technology is efficient and easy to use. It was exactly what I needed to simplify our hiring process", said Schneider [7].

3.3 MedSpeed: Improving Access to Underutilized Medical Equipment

Problem:
Medical equipment is underutilized and can be as low as 42% [3], with healthcare providers tending to view machines such as ultrasounds, MRIs, and CT scanners as assets depreciating in value, and belonging to a fixed location. Typically, a hospital might use an MRI scanner for five days a week between the hours of 8 a.m. and 2 p.m., meaning that the remaining time the machine is idle, gathering dust and accruing cost.

Solution:
In 2013, MedSpeed, a U.S.-based medical courier company noticed a gap in the market to increase by creating a network of healthcare organizations to share idle medical equipment, to increase utilization, reduce unnecessary purchases, and generate cost savings for both owner and user of the equipment [8].

Impact:
Asset sharing not only increases revenues for the owner and loanee of the equipment but also increases access to treatment, which ultimately contributes to improving outcomes and experience of care. By changing the mindset of how healthcare providers view their assets, MedSpeed has transformed the logistics arm of the company from making a loss, into a strategic asset. The company has partnered with 19 health systems in the U.S. and have seen a 350% revenue growth over the past five years, while also generating over $20m in savings per year for its customers. One partner based in the Great Plains saved more than $1.3M over five years through transportation optimization and better network utilization. In an interview with Modern Healthcare, MedSpeed CEO Jake Crampton says the real upside is the ability to share a wider variety of products and services, including at satellite facilities. "Plus," he adds, "it is easier to justify purchases of state-of-the-art equipment when it will be employed across multiple facilities." Patients also benefit through this increased access by providing evening and weekend appointments to fit around working patterns.

3.4 Connectus: On-Demand Dialysis Machines for Patients

Problem:
In 2007, Javier Artigas was diagnosed with chronic kidney disease, and by 2014, he required dialysis treatment three times a week to do the work his kidneys no longer could. However, Artigas's job involved significant travel to Latin American and African countries where it was not always easy finding somewhere safe that is, with uncontaminated waterto have dialysis treatment which prevents the buildup of dangerous toxins in the blood.

Solution:
In 2015, Artigas developed an app called Connectus for $1700, to connect kidney patients in South America with treatment centers when away from home [9].

Impact:
To date, Connectus serves 250,000 patients from nearly 150 countries, who need regular dialysis treatment and has 14,500 dialysis machines on its records. Since its launch, the app received the 2015 award for Best Innovative Healthcare Solution from the Massachusetts Institute of Technology (MIT), and Artigas has signed a contract to explore further business opportunities with Joe Gebbia, co-founder of AirBnB [9].

4 Three Sharing Models [10]

Sharing businesses are not all the same. There are at least three distinct models, which differ according to who owns the asset and who sets the price and other conditions.

- **Decentralized platforms**. An asset owner sets the terms and offers the asset directly to the user. The platform makes the match and facilitates the transaction in exchange for a small share of the fee. This is the Airbnb model. Upfront capital costs are low, but the platform must recruit providers to ensure adequate supply. E.g. Heal, who match patients to primary care physicians to make home visits.
- **Centralized platforms**. The platform itself owns the asset and sets the price. It has greater control over quality, availability, and standardization than a decentralized platform and collects a larger share of the transaction value, but costs to scale are much higher, too. This is the Zipcar and Rent the Runway model. It requires significant upfront capital and high utilization to be viable.
- **Hybrid platforms**. Asset owners offer a service with price and standards set by the platform. Ownership and risk are decentralized, while standardization and service level are centralized. This is the Uber and Lyft model. As with the decentralized model, upfront costs are low and provider recruitment is crucial. The platform must also carefully manage its relationship with providers since they have less control than they would under the decentralized model. E.g. MedSpeed, who facilitate the hiring of underutilized medical assets between one provider to another.

5 Benefits and Challenges of Applying Sharing Economics to Healthcare

Benefits

- **Patient-driven demand**: at its heart, the sharing economy highlights where there is immediate supply and demand for services. Because demand is driven by the consumer, it could be argued this could open up greater value for patients, and reduce unnecessary treatment and processes that are part of more deep-seated care delivery models.
- **Improved access**: through the facilitation more efficient utilization of assets (e.g. healthcare workers, medical equipment), this improves access to healthcare, while at the same time obtaining greater value out of existing resources for both the owner and loanee of any services or equipment.
- **Greater transparency**: by enabling users to rate the service provided, sharing economy can help patients select top providers and hence incentivize providers to continuously improve quality.

Challenges

- **Complexity of healthcare**: Providing medical treatment is more complex and high-risk than offering a taxi ride across town or renting out an apartment for a long weekend. Austin Frakt, at The Incidental Economist, is brutally honest in his assessment of healthcares difference: "Make a mistake with your choice of airline, lawyer, auto mechanic, or cell phone and you be inconvenienced for a time. Make a mistake with your healthcare, and you could pay for it with your life". Also, healthcare's web of different stakeholders from patients, providers, governments, insurers, life sciences and research, have interdependencies and, as a result, a change in how products and services are delivered or financed are likely to face substantial systemic, regulatory and commercial challenges.

- **Helping patients make informed decisions**: choice of healthcare provider is not always based on price. The quality and access to care, as well as the expected outcomes, are other important variables patients want to consider when making an informed choice about services they use. Offering a substantial database end-user reviews would help address this issue and give patients greater confidence when purchasing services digitally.
- **Who pays, who reviews?** One of the benefits of services that belong in the sharing economy is that consumers can leave reviews to rate their experiences. However, in healthcare, the end beneficiary of care is not always the person who pays for it.
- **Slower feedback loops**: the benefits of a treatment can take weeks, months and sometimes years before they can be properly assessed, which affects the ability to rate the service received instantaneously.
- **Scarcity of healthcare workers**: while there is vast potential for healthcare to expand further into the sharing economy—because everyone demands access to health services at some point—health workers are in short supply worldwide, which means that supply of some demand might be more challenging to fulfill.

6 Conclusion

Consumers have already spoken. They appreciate the convenience, variety, and cost-effectiveness the sharing economy has to offer. The economic and business rationale for sharing is strong, both for startups and for incumbents. The combination of these two dynamics is proof that the sharing economy is real, relevant, and a tangible opportunity rather than a temporary distraction or a passing fad. It can create new revenue streams and market opportunities.

That said, the sharing economy in healthcare is relatively young and underdeveloped. Most healthcare sharing businesses are at the beginning of the S curve, and the technological possibilities, regulatory environment, and consumer dynamics of sharing are still maturing. However, it is not too early to imagine how falling transaction costs, evolving business models, and changing consumer and patient behaviors will alter strategy and business models and deliver value to a range of stakeholders from patients, caregivers, operators, payors.

The sharing economy also offers solutions to bridge the disconnect between patients and providers of care, which has the potential to lead to greater efficiency and increased trust in the system, as patients become more empowered to make decisions about exactly what, where and how they access the services they need. For the sharing economy to successfully scale in healthcare, patients will need to be able to trust the platforms they use and for a wider economy of patient-generated reviews to give credibility to services on offer.

The sharing economy has already paved the way for numerous organizations to provide low-cost, high-value solutions to some major challenges in healthcare; and it still has the potential to offer more to tackle the challenge of meeting the increasing demand for healthcare, in the most efficient and patient-centric way possible.

In this context, being able to build at scale and first-mover advantages will arguably matter even more than before. Staying ahead of these changes and shaping the future of sharing in order to take advantage of the opportunities, and deal with the challenges that sharing will bring should be top of mind for healthcare professionals.

References

1. Investopedia (2018) What is the sharing economy? https://www.investopedia.com/terms/s/sharing-economy.asp
2. Wallenstein J, Shelat U (2017) The boston consulting group. Learning to love (or live with) the sharing economy. https://www.bcg.com/en-us/publications/2017/strategy-technology-digital-learning-love-live-sharing-economy.aspx
3. Horblyuk R et al (2012) GE healthcare. Out of control little-used clinical assets are draining healthcare budgets. Healthc Financ Manag 66(7):64–68. http://www3.gehealthcare.com.au/~/media/downloads/us/services/hospital%20operations%20management/white%20papers/out%20of%20control%20-%20asset%20management%20whitepaper.pdf?Parent=%7BD80E0C0C-461E-49BD-86FB-5FEF285DCA2C%7D
4. Petterson SM et al (2012) Projecting US primary care physician workforce needs: 2010–2025. Ann Family Med 10(6). http://www.annfammed.org/content/10/6/503.full.pdf+html
5. Gupta S (2014) The doctor will see you in a few weeks. EveryDay Health. https://www.everydayhealth.com/news/doctor-will-see-you-few-weeks/
6. Ramsey L (2017) Business insider US. A startup that's tackling a massive healthcare crisis is moving in on a $7 billion business. https://www.businessinsider.my/nomad-health-freelance-health-care-work-for-nurses-2017-7/
7. Nomad (2018) Nomad health expands into full-time healthcare jobs! https://blog.nomadhealth.com/nomad-health-expands-into-full-time-healthcare-doctor-jobs/
8. ModernHealthcare (2018) Is the sharing economy the disrupter healthcare needs? https://www.modernhealthcare.com/article/20180405/SPONSORED/180409955
9. Lipman N (2018) How saving a writer's life helped my dialysis app go global. BBC News. https://www.bbc.co.uk/news/stories-42778554
10. Wallenstein J, Shelat U (2017) The Boston consulting group, Hopping aboard the sharing economy. https://www.bcg.com/publications/2017/strategy-accelerating-growth-consumer-products-hopping-aboard-sharing-economy.aspx

Industry4.0

Edoardo Calia and Davide D'Aprile

Abstract The fourth industrial revolution is already showing its disruptive potential in many fields, primarily in the manufacturing, logistics and energy sectors. It is expected to revolutionize production processes, business models and IT infrastructures: eventually, the supply chain will turn into more efficient, adaptable and scalable workflows, ultimately driven by a nearly real-time and utterly bespoke demand of new or enhanced products and services. Science is already supporting this transformation, mostly through its recent developments in AI, finally resulting in an exponential availability of approaches, methods and tools to support and boost it. An ever increasing availability of data, coming from different sources, generated by humans and machines are being fruitfully integrated and will ignite the Industry4.0 paradigm. Machines will be smarter, they will make decisions and trade resources and services using virtual currencies in the emerging framework of the Machine Economy.

Keywords Industry4.0 · Digital manufacturing · Digital identity · Decentralized value chain · Artificial intelligence

1 Introduction

The digital transformation that we have witnessed in the past five decades took its first steps from a centralized scenario where a few large computers (mainframes) were located in military or academic premises, requiring entire rooms for their installation and operation, and evolved into todays world where a large number of small processing devices are pervasively embedded in the environment, from which they

E. Calia (✉)
Links Foundation, Via Pier Carlo Boggio 61, Torino 10138, Italy
e-mail: edoardo.calia@linksfoundation.com

D. D'Aprile
World Green Economy Organization, Happiness Street, Dubai, UAE
e-mail: davide.daprile@worldgreeneconomy.org

collect data and send it wirelessly to more powerful nodes able to store and process the data transforming it in useful information.

This relatively long process unfolded along several important milestones:

During the **80s**, computing machines became available on a personal scale, and at the same time some standards were developed for computer to computer communication. These included the Internet Protocol (IP) suite, which later became the official protocol stack of the network as we know it.

The **90s** were the time that saw Internet connectivity become available to the public: private providers stepped in, and the network rapidly grew to a worldwide infrastructure. Its hierarchical architecture was strengthened with the definition and adoption of Autonomous Systems and advanced routing protocols, and computing power started to be distributed at the edge with the introduction of Personal Computers and workstations. Thanks to its new organization in multimedia, hypertext format, digital information became available at a click of a mouse using browsers, a new family of applications introduced in these years (Mosaic, Netscape) that were set to change forever the Internet user experience.

The digital (r)evolution kept progressing even at a faster pace in the *first two decades* of the new century, with significant innovations both at the core and at the edge of the Internet. After the enthusiasm in favor of distributed computing paradigm, which cast a dark shadow over the popularity of large data centers, a new wave of computing resources centralization and management came with the introduction of the cloud paradigm in its many different implementations (SaaS, IaaS, PaaS,[1] etc.).

This evolution of the Internet core didn't stop the trend towards the massive distribution of (small) computing power at the periphery of the architecture, enabled by the availability of small embedded systems capable of performing limited computation and transmitting data over wireless links using low power technology. The *Internet of Things* was born, a paradigm where any object can be digitally enabled, paving the way to the close interconnection between the physical and the digital worlds represented by *cyber-physical systems*.

Finally, its worth mentioning one of the hottest and most disruptive topics of the current digital era: *decentralization* based on a modern version of *distributed ledgers*. This trend, started in 2008 [1] from the idea of building a system where financial transactions could be completed in a secure way without requiring trust and the intervention of central authorities, is spreading to almost all possible transaction-based services (not limited to financial applications). Built on the strong pillars of cybersecurity principles, algorithms and architectures, distributed ledgers became so popular in the last 10 years that at the date of writing more than 1200 services have seen their decentralized counterpart being introduced in the market.

The result of this technological progress is a scenario dominated by data, collected by small autonomous devices able to communicate wirelessly over a short distance. At the core of the architecture are the most powerful computing resources, connected

---

[1] *Software as a Service*, *Infrastructure as a Service*, and *Platform as a Service* are different categories of services offered by cloud computing providers.

to a smart periphery via a dense grid of high bandwidth cables and wireless channels. Its the architecture that makes the vision of Mark Weiser [2] come true.

For those who were lucky enough to witness this (r)evolution going on in the past decades it is easy to recognize similar (potentially faster) processes started at the beginning of the new century in sectors other than the Internet itself and the Web. Three of the most noticeable contexts where this is happening are *logistics, energy* and *manufacturing*: the IoT paradigm offers all the great advantages of the digital era also to these traditionally analog worlds.

Logistics takes care of moving physical goods from a source location to a destination, passing through storage hubs and a number of carriers with different performance. Most of the handling of packages up to now has been carried out with little digital information support, and—even more relevant—with very little cooperation between adjacent carriers or partners.

Optimizing the delivery time, monitoring the hop-by-hop path followed by packages and keeping all the interested parties informed about the different phases of the process are issues having strong analogies with the delivery of digital packets in Internet-like networks. The advent of digitization in this space is pushing the evolution of logistics into the so called Physical Internet [3]: the same principles that have been tested for decades in the forwarding of small amounts of digital data turned to be very beneficial to scientists and specialists studying how to better move around small or big amount of physical goods. Digitization in Logistics will also help operators offer a better service at a time where e-commerce brought to the scene new requirements and issues. Among them are short (and possibly certain) delivery times and (even more challenging) higher *back-route* traffic,[2] a relevant side effect of e-commerce for which the logistics networks were not prepared.

A similar evolving scenario is having strong impact on **energy** networks, traditionally built around a small number of large production plants connected to end users acting as passive consumers. Even in this case the analogy with the Internet is quite strong: the evolution of consumers into *prosumers* (consumers also able to actively generate energy at their premises, to be sent back to the network) is the equivalent of the distribution of computing power to everybody's desk when workstations, personal computers and lately mobile devices were introduced. Distributed power generation requires distributed intelligence to monitor the amount of energy generated at the periphery and to manage its local use or its re-introduction in the network.

Of the three mentioned above, **manufacturing** is perhaps the area where the digitization process is receiving the greatest level of attention, possibly because of the higher impact that it can have on the economy of each industrial country. *Industry4.0*, or the *Fourth Industrial Revolution*, will be the focus of this chapter in which we will highlight both its main technological aspects and its potential economic impact with the new business models it brings to the scene.

The fourth industrial revolution takes its name from the brave assumption that we are going to witness in the coming years a disruption comparable to the introduction

---

[2]Due to aggressive return policies designed to increase customer satisfaction.

of the steam or coal engine (1st industrial revolution), the internal combustion engine (2nd) and the atomic energy and information technology (3rd).

The next industrial revolution starts with the IoT paradigm brought to the shopfloor, bringing the physical and digital worlds closer to each other. It's the era of the *Cyber-Physical Systems* and of a new generation of services like personal fabrication, distributed factory, and a higher involvement of the end user in the production processes via better, timely, and bi-directional information flow. Machines will be smarter, they will make decisions and trade resources and services using virtual currencies in the emerging framework of the Machine Economy.

2 The Fundamental Building Block: Connected Machines

From a digital/technical standpoint digital manufacturing is just one of the many applications of the Internet of Things paradigm, where the things are represented by all the protagonists (including, but not limited to, human operators) of the manufacturing process.

Among these, robots and other CNC (Computer Numerical Control) machines play a fundamental role for which they must be properly equipped or upgraded.

CNC technology is quite old, much older than the Internet: the first numerical control machines date back in the '40s [4]. For several decades the improvements for such machines aimed at making the machine itself better, working on one side on the manufacturing process (mechanics, tools etc.) and on the other on the control electronics and algorithms. But the machine as a whole often remained mainly a standalone device, with no connection interface with the outside world except for a reader (initially punched cards, then floppy disks, and more recently USB memory sticks) used to manually upload a preprogrammed sequence of the operations required to complete a specific task.

Most manufacturing companies today (the majority of them, small or medium enterprises) still use even modern, sophisticated CNC machines as isolated tools; and machines configured in this way require slow, manual interventions to be programmed (in the best case this just requires manually copying files from CAD stations to the machine, but often the procedure requires hours or even days).

Most modern devices come equipped with a network interface similar to the one used by computers (Ethernet): they are ready—sometimes before their owners—to participate in the Industry4.0 challenge.

At the beginning of the 21st century, CNC equipment was—referring to network connectivity—at the stage where computers were twenty years before. In the mid 80s NICs (Network Interface Cards) were an option for personal computers, and so was the TCPIP software[3]: only users interested in accessing the Internet would buy

---

[3]The TCPIP is the suite of standard software protocols needed to exchange information over the Internet.

those as optional aftermarket components (something nobody would even consider today!).

So why should one want to connect a CNC machine to the Internet? The reasons are numerous, as the rest of this chapter will clarify. To name the most common:

- data collection from onboard sensors, to be sent to remote monitoring systems
- access to repository of G-code[4] files or templates, ready to use
- access to local (e.g. company intranet) repository of configuration files
- information exchange with other tools/machines/enterprises
- access to predictive maintenance services (see Sect. 6.3).

All these services are kind of basic requirements for the migration to digital manufacturing and Industry4.0.

3 Digital Manufacturing

The term digital manufacturing is sometimes erroneously associated to the process of building something using a CNC machine. Even if this is a necessary condition, it is not sufficient. Digital Manufacturing in the context of Industry4.0 refers to the entire production process becoming digital, ideally from the concept idea of the designer, engineer or architect to the actual construction of the target product. And becoming digital basically means becoming automated, requiring as little human intervention as possible from start to completion.

An example of actual digital manufacturing process is 3D printing: the object to be printed is first designed using a CAD program on a computing system (a personal computer or a web-accessible online service). The CAD program generates a file using an intermediate format that all 3D printers can interpret and convert in the appropriate G-code, so the file can be directly sent to any 3D printer equipped to match the specific printing requirements (material, size, accuracy etc.).[5]

3D printing had an important role in the evolution towards Industry4.0 because of the awareness (sometimes unconscious) it created about digital manufacturing. The excitement about 3D printing increased fast also thanks to the Maker Movement, started in 2005 by Dale Dougherty (founder of the Make: magazine): one of the most popular tools that makers started to build themselves were affordable 3D printers. Even though these were amateur version of professional machines, they could actually be included in a fully digital manufacturing process that anybody could put together at home or at school. When making the first moves along the path leading

---

[4]G-code is one of the most popular languages used to program CNC machines: it includes elementary commands to be executed in a specific order (move the tool to a certain point, drill a hole, set a specific speed, etc.).

[5]3D printing is an oversimplified example though, because it is essentially a 2D process iterated over a pre-determined number of layers. A more complex CNC machine operating in real 3D (like a multi-axis mill) often requires some—human-operated—fine tuning on the G-code before the actual object can be manufactured.

to digital manufacturing *the real disruption lies in the ICT/digital portion of the architecture more than in the quality of the CNC machine(s) used.*

3.1 Makers: Where It All Started

The wide adoption of digital technologies that started in the 90s, when the Internet and its services were made available to the public, remained for 15 years confined in the digital world: new devices, software tools and services were introduced in the market at a constantly growing pace, allowing everybody to participate in a revolution which brought new ways to instantly access and share information, starting a process set to disrupt the communication and information businesses forever.

At some time around 2005, when the adoption of digital technologies and tools had already become a reality (and in some cases given for granted) for a significant portion of the population, a new tendency emerged: extending the digital revolution to the physical world, adding a modern, digital soul to objects and machines. People already familiar with DIY, who were used to tinkering with objects, fixing them or turning them into new tools, received with great enthusiasm the possibility to go even further, exploiting the almost infinite range of opportunities opened by adding sensors, small motors, microcontrollers and other electronics to previously analog and dumb objects. They were now able to actually *make* new things: the *Maker movement* was born.

The makers community grew rapidly also due to their strong openness policy, supported by the possibility to easily share information: all projects were made public, including design files, construction instructions, HW and SW.[6] A significant boost to the spreading of the maker movement was the expiration of patents related to FDM (*Fused Deposition Modelling*) 3D printing in 2009, which made it possible for makers to start building their own (low cost) CNC devices to fabricate small objects at home. Information about these new technologies spread very fast, leading in a few years to a significantly increased general awareness on CNC equipment including robots, milling machines and many other devices that were sometimes designed and built for specific purposes. But 3D printers definitely played a great role in showing how a manufacturing process can be almost fully automated.

Even though profoundly pervaded with digital technologies, *making* still involved working with machines, tools, and raw materials (wood, metal, paper, Plexiglas etc.). The need quickly arose for having dedicated locations where carrying out the *making* activities: workshops where electronic boards, motors, controllers, started to show up besides milling machines, robots, drills and other more traditional manual tools. *Makerspaces* started to open in many cities worldwide, and soon they became open places where DIY experts (now called *makers*), amateurs, common people—including families with kids—could go and have a taste of what was coming up in

---

[6]The first open source project of a low cost 3D printer was RepRap, started in 2005 in England (University of Bath).

the modern manufacturing arena. Makerspaces were not the first workshops dedicated to making and open to the public: a few years earlier (1998) in Boston, at the Massachusetts Institute of Technology (MIT), the director of the Center for Bits and Atoms prof. Neil Gershenfeld started teaching the course "How to Make (Almost) Anything". The Center also became the birthplace of *FabLabs*, digital fabrication laboratories equipped in a pre-defined way with flexible manufacturing and rapid prototyping machines including 3D printers, laser cutters, milling machines and lots of more traditional tools. FabLabs left the academic labs in 2002, when they started becoming public workshops in support of dissemination initiatives aimed at spreading the word of modern manufacturing. MIT kept moving in the same direction, starting the Fab Foundation in 2009 and the academic program Fab Academy, based on remote teaching and practical experiences hosted by the many Fablabs available worldwide: at the end of 2017 the network counted more than 1200 labs over the world, a number that keeps increasing. The above concepts are no news for the big manufacturing companies, where robots have outnumbered people in the shopfloor for several years now. The services that can be implemented on top of a fully digital manufacturing process represent a unique opportunity also for SMEs to hop on this new industrial revolution wagon. The following subsections will give an overview of some of the new features enabled by Industry4.0.

3.2 Production Monitoring and Planning

Every entrepreneur in the manufacturing business knows how important it is to have a good and timely understanding of the production progress. Optimizing the use of the machine tools helps to lower and better distribute the general costs across the different lots, also reducing storing and inventory costs.

An accurate picture of the whole production plant(s) is key to enable planning tasks in the short term and to support decisions about future investments.

In order to be effective, the information about the production state and progress must be available in real time, constantly checking and monitoring the state of each production stage.

Sophisticated monitoring and decision support tools can be developed by leveraging on the wide set of (digital) data collected from the (physical) world, offering entrepreneurs, managers and technical staff the right information at the right time.

Real time data collected from machines and other elements of the production process can also be fed to existing MES (Manufacturing Enterprise Systems) to help optimize the allocation of tasks to the available working stations, and to planning and simulation tools, allowing to carry out more accurate *What If Scenario Analysis* (WISA) for the long term.

3.3 Personal Fabrication

When manufacturing automation was introduced in the 40s [4], the main innovation was the possibility to program a machine tool to execute repeatedly over and over the same instructions, therefore enabling the making of multiple identical copies of the same object.

Programming such a machine was (and sometimes still is today) a very time consuming task, requiring several trial-and-error iterations before the ideal sequence of elementary tasks was found. Moreover such programming phase could only be carried out by specialized personnel having the required programming competences.

But once the programming phase was completed, the machine could run continuously for years with just a few interruptions for maintenance, dramatically reducing the cost of the manufactured items. This was the beginning of the *mass production era*.

Mass production is competitive because the ratio between the time required to program the machine and the operation time during which that program is executed is very small, almost negligible. And that is because the operation time is much longer (hundreds of thousands of cycles, several months) than the programming time (a few hours or days).

The same ratio would remain small for smaller operation time—leaving the overall competitive/sustainable—if also the programming time was significantly reduced. Programming time would be close to zero if machines could fetch a ready-to-execute new configuration file from the company intranet or even from the Internet. And shorter operation time could translate in smaller production lot size (small number of cycles executed before a new program is needed). The combination of these two features leads to one of the major disruptions brought by Industry4.0: the possibility to offer *personal fabrication* (sometimes called *Lot size 1*).

Summing up the above description, a key requirement for setting up personal fabrication is the ability to automatically program the machines, starting from a project drawn using a CAD software tool. It is easy to see how in this case the cost of producing individual different (*personal* and *unique*) items is the same as the cost associated to the mass production of hundreds of thousands of identical objects. With a much higher perceived value for the customer.

This all-digital scenario also calls for a change in the role of customers, who become active participants in the manufacturing process by providing their personal requirements. The personal fabrication architecture then also requires tools made available to customers in order to put together their own, unique product order.

Even if a complete description of these features is out of the scope of this book, we can mention some examples of interaction tools that some advanced manufacturing companies are offering to their customers:

- A user could upload her own CAD file to a web portal, from which it would be transferred to the most appropriate machine[7] (for the reasons mentioned above this works particularly well for 3D printing services). Examples of companies offering 3D printing as a service include *shapeways, i.materialize. sculpteo* and others.
- A user could use a web interface to draw an object from scratch or modify/personalize a template of the desired object by changing its size, shape, color, etc. In this case the web interface hides most of the complexity associated with using an actual professional CAD tool, at the cost of limiting the available personalization options.[8]

Considering that the above mentioned features greatly simplify the programming phase of machine tools, they can also be leveraged by professional users (i.e. the design department of a manufacturing company).

3.4 Distributed Factories

In the manufacturing world as we know it, most (small or medium) enterprises are organized in a centralized way, with a few locations where the whole process is carried out, from design to actual manufacturing. The items built at these locations are then shipped to customers all over the world, implementing a paradigm and a business organization dominated by moving *atoms*,[9] or physical goods [5].

The adoption of digital processes in manufacturing brings the opportunity to shift to a different paradigm where *bits* are moved instead of atoms. The digital manufacturing process unfolds mainly in a virtual environment, and it reconnects to the physical world only when the actual manufacturing process takes place (that is when the configuration file is uploaded on a machine tool). This paradigm shift enables a more modern version of the manufacturing process, in which the *factory* becomes *distributed*.

The new model is based on the introduction of *digital manufacturing hubs* in different countries and locations, so to be closer to the customers. These centers would receive configuration files generated at the enterprises' design and engineering departments, and would build the requested objects or components. For relatively common production processes and materials the manufacturing hubs could represent a shared resource for design and engineering centers, bringing into the manufacturing arena the advantages typical of the *sharing economy*.

---

[7]This kind of interaction requires that users are able to use a CAD tool and generate the appropriate file to be uploaded.

[8]For this kind of service a technology known as parametric design, which allows to generate CAD files that can be easily personalized by simply changing some parameters, is gaining a lot of popularity.

[9]The use of the terms bits and atoms to indicate digital vs physical phenomena and activities was introduced by Prof. Neil Gershenfeld, director of the MIT Center for Bits and Atoms.

An amateur version of a (shared) distributed factory is represented by the already mentioned worldwide network of *FabLabs*: official FabLabs are all equipped with relatively similar CNC machines, so that projects put together by makers can be shared and replicated all over the world.

4 Digitization of Manufacturing

As highlighted in the previous sections, a key requirement for the implementation of Industry4.0 features and services is the ability to build a 'digital image' of the production process, constantly updated using real time data coming from the physical world.

Building such an information system is not an easy task, mainly due to the complexity of the corresponding process.

An approach that can be used to deal with such difficulty comes from the modelling and simulation experience: complex systems can be decomposed in (more or less) elementary modules, designed to interact among each other in the same way as the corresponding physical elements in the physical world.

In the case of manufacturing the choice of the elementary modules depends on the level of accuracy and detail most suited for the specific process. As a general rule, modules can be machine tools, manual working stations, vehicles or AGV (automatic guided vehicles) used for logistics, but also the parts to be assembled or built, and the expected final product.[10]

It is important to note that we are not talking of traditional modelling and simulation modules: what is needed for Industry4.0 is a digital replica of every interesting element of the process (much closer to the original than a simple model used for simulation), and that needs to be constantly in touch with the physical object it represents with a bi-directional data channel.

In the world of manufacturing this digital module has been given a name that says it all: the *digital twin*. Due to its generality and flexibility, the concept is quickly being adopted by many (almost all) other application domains like energy, social phenomena, economics, mobility, health, etc.

4.1 Digital Twins

A digital twin is a digital representation at physical and functional level of a real life component, process or service, augmented with operational data collected during its full lifecycle.

---

[10]The range of possible modules is very wide, and an accurate set can be defined only on a case by case basis.

By implementing digital twins of all the main elements of a complex system, the digital twin of the whole system is obtained. With reference to manufacturing, the digital twin of the production process would be the optimal underlying information system needed to enable most of the services at the base of the industry4.0 paradigm. Below are two relevant examples of digital twins of interest for a manufacturing process:

- The digital twin of a machine tool stores the machine functional description (useful for simulation and WISA), information referring to the machine purchasing (maker, model, date of purchase, cost, etc.), details on all the maintenance interventions carried out during its operational life, and all the data coming from onboard sensors, useful to enable real time monitoring, predictive maintenance services, etc.
- The digital twin of a component to be manufactured could store the initial requirements coming from the customer (see the section on personal fabrication above), and data generated during the production process (how many stages, which machine was used at each stage, how much each stage took to be completed, etc.). Once the product is put on the market or acquired by its original buyer, the digital twin ideally keeps storing information about the object location, its owner, its use (if applicable), and all relevant events up to the end of its life.

As mentioned in the examples above, data stored in a digital twin is quite heterogeneous: some information is static (details on the purchase of a machine), some is slowly variable in time (data referring to maintenance interventions), some is extremely dynamic (real time data coming from sensors).

This heterogeneity needs to be taken into account when designing the digital twin architecture, which doesn't need to be implemented in just one location or on one database using a specific technology. Digital twins are built around different data sets, which can be distributed on different information systems based on the requirements in terms of confidentiality, size, dynamic evolution, security, visibility or other. The resulting system can be quite complex, and particular attention needs to be paid to the definition of the permissions granted to the entities in charge of instantiating a digital twin, or to the entities that will read and write information to and from it: a set of cybersecurity issues *related to objects* similar to those studied and solved for humans and servers.

In this futuristic world of *cyber-physical systems* each element needs to be given a unique and certain digital identity, as well as a set of permissions to perform actions within the appropriate scope. These topics, as well as examples of sophisticated services enabled by digital twins, are covered in the remaining part of this chapter: in particular Sect. 5 is dedicated to digital identity of objects, and Sect. 7 deals with innovative business models and services.

5 Cybersecurity and Digital Identity for Objects, People and More

Readers lucky (and old) enough to have witnessed the exponential growth of the Internet in the 90s (when it left the military and academic labs to become a technology for all) will also remember how cybercrime boomed in the same period. The first registered Internet virus (the Morris Worm) was developed by a University student in 1988, when there were only 60,000 computers connected to the Internet. The wider adoption of Internet connectivity by the general public brought to a higher number of online information sites and databases. A direct consequence was a fast spread of cybercrime, perpetrated by hackers who found in the network (at the time very weak) a convenient vehicle to reach their victims.

Security and encryption were not included in the first specifications of TCPIP, making the early stage global network an easy target for illicit activities, including credential theft and unauthorized access to networked resources.

This past experience is very valuable now that access to the Internet is quickly being extended to entities other than humans. Attacks once limited to personal computers and servers today also target webcams, home gateways, smart appliances, robots, and—apparently soon—connected vehicles.

Extending security to objects like the ones mentioned above is not an easy task: we cannot simply apply to objects the same principles and technologies that have been developed for human users. A few examples:

- We are used to choosing strong passwords (or forced to do so) for each of the online services we access online, and to change them periodically. Choosing and remembering all these passwords is a pain for every Internet user. *How can we imagine to carry out similar actions for tens or hundreds of personal connected objects?*
- Connected objects, including laptops and servers, require constant software updates of their operating system and individual applications. These updates, even if highly automated, require from time to time the user's attention (easily obtained for a laptop or server, due to the frequency of interactions users have with them). *Who is in charge for updating the software of unattended appliances, like a webcam? (Many objects are plug and play, users buy and use them but have no knowledge about network and software maintenance)*
- Internet users carry a digital identity in order to obtain access to personal information and/or other online resources. Common strong digital identity architectures rely on cryptography, which is based on digital strings called keys. Identity-related keys are managed using *digital certificates*, issued by trusted *certification authorities* (CA). For security reasons, mainly related to counter brute force attacks, certificates have a limited validity, and they must be renewed—before the expiration date—interacting with the relevant CA. *How can digital certificates for objects be handled? Scalability is also an issue: the need will soon arise to manage the identity of several billions of objects.*

When humans are not in the picture, security and digital identity need to be set up with stronger requirements. The Internet of Things calls for a solution for digital ID of objects that matches (among others) the following specifications:

- Issued once, never expires/changes/is revoked/is updated
- Scalable up to multiple billions objects
- Uses/exchanges the minimum information required for the specific service to be accessed
- Full empowerment of the owner of the ID/object for the ID management.

Distributed Ledgers Technologies (DLTs) are emerging architectures that respond to most of these requirements:

- heavily based on strong cryptography
- able to store information in a secure, immutable way
- some of them are designed for scalability
- every information exchange (transaction) is digitally signed, in order to uniquely identify the originator and the destination.

The following subsections includes a short overview of the recent developments of Decentralized Identity (DID).

5.1 Decentralized, Self Sovereign Identity for All

The current management of credentials to access digital services must cope with the lack of an intrinsic identity management in the Internet architecture. Identity management was designed and included on top of existing, common layers or features instead of *being one* of them. The resulting scenario is a **centralized** one, where each service provider handles locally defined user credentials dedicated to access only its services, with two major disadvantages:

1. the service provider holds the right to open, restrict or cancel access to users without their consent
2. users have to memorize and use different passwords for different services, with a resulting very bad user experience.

In his paper "The path to self sovereign identity" [6], Christopher Allen identified four main steps of the evolution of digital identity management: *centralized, federated, user-centric, self-sovereign*.

Federated identity provides a partial solution to the issues listed above: it is based on *secure access delegation* standards such as OAuth, which allows users of large portals (e.g. social networks) the possibility to use their credentials to authenticate themselves with third party websites by automatically sharing the required information in a safe way.

While this fixes the second issue mentioned above, it leaves the management of users identity and credentials in the hands of a central entity.

If this situation is somewhat tolerated by human users, it is definitely not suited for the case in which access to services needs to be granted to objects and services.

The raise of the number of unanimated users connected to the Internet is pushing hard towards the definition of a completely different architecture for the management of users' identity, in which the users themselves (or the owners of the connected objects) have full control of the relevant digital identities.

The paradigm must also change: due to the annoying complexity of using the traditional username/password scheme, users do not take the necessary care, leaving behind them significant security vulnerabilities.

User-centric **identity** aims at putting users in full control of their own data and of its distribution to other parties. Information related to users' identity and other personal details must be exchanged only under users request. A good description of user-centric identity principles and operation can be found in [7].

Even though it gives users a good level of control over their identity and personal information, the user-centric identity management scheme still relies on the presence of digital identity providers who take care of the handling of personal information to grant access to all the services users subscribed to, with the associated security risks.

The end point of the identity management evolution is **Self-Sovereign Identity**, where also the role of the identity provider disappears, in the sense that every user is provider of her own identity. This implies the complete decoupling of identity information from specific services: user identity exists just because *user exists*, independently on any specific service.

Self-sovereign identity was also named "The Internet for Identity" by Phil Windley in the article he wrote in 2016 [8]: it represents the solution to the missing layer of the Internet architecture dedicated to digital identity management.

In order to turn these principles into a set of actual features, an infrastructure is needed which satisfies all the requirements of Self Sovereign Identity: *user control*, *security*, and *portability* (possibility to be used to access different services).

These features also need to be offered in a context where trust is guaranteed without the intervention of central authorities: this is why an ideal underlying architecture for the implementation of self-sovereign identity service is represented by the emerging distributed ledger technologies (DLTs).

The role of the different types of DLTs in the implementation of a self-sovereign identity architecture is described in depth in [9]. Most of the references cited in this section are also taken from the same white paper.

The Sovrin Identity Network is a global public utility, technically implemented as a *public permissioned* (hybrid) ledger governed by the Sovrin Foundation and built around open source, distributed ledger technologies with no central control entity.

6 Using AI for Innovative Services

Self-conscious technology is one fundamental paradigm of Industry4.0 that matches particularly well with what AI aims to achieve: hyper-aware systems able to sense their environments as well as take or suggest actions based not only on human commands but also on their own perceptions.

An evolutionary force is shaping the way conceptual models and related tools are empowering machines throughout a journey which will see human beings delegate to AI the task of making decisions (slowly but steadily today, but expected to grow exponentially in the near future).

Within the context of Industry4.0 there are two fundamental innovations that are spurring this paradigm shift:

- the traditional supply chain is changing into a new totally digitized and interconnected configuration, known as *Digital Supply Network* (DSN);
- the rise of a software/hardware artifact, called *Digital Twin* (see Sect. 4.1), providing a real-time connection between the physical and the digital worlds;

If we analyze the context trying to draw a cause-effect chain, it seems clear that the Industry4.0 revolution could only be triggered and sustained by the simultaneous availability of several different technologies. We are talking about *robotics* and *machine learning*, the *Internet of Things* and *smart sensors*, *cloud computing and software as a service*. These technologies, once thought of separately, nowadays are being combined in a rather holistic plan having the main objective of merging the physical and digital worlds. This will ultimately led to controlling the former by mean of the latter.

Rather than providing here a definition for AI, we prefer to stress the main role this discipline has been playing, quite as a fundamental attractive force, variously linking and combining the above cited technologies, finally sparking the Industry4.0 phenomenon.

Basically, what we expect from AI is the ability to work faster and more accurately than human intelligence does. AI needs to be better than human beings at:

- *recognizing patterns*
- *making sound predictions*.

Recognizing patterns Automatic text, voice and video recognition are the most immediate and pervasive examples.

Let's think of Siri, Alexa, and other tools able-first-to transform sound waves into a sequence of bytes, first, and then to map them into commands: in the autonomous vehicle vertical this can be applied to some effective interactions with the car (like asking for time and location of the next meeting, highlights of the day, etc.), while it is taking a passenger to the office. Within the same scenario, you could ask your car's AI-powered controller to let you drive; from that moment onwards a complex algorithm will start scanning your face, hundreds of time per second, ready to get the control back to its AI cruise control if it detects you are falling asleep or under the influence of drugs.

Automotive is not the only sector where these technologies are being applied. In the healthcare field, machines have already caught up to humans in reading X-ray and making diagnosis. In the security sector, an authentication service can be provided through an array of tools, including voice and image recognition.

In the government sector, chatbots can be deployed and used as concierge services. As a noticeable example, Smart Dubai, one of the Dubai's Government entities, released in 2017 a one stop concierge service called *Rashid*.[11] Citizens, residents and tourists can ask Rashid information related to government services, e.g. consumer rights, how to start a business, visa processes, renting, buying or selling a property, etc. Since the same year, a new version of *Dubai Now*[12] has been allowing users to access to over 50 government services all in one place and to initiate transactions using voice commands, after logging into the system using DubaiID.[13] In 2018 the plan is to link Rashid and Dubai Now, in order to provide the user with a seamless experience, from inquiring for some needed information to performing an actual transaction.

Making sound predictions is what we expect to be going on behind the scene of a weather forecast engine, or of a recommendation system running on the Amazon or Netflix websites.

In the first case we want the system, given a certain location, to be able to predict not only a qualitative weather attribute (e.g. sunny, cloudy, windy, etc.), but also numerical values (e.g. min and max temperature, percentage of humidity, probability of rain, etc.). The former is known as a *classification* problem, the latter a *regression* problem.

In the second case, we ask the system to suggest the customer with a basket of products ordered by the likelihood of being selected and bought.

In both cases, the underlying classes of problems can be seen and solved by applying the same strategy: *optimizing a cost function*, where the cost function assigns negative values to wrongly predicted events. Optimizing will take the form of finding the absolute minimum for the cost function.[14]

Machine learning is a field of computer science, built on sound foundations that use statistical techniques, which has proved itself to fit particularly well with the above cited features of pattern recognition and making sound predictions.

The term was coined in 1959 by Samuel [10], showing that this discipline is not new at all, but it is only in the last couple of decades that it has imposed itself as the main force revitalizing the adoption of AI-based techniques in almost every

---

[11] http://www.rashid/ae.

[12] https://dubainow.dubai.ae.

[13] DubaiID provides single identity (being linked to the Emirates Identity Authority) and single login to over 600 government services.

[14] To be noticed here that, in a more general scenario, what we want to optimized is an *objective function*; optimization will take the form of maximizing the posterior probability (e.g. in a naive Bayes model), maximizing a reward function (e.g. in reinforcement learning), etc.

imaginable sector, having successfully contributed to a renovated trust towards the still not completely uncovered potential and effective applications of AI.

An operational definition, defining an algorithm designed in the machine learning field, was given by Mitchell [11]:

> A computer program is said to learn from experience E with respect to some class of tasks T and performance measure P if its performance at tasks in T, as measured by P, improves with experience E

Machine learning algorithms use data inputs to learn a function, representing a model of a real world sub-system, relying on this learnt representation to classify an unknown object, predict values, finding patterns, etc.

Machine learning classes of problems are usually divided into two broad categories, depending on the availability of a feedback to the system.

- *Supervised learning*, where the algorithm is presented with inputs and their desired outputs and the goal is to learn a function mapping input to output. Classification and regression problems are within this class.
- *Unsupervised learning*, where no output is provided with the inputs, leaving the algorithm with the task of finding structures. *Clustering* and *Association rules* are examples of machine learning tasks belonging to this class and are commonly used in recommendation systems.

Deep learning is a subset of machine learning which is grounded on building and connecting multiple (hidden) layers of *Artificial Neural Networks* (ANNs), which are made of basic unit called *Neuron*, borrowing its denomination from the biological structure, whose activation and firing behavior is simulated by it.

ANNs were presented to the scientific community in the 60s [12]; initially welcomed with enthusiasm, the neural network research was almost abandoned two years later, after machine learning research pointed out some key issues with the computational machines that processed neural networks [13].

Different aspects have factored in the *renaissance* of deep learning (and AI) in the last two decades (apart from a key mechanism introduced in 1975, called *back propagation* [14]): the exponential growth of data availability, the massive reduction of costs for hardware equipment, the ever increasing computation power and storage capacity, and the usage of GPU with dedicated firmware/software libraries. Among these factors data availability seems to mark the difference between deep learning and other more traditional machine learning approaches: ANNs greatly and effectively benefit from huge amount of data, being less prone to overfitting and dealing well with non-linear tasks.

Considering all the above, it is not surprising that deep learning adoption is on the increase: in the coming years an exponential number of new projects in computer vision, information retrieval, marketing, medical diagnosis, voice recognition, online advertising, etc. is expected to be built on it.

AI for Industry4.0 is nowadays being recognized and perceived by all the G20 leaders as a key enabler for the next generation of digital manufacturing, leading to a real disruptive and transformative power in traditional workflows, supply chains, value creations and business models.

In a panel discussion at the "Digitizing Manufacturing in the G20" Conference,[15] the following nine priority areas have been identified and expected to be supported by the G20 countries with a joint and coordinated effort into research, development and deployment tasks on AI:

1. Hybrid teams of human workers and collaborative robots in Smart factories
2. Deep learning for state-based and predictive maintenance of networked production machines and for understanding human behaviors of shop floor workers
3. Semantic technologies for worldwide interoperability of Machine-to-Machine communication in Smart factories and logistics
4. Human-aware and real-time production planning and scheduling for multi-agent systems and dynamic plan revision
5. Intelligent industrial assistance systems for human workers: proactive and situation-aware on-line help and training on the shop floor
6. Trusted industrial data exchange hubs and machine learning for industrial process mining
7. Active digital product memories and digital twins for intelligent asset tracking and production cockpits
8. Security technologies for intelligent intrusion detection and penetration testing for Smart factories
9. Long-term autonomy and self-learning as well as self-healing capabilities of industrial components.

Furthermore, the panel stressed the importance of international collaboration for boosting the application of AI in manufacturing. Several urgent needs emerged, summed up in following list.

– An open standard for AI and Industry4.0 should be issued
– An AI on-demand platform and a large-scale AI infrastructure that offers open specifications and example implementations of basic AI components for Industry4.0 should be made available, supporting cloud-based computing and data services
– Reference models, Semantic representation languages, and simulation platforms must integrate the latest AI developments to ensure long-term impact
– Open and secure data exchange of production data should be supported so that advanced machine learning can be applied to these training data sets in order to reach a new level of productivity, efficiency and quality in manufacturing.
– A consensus on the social, legal, ethical and privacy implications of AI technology in manufacturing.

The following sections will investigate more about how AI is contributing to some of the most peculiar and innovative aspects linked to Industry4.0, namely the

---

[15] 16–17 March 2017, Berlin. Panelists: Prof. Wolfgang Wahlster, DFKI, Germany (Chair); Prof. Paolo Traverso, FBK-irst, Italy (Co-Chair); Sahin Albayrak, GT-ARC, Turkey; Dr. Philippe Beaudoin, Element AI, Canada; Dr. Satoshi Sekiguchi, AIST, Japan.

phenomenon of digital twins Sect. 6.1, the digitalization of the supply chain Sect. 6.2 and the very promising field of predictive maintenance Sect. 6.3.

6.1 AI for the Digital Twin

A digital twin may have many applications during the life-cycle of a product, thanks to the enormous amount of data it receives and can potentially analyze at every step of the manufacturing process.

The physical process is outfit with a multitude of sensors, performing operational measurements, sensing the environment and then transmitting data to the digital twin.

At this point data can be augmented by adding information sourced by external systems, like ERP and CRM data, CAD models, etc. Then, data is aggregated and stored into data repositories, ready to be analyzed.

Data scientists and analysts develop models, leveraging machine learning algorithms, and produce visualizations and insights.

Visualizations can be useful to find anomalies and areas that may need some further investigation.

Insights can take the form of actions to be transmitted back to the physical process.

This close the virtuous circle Physical-Digital-Physical.

It is important to notice that data can be analyzed either as a batch or in real time or as a combination, following an architectural pattern called *lambda architecture*. Large volumes of data collected over a period of time are more efficiently processed by mean of batch analytics, whereas real-time analytics are more indicated for creating insights through a continuous processing of incoming data generated by an asset.

6.2 AI for the Digital Supply Network

As it has been discussed in Sect. 6.1, a digital twin allows to implement a profitable loop where information is firstly created at the manufacturing plant, secondly sent to digital systems, augmented and made ready for being analyzed, finally transformed into visualizations, potentially triggering actions back to the physical process.

This physical-digital-physical loop also constitutes the intelligent information and communication backbone of an evolved form of supply chain, one more distinctive feature of Industry4.0, called *digital supply network*.

Supply chains, traditionally linear in nature with quite a fixed and immutable sequence of phases, from design to deliver, are nowadays being transformed into a dynamic, interconnected and open ecosystem of supply operations.

The turning point here is integrating and leveraging information collected from multiple sources and locations, not only from in-house traditional ERP and CRM systems (usually taking the form of *structured data*), but also from sensors (outfitting production plants, manufactured products, etc.), and even social media.[16] In these latter cases, the data information will most likely take the form of unstructured data.

What makes digital supply networks really different from a traditional supply chain is not in the way goods physically go from the design to the delivery phase, rather it is the continuous flow of a variety of pieces of information fully connecting a graph where the nodes are representatives of the typical supply chain functions.

Batch and real time data interconnect and feed processes and sub-processes, increasing efficiency through a constant and pervasive usage of data analytics, which generates insights supporting descriptive (what happened), predictive (what is likely to happen) and prescriptive (what actions should be taken) analysis.

AI is at the core here, the brain and heart of this living and pulsing intelligent network of supply functions, that can make decisions based on mathematically sound predictions in order to increase the manufacture firm' efficiency through cutting costs, increase revenues and potentially finding new revenue streams.

In the following some examples of areas where AI can foster and boost the transformation within the supply's main functions are listed.

- *Product design*: sensors embedded into products can reveal the way customers are using them, triggering design modifications or opening the way to new business opportunities. Furthermore, historical data analyses can reveal some emerging trend (in the shapes, colors, textures, ...) in the customers' choices, leading to the creation of products looking more appealing to them.
- *Planning and Inventory*: machine learning based predictions can be successfully applied to understand the demand, subsequently enabling may forms of optimization, like dynamic inventory, real-time inventory, automatic replenishment, etc.
- *Supplier networks*: from a sharing economy perspective means and sources of productions can be shared, too. Suppliers can be analytics-driven sourced and they can also share assets.
  To be noticed how this radical paradigm shift is involving a corresponding dramatic change in the way we see and build trustworthy (business) relationships and related transactions on the Internet. We will talk more profusely about this in Sect. 7.
- *Operations*: in this area, it is important to bring to the reader's attention some new approaches being introduced in the field of *robotics*, where new possibilities of automation can be boost by applying *reinforcement learning* techniques to train robots. By reinforcement learning a robot can learn by emulation, by making mistakes, autonomously understanding the rules of the game through a mechanism of reinforcement or disincentives via rewards and punishments. Using a reinforcement learning approach, instead of programming the set of actions using a imperative approach (explicitly listing $IF(condition)DO(action)$ clauses) seems to be undoubtedly faster and probably more reliable.

---

[16] In 2015 an interesting and promising proof of concept was developed at Retechnica Ltd. (a startup based in London) for a well-known British airline company, consisting in analyzing—through NLP techniques—tweets posted on the company's Twitter page and triggering actions to be taken by different functions (marketing and sales, customer relationship, on-ground assistance, etc.).

Predictive maintenance, given the high costs of downtimes in a production system, is another welcomed contribution given by AI to this field (see Sect. 6.3).
- *Logistics*: Warehouses digitally connected to other functions are already a reality. Biggest companies like Amazon and Alibaba rely on robots augmented with AI for drop-offs and pick-ups of goods. Their intelligence is not only in the way they can search and view a particular product but also in their routing algorithms.

  The same level of almost reached perfection and efficiency is now being introduce in the delivery phase: what can be easily forecast here is an ever increasing presence of intelligent and autonomous trucks driving our roads next decades.
- *Sales and After-sales*: Sales can benefit from AI, particularly from machine learning techniques to increase the revenue stream leveraging segmentation and association-rules algorithms. In addition, pricing can be dynamically set, depending on real-time demand and inventory data.

  Predictive after-sale maintenance is yet another not new concept, still being revitalized by a larger data availability to feed machine learning engines (see Sect. 6.3).

It is important to stress one more time here that, in all the examples given above, AI will be successful in adding real value and being a keystone in this epochal transformation of the supply chain into a digital network of supply functions if data will be massively collected and consistently shared across all the nodes.

6.3 Predictive Maintenance

A maintenance program is an effort to mitigate the risk of downtime in manufacturing facilities.

Usually a maintenance program falls into one of the following categories.

- *Reactive* maintenance: where the issue is fixed when it manifests itself. This leads to the maximum utilization of the machine when it is working correctly, but it implies an unplanned downtime, besides potentially bigger damages.
- *Planned* maintenance: leads to lesser unplanned downtime, but the machine is not exploited at its maximum possibilities.
- *Proactive* maintenance: it is a data-driven approach, aiming to spot and fix some potential causes that can head to a major failure (e.g. continuously sensing temperature and humidity conditions and dynamically adapting the production plant mode of work to some derived parameters). It combines the advantages of the reactive and planned approaches, but it needs some organizational changes (e.g. availability of sensors, IT systems, etc.).
- *Predictive* maintenance: it can be seen as an evolution of the proactive approach. Rather than measuring only parameters for which correlations with a possible failures are already known, it pushes the data-driven approach to its limits, collecting data from a myriad of sensors and analyzing them in real time, leveraging state-of-art machine learning algorithms, responsible to build up survival models based of the analysis of a massive amount of historical data.

It is evident, then, that the main advantage of predictive maintenance over the other paradigms lyes in its lifetime-optimal utilization of the machine, suggesting for a replacement of the machine or of a spare part on slightly before-needed.

The importance and promises of predictive maintenance is all but new. Already in 2010, a report by the U.S Department of Energy claimed that it could reduce maintenance costs by 30%, downtime by 45% and downtime by 75% [15]. According to Gartner [16] in 2013, much of the nearly 2 trillion USD in new value across the global economy would be accounted by predictive maintenance. Companies like GE, Cisco, IBM, and Intel believe that predictive maintenance capabilities will create an additional 100 billion USD in value for the energy and utilities industries by 2020 [17]. Finally, according to Deloitte:

> predictive maintenance can reduce the time required to plan maintenance by 2050 percent, increase equipment uptime and availability by 10, 20 percent, and reduce overall maintenance costs by 510 percent [18].

So, *why now*? Again, the main triggers for this late adoption is in the huge availability of sensors (implying collections of massive amount of data), computational power (for analyzing data using CPU/GPU greedy machine and deep learning algorithms) and bandwidth (for transporting data and feedbacks, finally realizing the physical-digital-physical loop) industry can take advantage of nowadays.

Predictive maintenance requires some fundamental interconnected pieces of technologies.

- *Sensors*
- *Network*
- *Integration tools*
- *AI for Advanced Analytics*
- *AI for Advanced Interface*.

Let's consider now the subject of maintenance not only as a way to increase firms' productivity, but also as an additional revenue stream in the after-sales phase.

As pointed out in [19], an additional perspective onto this subject is that the product's life-cycle does not end when it is released on the market, but it also continues within the after-sales phase: this possibility enables failures to be continuously monitored and evaluated. It is therefore of paramount importance to continuously measure the product defect tendency, to identify it at an early stage, allowing the endorsement of test improvements, as well as any appropriate engineering changes.

On the other hand, in the last few years customers have been led to demand better quality products with very personalized services, extending throughout the entire period of ownership. In this context, both remote and on-site assistance services should evolve into a full customer-oriented strategy, enabling the support to go well beyond the warranty, all the way into the sphere of the steady improvement of the installed base performance indexes.

A change in vision, indeed. *Assistance services are no longer considered as a source of costs, they become instead part of the enterprise strategies in order to increase the overall business turnover.*

7 Innovative Business Models and Services

In the previous sections we have gone through the main technological elements at the base of the digitization of industrial processes, which represent a fundamental component of the innovation behind the Industry4.0 paradigm.

The availability of accurate and timely digital information about the numerous phases and tasks of a complex process also enables the development and adoption of new business models and services where machines and objects play a new active role, evaluating the context and making informed autonomous decisions. Some of the most promising (in some cases already implemented) are included in the sections below.

7.1 Optimization of Production and Smart Scheduling

Planning the assignment of specific tasks to machines in a production plant is a critical issue because of the impact it can have on optimization of timing, warehouse stocks and other resources. Experts today make their decisions based on experience and assumptions about the average behavior of the overall process.

In an industry4.0-enabled environment, scheduling decisions could be dynamically made in an innovative scenario where the machines and the components to be machined jointly cooperate with the MES (Manufacturing Enterprise Software) in order to maximize the efficiency of the overall process.

The process could follow the steps listed below:

– When a user places an order for a specific object or component, the corresponding digital twin is instantiated, and initialized with all the user requirements (size, materials to be used, color, special finishing etc.)
– The digital twin of the product issues a request to the MES in order to be included in the pipeline. The MES identifies the correct sequence of machining operations needed (this decision can be taken starting from information found in the product digital twin: CAD files, material chosen by the user, etc.), and starts a negotiation with the digital twins of the candidate machines to identify the best match for that specific object (*personal fabrication*)
– The assignment of the product to specific machines can be followed, checked and updated at each step of the production, taking into account the real time state of the workstations (which could be different from what initially planned).

The ability of collecting data from sensors installed onboard each workstation also enables an accurate calculation of the lot (or even individual object) production costs, taking into account parameters such as the amount of raw material used, the energy consumption, the tools wear etc.: a feature which is greatly appreciated by entrepreneurs and production managers.

It is easy to extend the concept of optimal scheduling of manufacturing tasks to the distributed factory scenario described earlier in section a.b. Digital fabrication hubs started in numerous locations could be shared among designers and businesses.

7.2 Machine Economy

In the Industry4.0 paradigm machines become smart and connected objects, able to collect information both from the local progress of the machining task and from external sources.

Machines will be able to talk to each other as well as to a central decision making application (an advanced MES). The availability of detailed information about the requirements (e.g. collected from the product digital twin) gives a machine the possibility to calculate its own balance sheet, buying and selling resources and services.

If this is the case, security becomes a top priority issue: each object participating in the P2P (or M2M in this case) economy must be properly identified (hence the relevance of an accurate management of its *digital identity*), and it must be equipped with a digital wallet to send and receive transactions.

As we have seen in Sect. 5, the underlying architecture which could satisfy all these needs together (including the management of frequent, small payments) is represented by Distributed Ledgers. Some DLTs are emerging, like IOTA [20], with a specific orientation to the needs of the economy of things: low transaction fees, high scalability, high security (quantum-resistant in the case of IOTA) and compatibility with distributed identity management standards, like Distributed Identity (DID) and Distributed Public Key Infrastructure (DPKI).

References

1. Satoshi Nakamoto B (2008) A peer-to-peer electronic cash system. https://bitcoin.org/bitcoin.pdf
2. Weiser M (1991) The computer for the 21st century. Sci Am (265):94–104
3. Benoit M (2012) Physical internet foundation. In: 14th IFAP symposium on information control problems in manufacturing (INCOM), vol 45. Elsevier, Bucharest (Romania), pp 26–30
4. CNC.com, The history of computer numerical control. Retrieved from CNC.com: http://www.cnc.com/the-history-of-computer-numerical-control-cnc/
5. Gershenfeld N, Euchner J (2015) Atoms and bits: rethinking manufacturing: an interview with Neil Gershenfeld: Neil Gershenfeld talks with Jim Euchner about the internet of things and the coming revolution in manufacturing. Res Technol Manag J 58(5), September–October 2015. www.Questia.com
6. Allen C (2016) The path to self-sovereign identity. http://www.lifewithalacrity.com/2016/04/the-path-to-self-soverereign-identity.html
7. Cameron K, Posch R, Rannenberg K (2008) Proposal for a common identity framework: a user-centric identity metasystem
8. Windley P (2016) An internet for identity. http://www.windley.com/archives/2016/08/an_internet_for_identity.shtml

9. Reed D, Tobin A (2017) The inevitable rise of self-sovereign identity. Sovrin Foundation
10. Samuel A (1959) Some studies in machine learning using the game of checkers. IBM J Res Dev 3(3):210–229
11. Mitchell T (1997) Machine learning. McGraw Hill, p 2. ISBN: 978-0-07-042807-2
12. Ivakhnenko AG, Lapa G (1967) Cybernetics and forecasting techniques. American Elsevier Pub. Co
13. Minsky M, Papert S (1969) Perceptrons: an introduction to computational geometry. MIT Press. ISBN: 0-262-63022-2
14. Werbos PJ (1975) Beyond regression: new tools for prediction and analysis in the behavioral sciences
15. U.S. Department of Energy Federal Energy Management Program, Operations & Maintenance Best Practices: A Guide to Achieving Operational Efficiency, August 2010
16. Middleton P, Kieldsen P, Tully J (2013) Forecast: the internet of things, worldwide. Gartner Inc
17. Barlow M (2015) Predictive maintenance—a world of zero unplanned downtime. O'Reilly
18. Coleman C, Damodaran S, Chandramouli M, Deuel E (2016) Making maintenance smarter—predictive maintenance and the digital supply network. Deloitte University Press
19. D'Aprile D, Bergadano F (2013) An integrated service management system built on Microsoft dynamics CRM. In: Microsoft pre-conference CONVERGE
20. Popov S (2015) The tangle. https://iota.org/IOTA/Whitepaper.pdf
21. Boschert S, Heinrich C, Rosen R (2018) Next generation digital twin. In: TMCE conference 2018. Las Palmas de Gran Canarias, Spain

Industry 4.0 and The Sharing Economy

Robert Shorten, John Oliver, Deirdre Clayton, Ammar Malik and Hugo Lhachemi

Abstract We discuss the rise of manufacturing-at-the-edge in the context of the sharing economy. Some of the disruptive forces giving rise to edge manufacturing are discussed. As well as presenting some of the pressing open technical questions, many societal issues, and the unintended consequence of prosumer based societies are discussed.

Keywords Additive manufacturing · 3D-printing · Sharing economy

1 Shipping Data Instead of Items

When I (Shorten) was a boy, the world was a simple place. Market forces informed entrepreneurs of opportunity, *things* were manufactured, and then supplied to consumers via brokers (shops). For example, I recall buying my first bicycle. The choice was based on what was present in the local shop at the time and my input into the purchasing experience was limited to a *take-it* or *leave-it* choice. This picture, Fig. 1, of supplying goods to consumers has persisted since before the industrial revolution.

In Fig. 1, the role of the broker is to both inform consumers of product choice, to provide product access to customers, and to act as temporal storage buffers for manufacturing. While this model can be considered democratic in that it reflects the will of a society of consumers, it is certainly not fair, imposing consensus values

---

Based on lecture presented at EU Industry Day, Brussels, Feb 5th, 2019.

R. Shorten (✉)
Dyson School of Design Engineering, Imperial College London, London, UK
e-mail: r.shorten@imperial.ac.uk

J. Oliver · D. Clayton · A. Malik · H. Lhachemi
University College Dublin, Dublin, Ireland
e-mail: john.oliver@i-form.ie
e-mail: deirdre.clayton@ucd.ie
e-mail: ammar.malik@ucdconnect.ie
e-mail: hugo.lhachemi@ucd.ie

© Springer Nature Switzerland AG 2020
E. Crisostomi et al. (eds.), *Analytics for the Sharing Economy:
Mathematics, Engineering and Business Perspectives*,
https://doi.org/10.1007/978-3-030-35032-1_19

Fig. 1 Centralised manufacturing

on consumers, and is only able to address personalised choices of individuals in a limited manner.

This *centralised* picture of product delivery is barely recognisable today. Driven by internet innovation, and companies such as Google and Amazon, the brokerage element of the product delivery eco-system is evolving. Cities all over the world are reporting significantly reduced footfall, as the seemingly unstoppable rise of online shopping continues unabated. Instead of Fig. 1, the world is coming to more and more resemble an *on-demand* model as depicted in Fig. 2. Here the internet facilitates direct negotiation between consumers and manufacturers to create a model in which goods are delivered directly to consumers. Driven by these new forms of brokerage, and companies such as Amazon, the *on-demand* model is becoming ubiquitous. In this case, when I order my bicycle, I can choose from a range of frames, colours, optional extras and the internet is now the shop window, increasing choice significantly, and facilitating a certain degree of personalisation.

In the on-demand model, goods, even though they may be personalised, still have to be delivered from a central production site. The *delivery element* is hugely inefficient from many perspectives. As well as being costly for both consumer and manufacturer alike, it also can be costly from an environmental perspective (packaging and transport). Furthermore, confirming the veracity of centrally produced goods, remains largely a matter of trust in this model, with several recent notable examples of this trust being broken [1]. Driven by such concerns, and by technological advances in areas such as additive manufacturing, cyber-physics, and data science, and by a changing (more informed) demographic, a new model of product delivery is emerging; the model of the Prosumer Driven Society (Fig. 3).

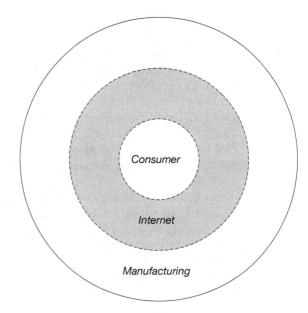

Fig. 2 Centralised manufacturing

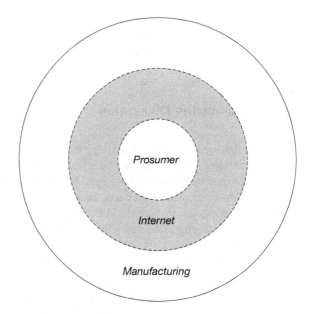

Fig. 3 Centralised manufacturing

In this model the consumer is replaced by a *prosumer*; that is a customer that both consumes and produces goods. In this model physical manufacturing takes place at the edge, and the role of the internet is to ship data (not parts) from the owner of the product intellectual property to the prosumer. Prosumers are already important players in the energy ecosystem (solar panels) [2, 3], but are also beginning to emerge in other markets due to the increasing availability of low cost 3D printing. In this model, when I am considering replacing my bicycle, the process begins with my generation of the bicycle design that is unique to my needs and desires. Manufacturers could bid for my custom, based on my design and specifications. As the prosumer, I am driving the design, cost, margin and manufacturing location based on my needs and beliefs.

The Edge Manufacturing/Prosumer model offers great advantages over traditional manufacturing models; including a potentially reduced environmental footprint; a manufacturing experience that is not consensus based, but rather bespoke, verifiable, and fair (both for customers and suppliers); greater societal resilience, as well as great convenience. In the not too distant future, citizens may be able to print their morning breakfast, while waiting for their new personalised running shoes to be printed using fairly sourced raw materials. Simultaneously, your elderly grandmother prints her morning prescription from the pharmacy, to aid her recovery after a personalised implant, and uses blockchain to verify to her doctors that the correct drugs were taken at the correct time, in the correct dosage, and in the correct order. After breakfast, your 3D printer is used by Amazon to deliver and print products for your neighbours, all for monetary reward for you. This is not utopia—this is the future world of additive manufacturing, cyber-physics and IoT, and the prosumer oriented society.

2 Manufacturing Disruption

So what is driving the trend toward edge manufacturing and the rise of prosumers? The past decade, in particular, has seen the emergence of new, hugely disruptive, technologies and business models, that are initiating change across a number of sectors and domains. Apart from low cost 3D printing, and ongoing developments in additive manufacturing, interest in making smart cities (and fairer), new business models based on sharing (rather than just sole ownership), and technological advances in areas such as internet of things (IoT) and cyber-physics, all are driving the search for new innovative services and applications of 3D printing technologies [4–9].

Smart Cities is all about making cities more efficient, and better places for citizens to live in. Clearly, the personalised nature of 3D printing, and the environmental and societal benefits of the technology, have a major role to play in making cities smarter and more resilient.

The Sharing Economy is all about the emergence of sharing expensive goods as an alternative to sole ownership. Using 3D printers as a sharing platform to empower communities of prosumers, and also machines, clearly has a big role to play in the emergence of a sharing economy eco-system.

Cyber-Physics is about orchestrating the interaction of humans and machines, and **IoT** is concerned with both ubiquitous connectivity, and actuation, and the societal opportunities that are afforded by such abilities. Both technologies are enabling 3D printed services.

Coupled with these we have the emergence of new, more connected, informed and altruistic, consumer groups. These consumer groups are interested in the environment, in fair trade, in the fair sourcing of products, in facilitating age-independent access to goods and services, and in on-demand access to products and services. Such consumers are well served by the functionality of 3D printing, for whom both the veracity and origin of products, and their environmental/carbon footprint, can be readily controlled.

Furthermore, worldwide demographic changes are also driving consumers to adopt new technologies and business models. Big driving forces behind both personalised and on-demand consumerism, and the sharing economy are: (i) a worldwide bulging of the middle classes where finite resources must be shared more efficiently; (ii) the needs of ageing populations and in particular, the provision of access to goods and services; and (iii) increased urbanisation and its effect on logistics and the environment.

3 Enablers of the Prosumer Society

3-D printing is a key enabling technology for a prosumer society. Despite much progress in the area of additive manufacturing, much research work remains to be done. The range of materials available, addressable product size, cost structure of the complete process (including post-processing) and regulations all need significant

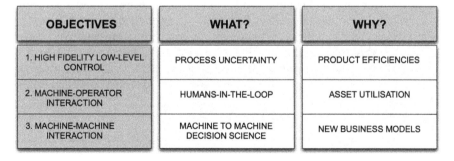

Fig. 4 Broad research directions

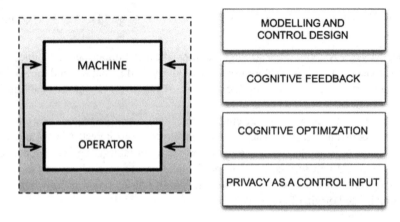

Fig. 5 Machine-operator interaction

advances in order to make the technology competitive.[1] The optimisation and efficient use of the machine themselves is also a key enabler and is one of the focus of our research in this area. Roughly speaking, three basic directions of work exist (Figs. 4 and 5).

1. **Making 3-D printing machines work better**: Roughly speaking, here we are interested in low-level aspects of 3D manufacturing, ranging from better understanding the chemistry of additive manufacturing and the PDE models that arise, to better making use of these models to realise high precision control systems.
2. **Making better use of 3D printing machines**: Here we are interested in helping consumers to use these machines in a better way than is currently possible. Designing 3D printed products is presently inefficient and difficult, and this is a major obstacle to widespread adoption and home use. Making this process better is a major research effort at the SFI funded Research Centre I-Form[2] and is driving research at the frontiers of cyber-physical systems, natural language processing, augmented reality, and control theory. For example, we are currently using augmented reality with haptic feedback to create virtual environments in which parts can interact with both real and virtual objects prior to printing (Fig. 6).
3. **Monetising networks of 3D printing**: A final frontier in developing a 3D-printing enabled prosumer society, is to develop the tools to allow individual machines to trade with one another, and to develop business models that efficiently exploit networks of such machines (Fig. 7).

---

[1] https://www2.deloitte.com/nl/nl/pages/consumer-industrial-products/articles/the-business-case-for-additive-manufacturing.html

[2] www.i-form.ie

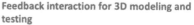

Classic work-flow for printing an object
- Computer-Aided Design (CAD)
- Printing of the model
- Iteration based on a trial and error procedure

3D model reconstruction in an augmented reality environment
- Layer by layer model reconstruction
- Model visualization in an augmented reality environment
- Augmented Reality glasses: Hololens

Feedback interaction for 3D modeling and testing
- Provide force feedback to the user to "feel" the 3D environment
- Gloves: Dexmo

Fig. 6 I-From hardware-in-loop platform

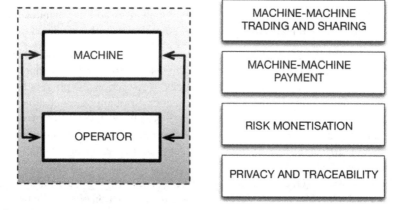

Fig. 7 Machine-to-machine decision science

4 The Implications of the Prosumer Society

Step-change advances in manufacturing have always been associated with fundamental change in other domains. One only has to think of advances in science, society, economics, even politics, that have been brought about by changes in manufacturing. It is not inconceivable that the rise of the prosumer society will similarly act as a catalyst for change in the coming decades (Fig. 8).

While some of the areas being transformed by contemporary advances in manufacturing are already evident, others are not. From a scientific perspective, as we have already mentioned, the area of cyber-physics, and more generally, data science

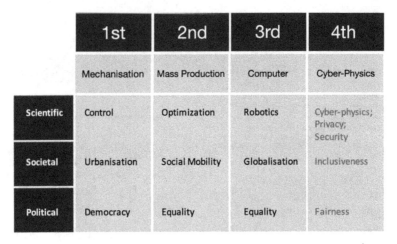

Fig. 8 Phases of manufacturing evolution, from 1.0 to 4.0, and societal/scientific change

and control, is benefiting enormously from developments in additive manufacturing. However, much work remains to be done. Important issues to be considered include:

(i) **The biology of 3-D printing**: One of the most important issues of our time concerns air-quality. For the most part, this interest has focussed on outdoor emissions from transportation and other sources. However, as 3D printers become more widespread in homes and offices, indoor air-quality may become a significant issue. Already, some studies have linked these devices to emissions of fine particulates, and VOCs [10]. A pressing requirement is to understand the effect of such emissions on humans so that these (networks of) printers can be safely certified.

(ii) **Privacy and IP**: One of the great advantages of 3D-printing is the ability to produce bespoke parts and products. However, the production of bespoke products has the potential to reveal very private information associated with individuals. In the medical domain, the concerns are evident. But also, a 3D-printer producing a breakfast croissant has the potential to reveal detailed information on eating habits of individuals. Much work therefore is required to obfuscate such information, so that individuals can trust these devices. Similar issues arise in the area of IP protection.

(iii) **Cyber-attacks**: While to the best of my-knowledge, no recorded examples of such attacks exist, 3D-printing for personal use opens a new frontier in the ability of adversaries to attack individuals. Imagine an attacker (remotely) tampering with a printer producing a drug or a food product?

(iv) **Fraud and standards compliance**: Centrally produced goods are strictly regulated to comply with regulatory standards. Locally produced goods offer a new possibility for individuals to cheat. Enforcement of standards (perhaps using blockchain) remains a major issue to be addressed by our data-science colleagues.

(v) **Enabling circular economies**: Will low cost, mass produced products remain the dominant forms of consumption for the general population? Resources are finite and acceleration of risks like climate change are not being adequately addressed. We need to consume less products and manufacture them in such a way that they can be refurbished, upgraded and repaired. To keep economies growing, the lifecycle cost and value of these products must be at least as high as the multiple throwaway goods they are replacing. Manufacture-as-a-service models need to include, repair, upgrade and refurbish as a service. We have to move to the mindset of consuming less volume of products, but each product being of higher value and cost, resulting in net economic growth from a monetary point of view, while reducing the pressure on finite resources. Policy can help, by developing incentives for the prosumer/manufacturing as a service model and enforcing well-informed carbon taxes on the traditional mass-produced products and supply chains.[3,4]

From a societal perspective, the issues are less clear. Innovation in many areas has been hampered by both the new challenges that arise as a result of this innovation, and by the unintended consequences of innovation. One only has to think of AirBnB, and the effect that it is having in property prices in many cities (Berlin, Dublin, Amsterdam) to realise that what is ostensibly a good thing (making use of under used capacity), may in fact be a bad thing. A less obvious, but more alarming example, concerns deposits on bottles/plastics to encourage recycling and more environmental behaviours. This seemingly good initiative has led to entire sub-economies emerging in many countries. Here armies of pensioners and homeless people sort through rubbish bins to recover bounty from discarded bottles, sometimes at the behest of organised syndicates. While such an initiative may do environmental good, it is extremely exploitative whereby the good of society is based on exploiting the existential needs of the few. Clearly, there is potential for such unintended consequences from the widespread deployment of 3D-printers. Other societal issues include more personal responsibility being placed on the individual (think of printing food and drugs), rethinking personal liability, enforcing regulatory standards, and a whole host of other social contracts associated with edge manufacturing.

Finally, the rise of the Prosumer will lead to new economic models and theories. Already Prosumer markets are being studied and designed as part of work on the sharing economy [11–15]. A fundamental feature of prosumer economies is the role of the individual. Networks of connected, well informed, and self-interested consumers, are at the heart of such markets. Such considerations should lead to a shift from macroeconomic analysis tools towards a more atomic (micro-economic) study of markets and their design. Perhaps the work of Adam Smith and the *self-interested man* will once again rule economics.

---

[3]Growth Within: A Circular Economy Vision for a Competitive Europe, Ellen MacArthur Foundation and McKinsey Center for Business and Environment, Available online: www.mckinsey.com.

[4]https://www.institutmontaigne.org/ressources/pdfs/publications/policy-paper-circular-economy.pdf.

5 Concluding Remarks

Imagine a world where your meals are printed while you wait for your new personalised running shoes to be printed. Your elderly grandmother prints her morning prescription from the drug store, to aid her recovery after a personalised implant, and uses blockchain to verify to her doctors that the correct drugs were taken at the correct time, in the correct dosage, and in the correct order. Your 3D-printer is then used by Amazon to deliver and print products for your neighbours, all for monetary reward for you. Now imagine a world where absolutely nothing is private. Your personalised food tastes are used to inform advertisers; and your personalised medical implant records reveal inner vulnerabilities. Imagine a world where your 3D-printer can reveal to adversaries what you are doing and when you are doing it, where indoor air-quality is the pressing health issue of the day, where product liability is the sole responsibility of the individual, and where an army of homeless people work to control the sorting of 3D printing cartridges that are discarded by the well-off. These are the two faces of the future world of additive manufacturing, cyber-physics and IoT. 3D-printing will be a catalyst for change and good, both in society and science. It will empower entire communities and lead more inclusive societies. The scientific questions stimulated by the area will drive new innovation in manufacturing, and will be highly beneficial in other domains. The unintended consequences of this new technology remain undiscovered.

Acknowledgements This work is supported in part by a research grant from Science Foundation Ireland (SFI) under Grant Number 16/RC/3872 and is co-funded under the European Regional Development Fund. RS was also partially supported by SFI grant—16/IA/4610.

References

1. Bachmann R, Ehrlich G, Ruzic D (2017) Firms and collective reputation: the Volkswagen emissions scandal as a case study. CEPR Discussion Paper No. DP12504. Available at SSRN: https://ssrn.com/abstract=3089760
2. Agnew S, Dargusch S (2015) Effect of residential solar and storage on centralized electricity supply systems. Nat Clim Chang 5(315):
3. Parag Y, Sovacool BK (2016) Electricity market design for the prosumer era. Nat Energy 1
4. Crisostomi E, Shorten R, Studli S, Wirth F (2017) Electric and plug-in hybrid vehicle networks: optimization and control. CRC Press, Taylor and Francis Group
5. Fioravanti A, Marecek J, Shorten R, Souza M, Wirth F (2017) On classical control and smart cities. arXiv:1703.07308
6. Narasimhan C, Papatla P, Jiang B, Kopalle PK, Messinger PR, Moorthy S, Proserpio D, Subramanian U, Wu C, Zhu T (2018) Sharing economy: review of current research and future directions. Cust Needs Solut 5(1):93–106
7. Hamari J, Sjoklint M, Ukkonen A (2016) The sharing economy: why people participate in collaborative consumption. J Assoc Inf Sci Technol 67(9):2047–2059
8. Lan J, Ma Y, Zhu D, Mangalagiu D, Thornton TF (2017) Enabling value co-creation in the sharing economy: the case of mobike. Sustainability 9(9):1–20
9. Goudin P (2016) The cost of non-Europe in the sharing economy: economic, social and legal challenges and opportunities. Technical Report, European Parliamentary Research Service

10. Stefaniak A, Lebouf R, Duling MG, Yi J, Abukabda A, McBride CR, Nurkiewicz TR (2017) Inhalation exposure to three-dimensional printer emissions stimulates acute hypertension and microvascular dysfunction. 335:1–5
11. Moret F, Pinson P (2018) Energy collectives: a community and fairness based approach to future electricity markets. IEEE Trans Power Syst
12. Iosifidis G, Tassiulas L (2017) Dynamic policies for cooperative networked systems. In: Workshop on the economics of networks, systems and computation, Series NetEcon, pp 1–6
13. Einav L, Farronato C, Levin J (2016) Peer-to-peer markets. Ann Rev Econ 8:615–635
14. Ritzer G, Jurgenson N (2010) Production, consumption, prosumption: the nature of capitalism in the age of the digital 'prosumer'. J Consum Cult 10(1):13–36
15. Patel S, Rajagopal R (2017) The value of distributed energy resources for heterogeneous residential consumers. arXiv:1709.08140 [cs.SY]

Conclusion

Emanuele Crisostomi, Bissan Ghaddar, Florian Häusler, Joe Naoum-Sawaya, Giovanni Russo and Robert Shorten

The sharing economy is nowadays a major sector in the global economy and companies embracing the new trends of collaborative consumption have been growing at very rapid rates across the globe. The newly created markets that are already valued at multi-billion dollars confirm that the sharing economy is not a temporary fashion but rather a growing trend in changing consumers behaviors that will continue to create business opportunities, many of which are already threatening traditional industries such as transportation and hospitality.

The book overviewed the fundamentals of the sharing economy which are based on three main pillars: mathematics, engineering, and business. This book first reviewed the mathematics of sharing which provides a theoretical foundation to enable the proper design and dimensioning of sharing systems. While several mathematical models have been developed to study sharing systems, several challenges remain including developing models that scale to realistic sizes, enable real time decisions,

---

E. Crisostomi (✉)
University of Pisa, Pisa, Italy
e-mail: emanuele.crisostomi@unipi.it

B. Ghaddar · J. Naoum-Sawaya
Ivey Business School, University of Western Ontario, London, Canada
e-mail: bghaddar@ivey.ca
e-mail: jnaoum-sawaya@ivey.ca

F. Häusler
Moovel, Stuttgart, Germany
e-mail: florian.haeusler@gmail.com

G. Russo
Department of Information & Electrical Engineering and Applied Mathematics,
University of Salerno, Fisciano, Salerno, Italy
e-mail: giovarusso@unisa.it

R. Shorten
Dyson School of Design Engineering, Imperial College London, London, UK
e-mail: r.shorten@imperial.ac.uk

© Springer Nature Switzerland AG 2020
E. Crisostomi et al. (eds.), *Analytics for the Sharing Economy:*
Mathematics, Engineering and Business Perspectives,
https://doi.org/10.1007/978-3-030-35032-1_20

adapt to changing behaviors, and incorporate multiple competing and decentralized decision makers.

Then the book presented several recent technological developments that enable the sharing economy. The widespread adoption of the mobile phone has certainly provided the launch pad for the early adopters of the sharing economy. These technological developments continue to be an innovation platform for providing new services and user experiences. The recent innovations have been focusing on enabling fully decentralized systems that can autonomously reallocate resources to adapt to the continuously changing behaviors of consumers and resource owners. The underlying challenges also include security of massive scales of data, optimal coordination of decentralized peer-to-peer networks, and signaling to induce behavior change in a decentralized approach.

The final section of the book highlighted promising use cases of the sharing economy mostly focusing on transportation and energy. Furthermore, sharing data in itself is growing as a platform that will empower many new services such as future connected autonomous vehicles, and manufacturing (Industry 4.0). The next generation sharing platforms will include continuous sensing, reporting, and conditioning of the shared resources, autonomous operation, assignment, and adaptability of the resources to the users, and the adoption of secure decentralized transaction processing.

The sharing economy is still at its early stages where the technology as well as the prosumers base are developing and maturing. The current obstacles for the growth of the sharing economy continue to be the regulatory resistance and the fear of the traditional players that are threatened by the emerging business models. The continued research and development in the mathematics, engineering, and business aspects of the sharing economy will help dissipate these obstacles and produce a long-term transformation in collaborative consumption of resources.

CPSIA information can be obtained
at www.ICGtesting.com
Printed in the USA
LVHW051145150321
681577LV00001B/31